The Handbook of Environmental Chemistry

Volume 1 Part D

Edited by O. Hutzinger

The Natural Environment and the Biogeochemical Cycles

With Contributions by
H. J. M. Bowen T. Frevert W. D. Grant
G. Kratz P. E. Long

With 58 Figures

Springer-Verlag
Berlin Heidelberg GmbH 1985

Prof. Dr. Otto Hutzinger

University of Bayreuth
Chair of Ecological Chemistry and Geochemistry
Postfach 3008, D-8580 Bayreuth
Federal Republic of Germany

ISBN 978-3-662-15263-8 ISBN 978-3-540-39209-5 (eBook)
DOI 10.1007/978-3-540-39209-5

Originally published by Springer-Verlag Berlin Heidelberg New York in 1985.

Softcover reprint of the hardcover 1st edition 1985

2154/3140-543210

Preface

Environmental Chemistry is a relatively young science. Interest in this subject, however, is growing very rapidly and, although no agreement has been reached as yet about the exact content and limits of this interdisciplinary discipline, there appears to be increasing interest in seeing environmental topics which are based on chemistry embodied in this subject. One of the first objectives of Environmental Chemistry must be the study of the environment and of natural chemical processes which occur in the environment. A major purpose of this series on Environmental Chemistry, therefore, is to present a reasonably uniform view of various aspects of the chemistry of the environment and chemical reactions occurring in the environment.

The industrial activities of man have given a new dimension to Environmental Chemistry. We have now synthesized and described over five million chemical compounds and chemical industry produces about hundred and fifty million tons of synthetic chemicals annually. We ship billions of tons of oil per year and through mining operations and other geophysical modifications, large quantities of inorganic and organic materials are released from their natural deposits. Cities and metropolitan areas of up to 15 million inhabitants produce large quantities of waste in relatively small and confined areas. Much of the chemical products and waste products of modern society are released into the environment either during production, storage, transport, use or ultimate disposal. These released materials participate in natural cycles and reactions and frequently lead to interference and disturbance of natural systems.

Environmental Chemistry is concerned with *reactions in the environment*. It is about distribution and equilibria between environmental compartments. It is about reactions, pathways, thermodynamics and kinetics. An important purpose of this Handbook is to aid understanding of the basic distribution and chemical reaction processes which occur in the environment.

Laws regulating toxic substances in various countries are designed to assess and control risk of chemicals to man and his environment. Science can contribute in two areas to this assessment; firstly in the area to toxicology and secondly in the area of chemical exposure. The available concentration ("environmental exposure concentration") depends on the fate of chemical compounds in the environment and thus their distribution and reaction behaviour in the environment. One very important contribution of Environmental Chemistry to the above mentioned toxic substances laws is to develop laboratory test methods, or mathematical correlations and models that predict the environmental fate of new chemical compounds. The third purpose of this Handbook is to help in the basic understanding and development of such test methods and models.

The last explicit purpose of the Handbook is to present, in concise form, the most important properties relating to environmental chemistry and hazard assessment for the most important series of chemical compounds.

At the moment three volumes of the Handbook are planned. Volume 1 deals with the natural environment and the biogeochemical cycles therein, including some background information such as energetics and ecology. Volume 2 is concerned with reactions and processes in the environment and deals with physical factors such as transport and adsorption, and chemical, photochemical and biochemical reactions in the environment, as well as some aspects of pharmacokinetics and metabolism within organisms. Volume 3 deals with anthropogenic compounds, their chemical backgrounds, production methods and information about their use, their environmental behaviour, analytical methodology and some important aspects of their toxic effects. The material for volume 1, 2 and 3 was each more than could easily be fitted into a single volume, and for this reason, as well as for the purpose of rapid publication of available manuscripts, all three volumes are published as volume series (see Preface to Parts C and D of the Handbook). Publisher and editor hope to keep materials of the volumes one to three up to date and to extend coverage in the subject areas by publishing further parts in the future. Readers are encouraged to offer suggestions and advice as to future editions of "The Handbook of Environmental Chemistry".

Most chapters in the Handbook are written to a fairly advanced level and should be of interest to the graduate student and practising scientist. I also hope that the subject matter treated will be of interest to people outside chemistry and to scientists in industry as well as government and regulatory bodies. It would be very satisfying for me to see the books used as a basis for developing graduate courses on Environmental Chemistry.

Due to the breadth of the subject matter, it was not easy to edit this Handbook. Specialists had to be found in quite different areas of science who were willing to contribute a chapter within the prescribed schedule. It is with great satisfaction that I thank all authors for their understanding and for devoting their time to this effort. Special thanks are due to the Springer publishing house and finally I like to thank my family, students and colleagues for being so patient with me during several critical phases of preparation for the Handbook, and to some colleagues and the secretaries for technical help.

I consider it a privilege to see my chosen subject grow. My interest in Environmental Chemistry dates back to my early college days in Vienna. I received significant impulses during my postdoctoral period at the University of California and my interest slowly developed during my time with the National Research Council of Canada, before I could devote my full time to Environmental Chemistry in Amsterdam. I hope this Handbook may help deepen the interest of other scientists in this subject.

Otto Hutzinger

Preface to Parts D of the Handbook

Parts D of the three series
- The Natural Environment and the Biogeochemical Cycles (Vol. 1)
- Reactions and Processes (Vol. 2)
- Anthropogenic Compounds (Vol. 3)

are now either available or in press. During their preparation it became obvious that further parts will have to follow to present the respective subject matters in reasonably complete form.

The publisher and editor have further agreed to expand the Handbook by three new series: Air Pollution, Water Pollution and Environmental Trace Analysis.

Again, I thank all authors as well as collaborators at the Springer Publishing House for their cooperation and help. Thanks are also due to many environmental chemists and reviewers in particular for their critical comments and their positive reception of the Handbook.

Bayreuth, December 1984 O. Hutzinger

Contents

List of Contributors

Dr. H. J. M. Bowen
The University of Reading
Department of Chemistry
Whiteknights
Reading, RG6 2AD
England

Dr. T. Frevert
Lehrstuhl für Hydrologie
Universität Bayreuth
Postfach 3008
D-8580 Bayreuth
Federal Republic of Germany

Dr. G. Kratz
Mainstraße 28
D-6370 Oberursel
Federal Republic of Germany

Dr. W. D. Grant
Dr. P. E. Long
Department of Microbiology
University of Leicester
University Road
Leicester LE1 7RH
England

The Cycles of Copper, Silver and Gold

H. J. M. Bowen

The University of Reading, Department of Chemistry
Whiteknights, Reading RG6 2AD, United Kingdom

Summary

Concentrations of copper, silver and gold in rocks, and the types of minerals in which these elements occur, are briefly reviewed. Natural dissolution processes are mostly speculative. Concentrations of copper and gold in natural waters, but not those of silver, are fairly well known. The concentrations of copper in the ocean varies consistently with depth, but measurements of silver and gold are too difficult to establish a concentration profile. The main species present in natural waters are uncertain but may be humates. All three elements are enriched with respect to aluminium in aerosols, and a marine source is suspected. Concentrations in soils are known but the chemical species present are not. Marked biochemical accumulation of these elements is rare. The masses of the elements in the main environmental reservoirs can be estimated, but the fluxes between the reservoirs need further study. Recent industrial perturbations of these fluxes have caused global contamination of the atmosphere and local contamination of rivers, estuaries, lake sediments and soils. Biological effects of copper contamination of the environment are rare.

Introduction

The three metals of group IB all occur native and have probably been known to man longer than any other free metals. Their environmental chemistries differ be-

Table 1. Radii of Group IB metals/pm [1–3]

Element	Radius of M(0) [a]	Radius of M(I) [b]	Radius of M(II) [c]	Radius of M(III)
Cu	127.6	96	81	N.D.
Ag	144.5	123	N.D.	N.D.
Au	144.2	137	N.D.	85

[a] Half the shortest internuclear distance in the solid metal
[b] Octahedral radius
[c] Owing to Jahn-Teller effect, octahedral Cu(II) has 4 short and 2 long bonds

cause their most stable valency states are all different. Copper occurs mostly as Cu(II), in which form it behaves like a transition metal and has blue or green hydrated salts; only a few minerals contain Cu(I) or Cu(0). Silver occurs mostly as Ag(I), which is colourless, or as Ag(0) in the native metal and its alloys; natural Ag(II) and Ag(III) are unknown. Gold occurs either as Au(0), or as tellurides such as calaverite($AuTe_2$) where the valency cannot be assigned unambiguously; Au(I) and Au(III) may occur in natural waters.

The radii of the environmental species of these three metals are given in Table 1. Despite the fact that the radius of Cu(II) is close to those of Mg^{2+} (80 pm) and Fe^{2+} (86 pm), neither this cation nor the other metals fit easily into the magnesium silicate lattice of olivine or into aluminosilicate lattices such as feldspars.

Natural Occurrence

Copper and silver are thiophilic, while gold is siderophilic in the earth's crust [4]. The known mineral species are listed by Wedepohl [5] and can be classified as in Table 2. The thiophilic nature of Cu and Ag, and the markedly oxyphobic nature of Ag and Au, are shown in this table. In contrast, copper sulphide minerals readily weather to secondary minerals which are basic salts of a variety of oxyanions.

Copper occurs mainly as chalcopyrite($CuFeS_2$) and less often as chalcocite (Cu_2S) and covellite(CuS). The last-named mineral is better formulated as $Cu_2(I)Cu_2(II)S_2(S_2)$ since it contains copper in two valency states and both sulphide and disulphide groups. Silver occurs mainly as argentite(Ag_2S), while gold occurs mostly native, or as alloys, or as the telluride calaverite. Common rocks

Table 2. Classification of known minerals of Group IB Metals [5]

Element	Alloys	Sulphides, Selenides, Tellurides	Oxides + Hydroxides	Halides	Silicates	Basic oxysalts
Cu	8	91	11	18	11	96
Ag	6	47	0	7	0	2
Au	10	11	0	0	0	0

Table 3. Mean concentration of Cu, Ag and Au in various rocks (in ppm)

Rock	Cu	Ag	Au	References
Iron Meteorites (type Om)	167	0.02	1.1	[5]
Carbonaceous Chondrites (type CC_1)	182	0.39	0.17	[5]
Achondrites	8.1	0.033	0.014	[5]
Moon (Apollo Breccias)	15	0.0027	1.7	[5]
Moon (Apollo Basalts)	10	0.001	0.042–0.23	[5]
Dunite DTS-1	7	0.008	0.0008	[6]
Basalt	90	0.1	0.0005	[5, 7]
Granite	13	0.04	0.0018	[5, 8, 9]
Shale	39	0.07	0.0025	[5]
Limestone	5.5	0.12	0.002	[5]
Sandstone	30	0.25	0.003	[5]
Coal	15	0.05	0.0004	[10]
Manganese Nodules	2,600	6	0.0025	[11, 12]

such as basalt and granite contain relatively little of any of these minerals, which tend to occur in segregated ore bodies associated with other sulphides such as pyrite or galena. The mean contents of the three metals in various rocks are given in Table 3.

Ore deposits of copper, silver and gold are markedly localised. They tend to occur in acidic igneous rocks, especially in places where the earth's crust is buckling near plate boundaries, such as the Andes and the Rocky Mountains. Some rich localities are given in Table 4.

Gold is so insoluble and unreactive that it is found with other heavy metals in rock debris of certain rivers.

Table 4. Areas where ores of Group IB occur [13, 14]

Continent	Cu	Ag	Au
Asia	Caucasus, Urals, Japan, N. Korea, Philippines	S. Russia, Japan, Korea, Philippines, W. Java, Burma, Udaipur, India	Siberia, Japan, N. Philippines, S. India
Europe	N. Norway, S. Spain, German Democratic Republic, Bulgaria, Cyprus	S. France, N. Spain, Germany, Czechoslovakia, Yugoslavia, Sardinia, E. Greece	N. Spain, Yugoslavia
Africa	Zaire, Zambia	Algeria, Katanga, Angola, Transvaal	S. Africa
Australia	Mt Isa, Queensland	Mt Isa, Broken Hill, Tasmania	Queensland, W. Kalgoorlie, N. Tennent Creek
N. America	Arizona, Utah, Rockies, Sudbury	Rockies, Alaska, Sudbury, Newfoundland, Labrador	Rockies, Alaska, Sudbury
S. America	Andes, especially Chile and Peru	Andes, Mexico, Peru	Andes, Mexico, Columbia, Minas Gerais, Brazil

Garrels and Christ [15] have constructed a pH/E_h diagram for the CuFeOSH system which accounts for the coexistence of reduced and oxidised minerals such as $CuFeS_2$ and either Fe_2O_3 or Fe_3O_4. Their diagram would be both more precise and more complicated if free energy data were available for $CuFeS_2$, bornite (Cu_5FeS_4) and other mixed sulphides of iron and copper. Chalcopyrite appears to be the most stable ore of copper, but is converted to bornite at high activities of either oxygen or sulphur.

Cox [16] has reviewed the occurrence of copper ores and finds they present a complex picture. Primary chalcopyrite ores are thought to crystallise from cooling magmas when the residual liquid becomes saturated in sulphur, at a concentration of about 0.3% S. The three main types of chalcopyrite deposits in igneous rocks are:

1. Lens-shaped masses, often associated with silver and gold, which are formed from undersea volcanoes and later exposed, e.g. in Cyprus and at Kidd Creek, Ontario.
2. Layers in intrusive mafic rocks, associated with millerite(NiS), as at Sudbury, Ontario.
3. Haloes in porphyries surrounding intrusive granites, especially at converging plate boundaries as in Chile and Butte, Montana. Enrichments in the concentric haloes often follow the order: Granite; Cu; Pb, Zn, Ag; Mn, Ag, Au, Cu, As, Sb. Porphyritic copper deposits are localised in the earth's crust in suggestive way as shown in Fig. 1 [17].

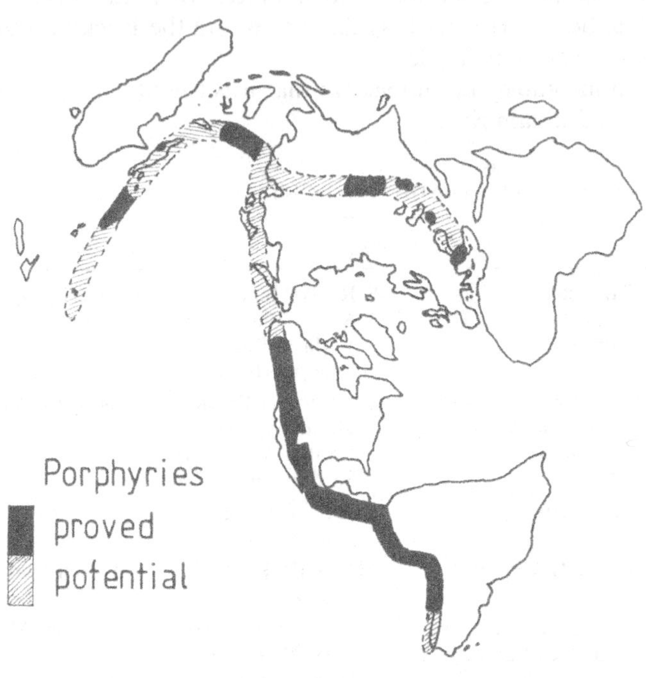

Fig. 1. Worldwide distribution of porphyry copper [17]. This polar view shows the localisation of porphyries in belts along the Pacific coast of the Americas, across central Asia and round the Pacific "Fire belt".

Secondary alterations include dissolution of the chalcopyrite to form dilute copper solutions, which may then react with pyrite to form deposits of chalcocite(Cu_2S) or form various insoluble basic salts, including silicates. The origin of substantial copper deposits in sedimentary rocks, as in Zambia, is still controversial.

Silver and gold ores are believed to form by deposition from solution, probably at high temperatures. Boyle [18] estimates that silver could diffuse 20 m/My through sandstone saturated with water at 25 °C.

Natural Dissolution Processes

The solubilities of some of the Group IB minerals are given in Table 5. The contrast between the extreme insolubility of metallic gold, copper sulphides and silver sulphide with the low solubilities of the other minerals is marked.

Copper and basic copper salts are soluble in many natural environmental acids, which do not dissolve sulphides, silver or gold. Basic copper salts are much more soluble in water containing carbon dioxide than in pure water [19]. Silver chloride is soluble in dilute ammonia. Elemental silver and gold are soluble in cyanide solutions which may be produced from certain plants [20, 21]. Although the sulphides are very insoluble in water and alkali sulphide solutions at room temperature, they become much more soluble at high temperatures, or in the presence of superheated brine. Sulphides may also dissolve under oxidising conditions, notably in acid solutions where the bacterium *Thiobacillus ferrooxidans* is present [22]. Metallic gold can be dissolved to the extent of 220 mg/L by sodium bisulphide solutions at 300 °C and 1,000 bar, possibly as $Au(SH)_2S^-$, and corresponding telluride complexes may exist [23, 24]. Another possibility is that gold may be dissolved by superhot brines in the presence of oxidants such as manganese dioxide to form $AuCl_4^-$ [5, 21], or by humic acids which can dissolve 0.3 mg Au/l [25]. Humic acids are known to be strong complexing ligands for both copper and silver in natural waters. Wedepohl [5] considers that the solubilities of copper and perhaps the other metals in natural waters are likely to be controlled by the availability of complexing ligands. Boyle [26] contests this and considers the concentrations are more likely to be determined by the kinetics of input and removal.

Table 5. Solubilities of some minerals of Group IB (M = Cu, Ag, Au) [19]

Mineral	Solubility (mg L^{-1})	Solubility (mg M L^{-1})
Ag	0.028	0.028
AgCl	1.55	1.17
Ag_2S	1.07×10^{-10}	0.93×10^{-10}
Au	4×10^{-7}	4×10^{-7}
Cu	0.166	0.166
CuS	2.44×10^{-10}	1.63×10^{-10}
Cu_2S	1.89×10^{-9}	1.51×10^{-9}
Malachite, $Cu_2(OH)_2CO_3$		0.0013

Occurrence and Speciation in Fresh Waters

Many of the older determinations of copper in waters are erroneous through contamination or inadequate sensitivity or precision [26]. Determinations of silver are probably still less reliable, judging by the poor precision of results for reference materials. Determinations of gold are rarely made and only neutron activation has the sensitivity needed. Some recent determinations are collected in Table 6. The median concentrations of soluble copper, silver and gold in fresh waters are 3, 0.3 and 0.002 μg/l respectively, but the value for silver may be found to be lower in the future. Wedepohl [5] suggests that the value for copper may be a reflection of the solubility of malachite (Table 5), but this is unlikely as the latter varies widely with [CO_2] in solution. We need more data on the chemical species present as well as on solubilities and ligand-binding before we can understand why the median values are what they are.

In the case of copper, little if any of the soluble material exists as the hydrated Cu^{2+} ion. The latter gives a potentiometric signal at a copper ion-specific electrode, and is completely absorbed by passing the water through a column of Chelex-100 [44]. Copper species in fresh waters do not have these properties, but some of them are extractable into hexanol [45] or chloroform [44], or are absorbed by Sephadex gels specific for compounds of high molar mass. At least some of these high molecular weight complexes are humates or fulvates. Two such complexes have been found in sewage effluents [44, 46], and others have been reported in N. America [47, 48], Mississippi water [30], Japan [49] and the water of Luce, Scot-

Table 6. Concentrations of dissolved Cu, Ag and Au in fresh waters

Locality	μg Cu L^{-1}	μg Ag L^{-1}	μg Au L^{-1}	Method	References
World	3 (0.2–30)	0.3 (0.01–3.5)	0.002		[10]
World	1.5	0.3	0.002		[27]
R. Amazon	1.8	0.23		AAS	[28, 29]
R. Yukon	2.0			AAS	[28]
R. Mississippi	1.2–11.5	0.24		AAS, NAA	[29, 30]
R. Mackenzie	< 1–4			AAS	[31]
USA	2.8–6.3			ASV	[32]
New Mexico	< 0.2–10	<5		AAS	[33]
Britain	7 (0.5–50)	0.1 (0.01–1)			[34]
Belgium	3.7–135	0.5–4.5		SSMS	[35]
R. Rhine	16	1.15		AAS	[36]
L. Nasser, Egypt	52	0.09	0.0009	NAA	[37]
5 Japanese rivers	1.6	0.081	0.00046	NAA	[38]
Latvia		0.25		AES	[39]
R. Capvern, France		0.28–0.6			[40]
10 US rivers		0.3 (0.23–0.38)		NAA	[29]
R. Rhine, France		0.38		NAA	[29]
New Zealand (hot)		0.6	0.04	AES	[41]
R. Thames, UK			0.0011	NAA	[42]
Auriferous rivers			0.002–0.13	NAA	[43]

Notes: AAS = Atomic Absorption Spectrometry, AES = Atomic Emission Spectrometry, NAA = Neutron Activation, SSMS = Spark Source Mass Spectrometry, ASV = Anodic stripping Voltammetry

land [50]. Humate and fulvate complexes of copper have stability constants in the range 10^4–10^{15} depending on pH etc [51] but are usually around 10^8–10^{10} [52]; they are characteristically destroyed by acidification to pH < 4 or by prolonged exposure to ultraviolet radiation. Evidence for other complexes involving hydroxo or carbonato-ligands is largely speculative. It is known that ionic copper(II) added to small Canadian lakes is lost with first order kinetics and a half-life of 1–2 days, but the ligands responsible have not been identified [53]. In polluted waters, the excess copper appears in the suspended matter or sediments rather than the soluble fraction. Copper(II) is strongly absorbed by peat, by hydrous ferric and manganese oxides and by clay minerals such as montmorillonite [5]. In sewage treatment, soluble copper is 33–50% removed by activated aerobic sludges and 97% removed by anaerobic sludges which contain sulphides [54].

Virtually no information is available on the speciation of silver and gold in fresh waters, but it has been shown that both silver ions and silver chloride react with humates to form soluble complexes [55]. Humates are able to dissolve gold powder at pH 3.5, and are also able to reduce tetrachloraurates to colloidal gold in solution [25]. Both silver and gold co-precipitate with antimony(III) sulphide deposited from hot spring waters in New Zealand [41].

When $^{110m}Ag^+$ in distilled water is treated with illite, montmorillonite, hydrous ferric or manganese oxides, 60–90% is absorbed onto the solid phase [29]. Murray and Meinke [56] added $^{110m}Ag^+$ to fresh water and found that about 65% was absorbed by suspended sediments between pH 3 and 7, increasing to 85% at pH 9. These workers also attempted to estimate how much tracer silver is desorbed when fresh and salt waters meet. In estuaries, colloidal humates tend to flocculate at high salt concentrations which can mobilise copper [57–9] and silver. In the Satilla estuary, Windom et al. [60] found no relation between copper concentration and salinity. Divergent results have been reported [61]. However, copper is known to accumulate in sediments off California around major sewage outfall pipes, although it is estimated that 90% of the copper in the sewage escapes into the ocean [62]. Estuarine sediments accumulate copper from polluted rivers such as the Rhine [63] and Severn [64], and silver and gold in the R. Thames [65].

Occurrence and Speciation in Sea Water

Earlier work on the concentration of copper in the open ocean has been rendered obsolete by the finding of Boyle et al. [66] that there is a well-defined variation of concentration with depth, which has been confirmed by others [67, 68]. Results of some recent determinations are given in Table 7, and earlier values have been reviewed [69, 70].

Surface sea water near coasts is often contaminated with industrial copper and has been found to contain between 0.3 and 3.8 μg Cu L^{-1} [75]. In uncontaminated sea water, surface concentrations are always lower than in deep water. A typical depth profile of copper concentration in ocean water is given in Fig. 2. The concentration is highest in deep water and lowest at about 500 m depth, with a small rise often found in the upper 100 m. The current interpretation is that two com-

Table 7. Concentration of copper in sea water in μg L^{-1}

Ocean	Surface [Cu]	[Cu] at 5 km depth	Reference
Pacific	0.1–0.22	0.37–0.68	[66]
Pacific	0.034–0.1	0.22–0.33	[68]
Indian	0.23		[71]
Indian	0.59 (0.23–1.02)		[72]
Atlantic	0.09	0.69	[67]
Atlantic	0.2		[73]
Atlantic		0.003 [a]	[74]

[a] in particulate matter only, constituting 15 μg L^{-1}

peting processes occur in the deep ocean, copper release from sedimenting or sedimented particles and scavenging of copper by other particles. Brewer at al. [76] have shown that the copper content of oceanic particles near Barbados increases from 29 ppm at 988 m to 129 ppm at 5,086 m, presumably by scavenging. In addition aerial input of copper is large enough to cause local rises in surface concentration. Particulates from the North Atlantic have been shown to include tenorite(CuO), malachite and brass, of which the last must be from industrial sources [77]. The molar mass range of copper species in marine sediments is 700 to 10,000 [78].

The actual forms of copper in sea water could include ionic copper, inorganic species, copper humates, colloidal copper as well as copper in particulates and liv-

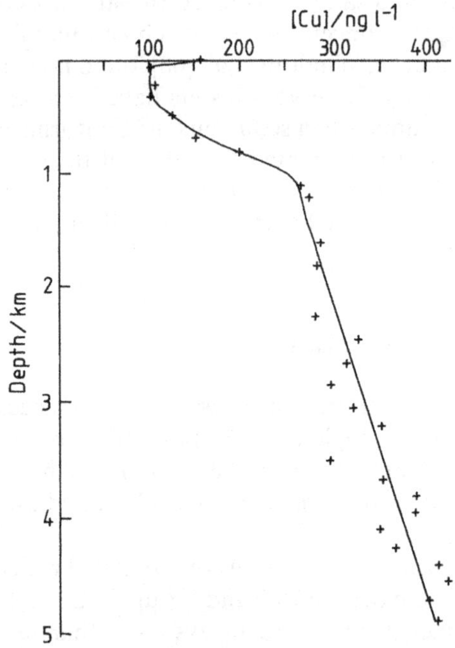

Fig. 2. Depth profile of copper in the Pacific Ocean at 33° 66′N, 139° 34′ W: redrawn from [66]. The reduced concentration in the upper kilometre compared with the deep ocean is characteristic.

Table 8. Surface concentrations of silver and gold in sea water in ng/l

Location	Ag	Au	Method	Reference
All oceans	150–2,900		Several	[18]
Pacific	8		NAA	[84]
Pacific		1.1 (0.5–8)		[85]
Pacific		0.9	NAA	[86]
Atlantic	4.9 (0.4–20)		NAA	[87]
Atlantic	40–2,700	0.5–7.8	NAA	[88]
Atlantic		4	FA	[89]
Atlantic		8	AES	[90]
Atlantic		14 (10–506)	NAA	[91]
Atlantic		9	AES	[92]
Atlantic	0.14[a]	0.02[a]	NAA	[74]
North Sea	100	0.21	NAA	[93]
Irish Sea	26–40		NAA	[94]
Aegean Sea	176		AAS	[95]
Black Sea	89 (1–370)		AAS	[96]

[a] particulates only; FA = Fire Assay

ing organisms. Ionic copper is thought to be very low as little is absorbed by Chelex-100 resin; also 10–60% of copper in sea water is extractable into chloroform [44, 79]. No two authors agree in calculating the inorganic speciation of copper in sea water, although $CuCO_3$ may be important [80]. Organic complexes of copper in sea water have higher stability constants than CuEDTA [80–82], and it has been suggested that humates of molar mass > 50,000 are responsible for complexing copper, silver and gold [83]. Soluble copper could be lost as insoluble sulphides at the mid-ocean ridges, where sea water contacts hot basalt. In anoxic waters containing free SH^-, such as the Black Sea deeps, copper sulphide precipitates and the concentration of soluble copper falls below measurable limits [26].

There is poor agreement between different determinations of silver in sea water (Table 8). Schutz and Turekian [97] list many determinations which do not differ significantly with depth, with a mean of about 300 ng Ag L^{-1}. Robertson [87] gives a much lower mean of 5 ng Ag/l, and also finds no consistent variation with depth. Silver is readily lost by adsorption onto glass or plastic containers during storage [98]. The determination of silver in sea water without contamination or loss is difficult, and until better data are available it is pointless to discuss whether silver has a concentration profile like copper. The chemical species present are unknown, calculations favouring di-, tri-, and tetrachloroargentate anions [99] and experiments suggesting humate complexes [83]. Silver is probably precipitated from the ocean as the sulphide in the same regions as copper.

Most workers agree with the classic finding of Haber [89] that the oceanic concentration of gold is about 4 ng Au/l with higher values in some coastal waters. Heuss and Lieser [93] find an order of magnitude less gold in the North Sea, with 70% particulate, in disagreement with the low values of particulate gold in the ocean [74] (Table 8). Calculations making no allowance for colloidal or organically bound gold predict that the main species should be $AuCl_2^-$ [99]. Gold atoms formed by reduction are readily lost by adsorption on surfaces [91], and tracer studies show that gold is more likely to be lost by sorption from acid than from alkaline solutions [100].

Present evidence suggests that copper and silver, but not gold, are strongly scavenged from the ocean by manganese nodules, where their concentrations are two orders of magnitude higher than those found in common marine sediments (Table 3) [11, 12, 16]. Copper and silver are also enriched in the hot brines found in deep waters near cracks in the earth's crust, such as the Red Sea and Salton Sea, California. These brines contain up to 5 ppm Cu and 2 ppm Ag, probably as chlorocomplexes [5].

Volatile Compounds and Atmospheric Cycling

Copper, silver, and gold have boiling points of 2,595 °C, 2,212 °C, and 2,296 °C respectively and are not normally thought to be volatile elements. Nevertheless fumarole gases are known to contain copper(I) chloride [101, 102] and copper chlorides and sulphides occur as minerals around gas vents on Vesuvius and Alaskan volcanoes. Krauskopf [103] showed that copper sulphides are involatile but that copper(I) chloride could volatilise above 600 °C in the absence of hydrogen sulphide. The vapour pressure of silver chloride is about 10^{-6} bar at 600 °C, so that it may also volatilise slowly at high temperatures. Other volatile compounds of the metals are unlikely to occur in nature (e.g. acetonylacetates), or decompose when heated (e.g. alkyls) or are destroyed by water (e.g. copper formate). Gold appears to be enriched round a few calderas and volcanic vents. Its most volatile compounds are Au_2Cl_6 (subl. 265 °C) and AuF_3 (subl. 300 °C *in vacuo*), and the former may occur in traces in volcanic gases. Herron [104] claims that major volcanic eruptions such as that of Krakatoa in 1883 gave rise to increased deposition of copper and other metals in ice cores in Greenland. Table 9 gives some data for the three elements and scandium in volcanic aerosols. This data suggests that all three elements are concentrated relative to the non-volatile scandium in the hot vents of Mount Etna, but not in the other three volcanoes studied. Etna is estimated to emit 365 tonnes Cu, 3.3 tonnes Ag and 0.88 tonnes Au every year [109].

Table 9. Cu, Ag and Au in volcanic aerosols in ng/m^3 gas

Source	Cu	Ag	Au	Sc	Reference
Showashinzan, Japan	3–50[a]	<0.1–4[a]			[105]
Kilauea, Hawaii	200				[106]
Augustine, Alaska	15–516		0.00087–0.05	0.144–29	[107]
Mt St. Helens, USA		1.9		39	[108]
Mt Etna, Italy	3,200	0.03	0.008	0.0018	[109]
Mt Etna, Italy hot vents	243,000	0.69	0.05	0.0016	[109]

[a] In µg/l condensed water

Table 10. Atmospheric Cu, Ag, Au, Al and Sc in ng/m^3 air from various localities

Source	Ref.	Class	Cu	Ag	Au	Al	Sc	10^3 Cu/Al	Cu/Sc	10^3 Ag/Al	Ag/Sc	10^6 Au/Al	10^3 Au/Sc
Atlantic Ocean	[74]	N, U	0.79	0.027	0.003	160	0.035	5	23	0.17	77	19	86
Jungfraujoch	[110]	N, U	0.88	0.01	0.003	51	0.0077	17	114	0.2	1.3	59	390
Novaya Zemlya	[111]	N, U	0.61			1.65		370					
Oahu, Hawaii	[112]	N, U	2			6.7		300					
Chacaltaya, Bolivia	[113]	S, U	0.5	0.17	0.001	720	0.12	0.7	4.2	0.24	1.4	1.4	8.3
Antarctic, winter	[114]	S, U	0.079	0.001	0.00009	0.3	0.000037	260	2140	3.3	2.7	31	2500
Antarctic, summer	[114]	S, U	0.059	0.003	0.00008	0.83	0.00018	71	330	3.6	17	9.6	440
U.S.A. mean	[115]	N, P	500	1	0.06	3000	1	170	500	0.33	1	20	60
Buffalo, N.Y.	[116]	N. P.	70			2200		32					
San Francisco, CA	[117,118]	N, P	34		0.27	863	1.4	39	24			310	190
La Jolla, CA	[118]	N, P	16	0.05		900		18		0.056			
Ensenada, Mexico	[118]	N, P	20	0.03		3380		5.9		0.0089			
Niles, MI	[119]	N, P	290			1200	1.2	240	240				
Chilton, U.K.	[120]	N, P	9.8	0.185	0.0037	228	0.057	43	170	0.81	3.3	16	65
Belgium	[121]	N, P	5			104	0.008	48	620				
Japan mean	[122]	N, P	69	1.8		834	0.112	83	620	2.2	16		
Pretoria, S.A.	[123]	S, P	60			1720	0.39	35	150				
Mean Soil	[10]		30^d	0.05^d	0.001^d	71000^d	7^d	0.42	4.3	0.0007	0.007	0.014	0.14

Notes: N = North hemisphere, S = South hemisphere, U = unpolluted, P = polluted, ^{d} = ppm

Some values for the metals associated with particles in ambient air are given in Table 10, together with values for the non-volatile aluminium and scandium where available. Despite the wide variation in absolute values, the ratios Cu/Al etc. are much less variable and support the contention that aerosol composition is fairly uniform over the planet, though different samples of air contain very different amounts of aerosol. If we compare the ratios in aerosols with those in average soil, as suggested by Rahn [124], we find large enrichment factors as follows:

Cu/Al	12– 880
Cu/Sc	5– 500
Ag/Al	240– 5,100
Ag/Sc	140– 2,400
Au/Al	690– 22,000
Au/Sc	460– 18,000

The reasons for these high enrichment factors are not clear at present, but marine sources may be important. Cattell and Scott [125] measured copper in the air above the Indian Ocean near Tasmania and found values between 0.3 and 190 ng Cu/m^3. The copper content of the air declined with altitude and below 2 km was well correlated with magnesium and sulphate originating from sea water. However the enrichment factor $E = [Cu]_{air}[Mg]_{sea}/[Cu]_{sea}[Mg]_{air}$ had a value of about 20,000, implying that the ocean is an important source of atmospheric copper.

Since particulate matter has a lifetime of about 2–4 weeks in the atmosphere, it is not possible to investigate historical changes in composition by direct anal-

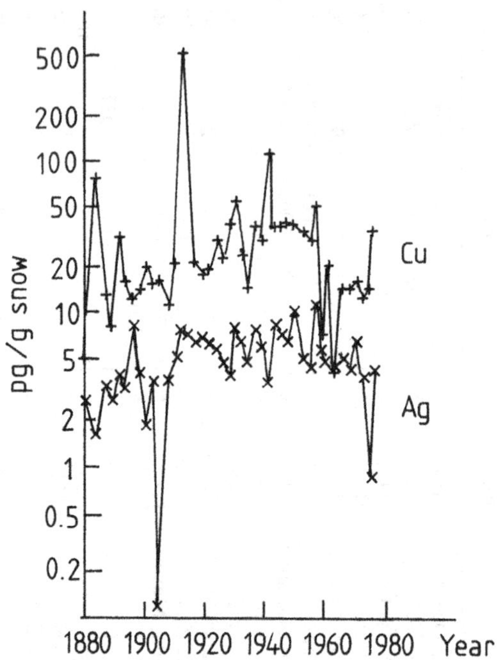

Fig. 3. Copper and silver in Antarctic snow, drawn from data in [126]. The seemingly random variations are not due to industrial pollution but may reflect volcanic inputs.

ysis, but indirect studies can be made by analysing layers in permanent snowfields or lake sediments. Measurements of copper and silver in antarctic snow are shown in Fig. 3 [126]. The variations over the last century are seemingly random and may be due to variations in volcanic inputs, but are not related to marine salt or terrestrial soil sources [127]. Mean values of deposition over this period are 38 pg Cu and 5.2 pg Ag per g snow, with Cu/Al and Ag/Al ratios of 0.033 and 0.0045 and enrichment factors over soil of 78 and 6,400 respectively.

Behaviour of the Metals in Soils

Selected reviews of the concentrations of the three metals in soils are summarised in Table 11. The values for silver and gold are uncertain.

Table 11. Concentrations of Cu, Ag and Au in soils in mg/kg

Cu	Ag	Au	Reference
30 (2–250)	0.05 (0.01–8)	0.001 (−0.02)	[10]
26 (1–389)	0.4	0.001?	[128]
(3–29)	(1–2.4)	(0.004–0.06)	[129]
28 (2–660)			[130]

Copper in soils has been reviewed by Reaves and Berrow [131], and the general behaviour of heavy metals in soils by Harmsen [132] and Tyler [133]. Cycling of copper has been investigated in oak forest [134] and salt marsh ecosystems [135]. In the former case the determinations were at the limits of their sensitivity, and the figures indicate that copper inputs to soil are about 1.5 mg/m^2 y from rainfall and 2.8 mg/m^2 y from forest litter. In the latter case tidal import and export of copper approximately balance, but the authors emphasize the involvemt of a surface microlayer of about 1.5 μm thickness on the seawater in the exchange of copper. Where inputs and outputs from soils can be measured, it is possible to calculate a residence time for an element in soil. This has been attempted for copper in soils from the Thames Valley, England, where the residence time is about 860 years. In uncontaminated, uncultivated soils the rate of biological cycling between soil and vegetation greatly exceeds the rate of geological cycling measured from inputs and outputs [10].

The chemical species present in soils are uncertain. In young soils copper (and silver) occur as sulphide inclusions in many types of rock. After weathering or microbiological attack copper ions are released and are then absorbed by clay minerals, hydrous iron and manganese oxides and humates. Hydrous manganese oxides and humates are believed to be the main absorbents [136]. Copper porphyrin complexes have also been found in soils [137]. Silver is also strongly absorbed by hydrous manganese oxides [138] and by humates. In some types of soils copper, silver and gold accumulate in the upper horizons when these are rich in organic matter [128].

Biogeochemical Behaviour

Marked concentration of copper, silver, and gold by organisms is uncommon. Copper differs from the other two metals both in its greater abundance and in the fact that it is essential for all groups of organisms. Its essentiality arises from the existence of at least thirty metalloenzymes which need copper to function. Virtually all these enzymes are redox catalysts and a few are dioxygen carriers; they contain copper bound in several different ways, whose nature is currently being elucidated. Copper(II) is coupled to iron(III) in cytochromes and to zinc(II) in some hyperoxide dismutases [10, 139]. Copper reacts readily *in vitro* with aminoacids and proteins, so it is not surprising that it forms metalloenzymes. It also reacts with some antibiotics and acts as the central atom in one natural porphyrin, the red pigment turacin. The relatively high copper content in enzymes could account for the observation that mitochondria are richer in the element than are nuclei, but such findings are subject to artefacts due to the isolation procedures used. Organisms exposed to high copper concentrations can form intracellular solid inclusions containing copper and sulphur, which may be a mode of detoxification. Such inclusions have been noted in algae [140] and in barnacles [141]. Silver and gold salts react with skin proteins to form deposits of colloidal metal which are black and purple respectively.

The uptake of the elements by terrestrial plants from soils is not well understood. Plant contents in μg/g dry matter are of the same order as soil contents, which suggests that no concentration takes place. However, copper is believed to be higher in plant sap than in soil solutions, but whether metabolic energy is needed to effect this is uncertain. The chemical form of copper in plant sap is unknown, but is probably not Cu^{2+}. The free ions Cu^{2+}, Ag^{+} and $AuCl_4^-$ are notably toxic towards algae, fungi, higher plants and to a lesser extent bacteria, as shown in Table 12. Cu(II) is used to control algal growth in lakes and fungi in vineyards. The toxicity may be due to the reactions with free -SH groups in cell membranes, or to binding to heterocyclic bases in DNA which alters the molecular conformation [148].

In regions of the world where copper is high in soils, three effects are observed:

a) Elimination of many susceptible species, including most woody plants.
b) Survival of a few tolerant species, some being copper-indicators.
c) Accumulation of copper in the tissues of some tolerant species.

a) Has been noticed in Norway, where copper-rich sites have no forest cover of pine, spruce or birch. b) Is marked in Katanga [149], but has been seen in many other places such as Wales [150]. c) Accumulator species for copper include fungi (*Agaricus* spp and *Boletus edulis*) [151, 152], lichens (*Acarospora sinopica*) [149, 153], bryophytes (*Merceya and Mielichhoferia* spp) [149] and many seed plants [154, 155]. It is not clear whether this specialised vegetation can cause accumulation of copper in the upper layers of the soil. Effects a) and b) are not known for silver or gold, but both metals are accumulated by the fungi which accumulate copper [151]. *Bacillus megaterium* is a bacterium which is able to dissolve metallic gold, possibly by excreting aminoacids as ligands to form anionic complexes [156, 157]. Gold is concentrated by seed plants which contain cyanogenic glucosides.

Table 12. Toxicity of Cu^{2+} and Ag^{+} to organisms grown in nutrient solutions

Group	Organism	Metal	Toxicity/mg L^{-1}	Reference
Bacteria		Cu	6.3	[142]
Green alga	*Chlorella*	Cu	0.07	[143]
Brown alga	*Macrocystis*	Cu	0.1	[144]
Bog moss	*Sphagnum*	Cu	0.64	[145]
Seed plant	*Oryza* (Rice)	Cu	0.1	[146]
Seed plants		Cu	0.5	[10]
Bacteria		Ag	3.8	[142]
Seed plant	*Phaseolus*	Ag	0.1	[147]

These plants release cyanides to the soil, which react with gold and perhaps also silver, to form a mobile cyano-complex [20, 158]. Hence some plants can be used in biogeochemical prospecting for gold [159].

The biological importance of copper in the ocean has been well reviewed by Lewis and Cave [70]. Almost all organisms accumulate copper and silver, but apparently not gold, from sea water. Accumulation factors of up to 180,000 for copper by benthic algae [160] and 51,000 for silver by diatoms [161] have been reported. Particularly high concentrations of copper are found in the blood and hepatopancreas of marine crustacea and mollusca, and in the jaws of polychaete worms [162]. The blood of these organisms contains the copper protein haemocyanin which functions in the transport of dioxygen. When the organisms die, the copper appears to be returned to solution in sea water, and no high concentrations are found in the upper layers of marine sediments. However, faecal pellets may constitute a major pathway for removal of copper and silver from sea water. The zooplanktonic euphausiid *Meganyctiphanes* contains 48 ppm Cu and 0.71 ppm Ag, but its faecal pellets contain 226 ppm Cu and 2.1 ppm Ag [163].

The uptake and excretion of copper, silver, and gold by humans has been briefly reviewed [164]. The daily diet contains on average about 2.5 mg Cu, 0.4–40 μg Ag and 10–1,000 ng Au, and the daily urinary excretion is about 0.05 mg Cu, 0.6–10 μg Ag and 1–80 ng Au. From these data and the body content of copper, one can estimate the residence time of copper in man to be about 4 years. None of the elements is markedly concentrated in any organ, though silver and gold may be enriched in hair, probably through external contamination. After absorption through the gut wall, the elements associate with albumins in the bloodstream, and are removed partly by the liver and partly by the sulphur-rich protein metallothionein in the kidney [165, 167].

Cycling of Copper, Silver, and Gold and their Industrial Perturbations

The study of elemental cycles is simplified in model systems by distinguishing a number of environmental reservoirs, each containing a mass m_i of metal. Attempts are made of estimate the pre-industrial fluxes between reservoirs, f_{ij} or $(dm/dt)_{ij}$, which can be used to calculate residence times in the various reservoirs from the relation $T_i = m_i/f_i$ where f_i is the total input (or output) flux.

Industrial perturbations are changes in the natural fluxes, $(d^2m/dt^2)_{ij}$. As will be seen below, only approximate values can be assigned to masses of metals in reservoirs, fluxes and industrial perturbations.

Masses of Metals in Environmental Reservoirs

Here we use a simple model consisting of only five reservoirs; Air, Ocean, Lithosphere, Ocean sediments and Soil + Vegetation. Estimates of the masses of copper, silver, and gold in these five global reservoirs are given in Table 13.

Apart from the atmosphere, the estimates of the masses of copper in various reservoirs agree fairly well with those of Nriagu [168], who used somewhat different assumptions. The lithosphere is the dominant reservoir for all three metals. Using present technology, world ore reserves are estimated to be 2×10^{20} g copper and 2×10^{17} g silver [171], which are both high proportions of the estimated content of the lithosphere.

Table 13. Estimated masses of Cu, Ag and Au in five environmental reservoirs

Reservoir	Index No.	m(Cu)/g[a]	m(Cu)/g[b]	m(Ag)/g[b]	m(Au)/g[b]
Atmosphere	1	2.6×10^9	2.9×10^8	8.2×10^6	3.7×10^5
Hydrosphere	2	2.1×10^{14}	3.5×10^{14}	7.0×10^{12}	5.6×10^{12}
Lithosphere	3	1.47×10^{21}	1.3×10^{21}	1.7×10^{18}	2.6×10^{16}
Ocean sediments	4		6.9×10^{19}	2.0×10^{17}	5.2×10^{15}
Soil + vegetation	5	6.8×10^{15}	6.0×10^{15}	1.0×10^{13}	2.0×10^{11}

[a] Nriagu [168]
[b] Bowen [10] & this work

Notes on Table 13:

1. The atmosphere reduced to stp would contain 4.1×10^{18} m^3. Table 10 shows that its metallic content is variable, and the values used here are based on the composition of the least contaminated site on Earth, at the South Pole. Preindustrial values may well have been an order of magnitude higher.
2. The ocean weighs 1.39×10^{24} g. Tables 7 and 8 show that concentrations of copper, silver, and gold are not well established and vary with depth; here they are assumed to be 250,5 and 4 ng/kg respectively.
3. The mass of continental crust, or lithosphere, is uncertain, but is taken here to be 2.4×10^{25} g igneous rock $+3.2 \times 10^{24}$ g sedimentary rock, with concentrations of the metals taken from Table 3.
4. The mass of submarine sediments is controversial, but Garrels and Mackenzie [169] give a value of 2.1×10^{24} g, of which 42% is continental shelf, 36% oceanic slope and 22% deep ocean. These sediments are considered to be 74% shale, 14% limestone and 11% sandstone, whose metal contents are given in Table 3.
5. The mass of world soils is taken to be 2×10^{20} g [170], and the concentration of metals in soils is taken from Table 11. The total content of metals in vegetation is negligible compared to that in soils.

Table 14. Estimates of nine environmental fluxes of Cu, Ag and Au before industrialisation

Source	Sink	Index nos.	$f(Cu)/Gg\,y^{-1}$	$f(Ag)/Mg\,y^{-1}$	$f(Au)/Mg\,y^{-1}$
Air	Ocean	1,2	13	97	45
Air	Soil	1,5	6	45	21
Ocean	Air	2,1	3.3	82	66
Ocean	Sediments	2,4	833	14,200	99
Lithosphere	Air	3,1	9.3	10	0.002
Lithosphere	Soil	3,5	824	12,400	99
Sediments	Lithosphere	4,3	833	12,400	99
Soil	Air	5,1	10	50	0.33
Soil	Ocean	5,2	820	12,400	120

Notes: Dynamic equilibrium is assumed, which implies the following relations:
$f_{12}+f_{15}=f_{21}+f_{31}+f_{51}$
$f_{21}+f_{24}=f_{12}+f_{52}$
$f_{31}+f_{35}=f_{43}+f_{24}$
$f_{51}+f_{52}=f_{15}+f_{35}$

Fluxes Between Reservoirs

There are twenty possible fluxes between the five reservoirs, but eleven of these are zero or negligible. Estimates of the preindustrial fluxes of copper, silver, and gold are given in Table 14.

Since fall-out over land and sea should be proportional to area, $148f_{12}=316f_{15}$. The dominant flux f_{52}, representing the transfer from soil to the ocean by rivers is calculated as $3.7\times10^{19}[M]_{river}+1.83\times10^{16}[M]_{shale}$, assuming that rivers carry to the sea each year 3.7×10^{19} g water and 1.83×10^{16} g sediment [10, 172]. f_{12} and f_{15} are calculated values of fall-out. f_{21} (Cu) is taken from Lantzy and Mackenzie [172], and used to estimate values for Ag and Au by assuming that surface ocean water contains 200,5 and 4 ng/kg of Cu, Ag, and Au respectively. f_{24} is calculated, and is equal to f_{43}. f_{31}, representing volcanic emissions, is taken from [172] for Cu and Ag, and the corresponding value for Au is estimated from limited data in Table 9. Values for f_{35}, the weathering of rock to soil, are calculated and disagree with earlier estimates by Bowen [10]. f_{51} values for Cu and Ag are taken from [172], and the value for Au is interpolated using the ratio of Au/Cu in soil from Table 11.

Our estimate for the important flux f_{52} (Cu) is of the same order as those of Mackenzie et al. [173] and Li [174], but much lower than that of Nriagu [168]; it is difficult to tell how much of this flux is influenced by industrialisation. The remarkably high flux f_{21} from ocean surface to atmosphere for gold needs confirmation.

Residence Times

Residence times for the metals in the reservoirs can be calculated from the estimates in Tables 13 and 14, but can be taken as orders of magnitude only. In the atmosphere, residence times of copper, silver, and gold are 6, 21, and 2 days respectively. In the ocean, they are 420, 560, and 36,000 years, while in soils they are 7,000, 800, and 1,700 years respectively. All the metals have residence times of the order of 10^7 years in ocean sediments and 10^8–10^9 years in the lithosphere.

Industrial Perturbations

Human activities have modified natural fluxes of metals by mining, transporting and using rich ore deposits. The annual amounts of copper, silver, and gold mined during the last hundred years are shown in Fig. 4–6, and the current annual production is 8.3×10^{12} g copper, 1.15×10^{10} g silver and 1.3×10^{9} g gold [175]. 99.4% of this copper, 99.9% of the gold but only 59% of the silver are retained for use as metal or alloys, whose rates of erosion and corrosion are uncertain. Silver is exceptional because of its extensive use in photography. Environmental

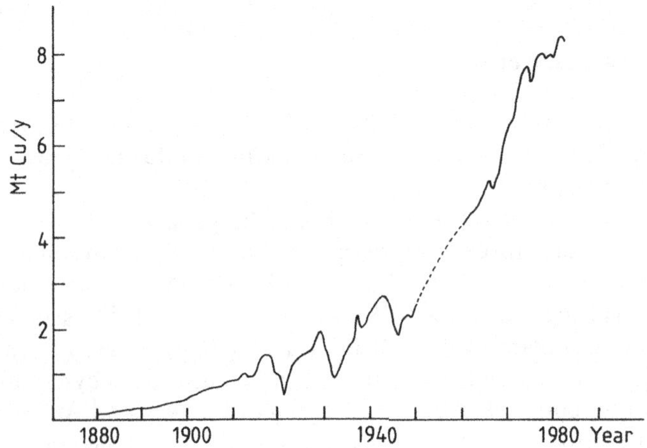

Fig. 4. World production of copper since 1880. The general trend is an exponential increase, with minor peaks caused by world wars and troughs during slumps.

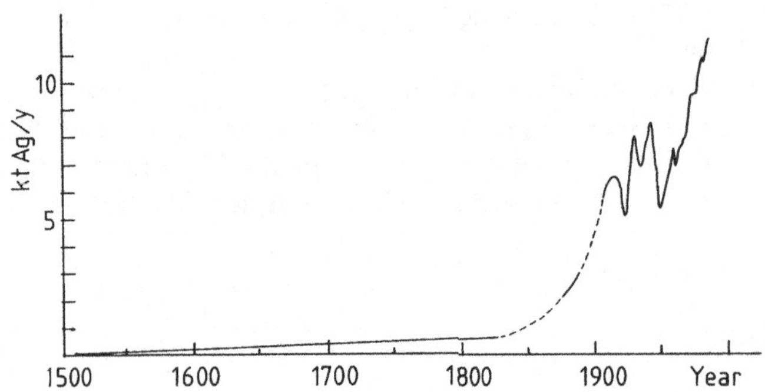

Fig. 5. World production of silver since 1500. The general trend is exponential, with fluctuations which may have been caused by the replacement of silver as a coinage metal in the present century.

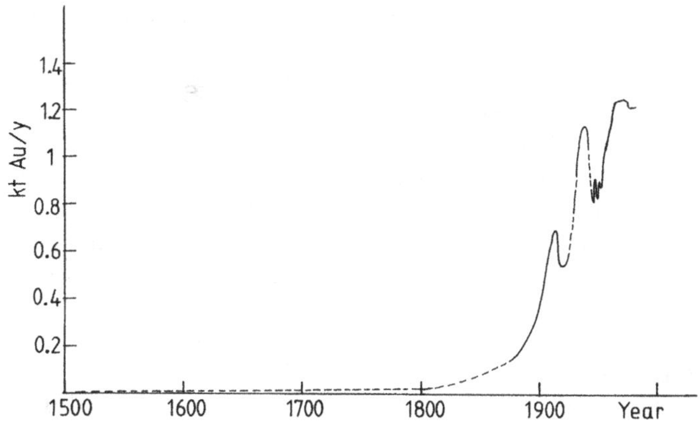

Fig. 6. World production of gold since 1500. Note the close parallelism to the curve for silver in Fig. 5.

fluxes directly affected by human activities are f_{31}, f_{52}, and f_{35}. Indirect perturbations of other fluxes, such as f_{24}, may occur.

Perturbation of Flux f_{31}, from Lithosphere to Atmosphere

This arises from smokes produced by burning fossil fuels and other industrial processes. Coal-fired power plants produce fly ash and small particles which are enriched in copper with respect to bottom ash; typical fly ash contains 300–1,000 ppm Cu [176]. The amount of carbon burnt as coal in the 1970's was 2.24×10^{15} g/y [10], but the amount of fly ash emitted is uncertain; recent estimates are 5% [177], 10% [10] and 90% [172]. Other industrial dusts are estimated to amount to 2×10^{14} g/y [172], but their composition is not well established, nor is it known what percentage of this dust falls out locally and how much contributes to the global aerosol. The burning of town refuse and sewage sludge may be an important source; in Japan particles released from these two wastes were found to contain on average 3,800 ppm Cu + 150 ppm Ag and 4,400 ppm Cu + 740 ppm Ag respectively [122]. All estimates agree that industrial perturbations to flux f_{31} are comparable to or much greater than natural fluxes. As a result, atmospheric copper rises from 5–50 ng/m^3 in rural areas to 30–200 ng/m^3 in urban areas, or even higher near smelters [168]; see Table 10. The evidence for the enrichment of silver and gold in urban air is more tenuous, but is probably real.

Perturbation of Flux f_{52}, from Soil to the Ocean

Emissions from the land surface to rivers and the ocean involve discharge of sewage and other wastes, the corrosion and erosion of copper pipes, wires, and roofs, and the loss of copper from algicides and marine paints. Quantitative estimates of these emissions are not easily made, but their orders of magnitude may be obtained as follows:

Sewage: In industrial countries sewage sludge production is 180 g dry matter/man day [178]. Assuming a world population of 4.4×10^9, 25% industrialised, sludge production is 7.2×10^{13} g/y. In Britain, representing 5% of the world's industrialised population, the median composition of sludge is 800 ppm Cu, 20 ppm Ag and <1 ppm Au [179]; 20% of sludge is discharged to sea, 40% to soils and 40% is buried as landfill. Using these figures we calculate annual emissions to be 1.2×10^{10} g Cu, 2.9×10^8 g Ag, and $<1.4 \times 10^7$ g Au.

Erosion and Corrosion: Assuming a loss of 1% of the annual production of Cu and Ag, and 0.1% for Au, the annual emissions would be 8.3×10^{10} g Cu, 1.1×10^8 g Ag, and 1.3×10^6 g Au.

Miscellaneous: 2.1×10^{10} g Cu from marine paints [168]; 5.7×10^9 g Ag from use in photography [180].

These perturbations may amount to 50% or more of the natural fluxes. However, as the fluxes in Table 14 are based on recent analyses of river waters, they presumably include the industrial contributions.

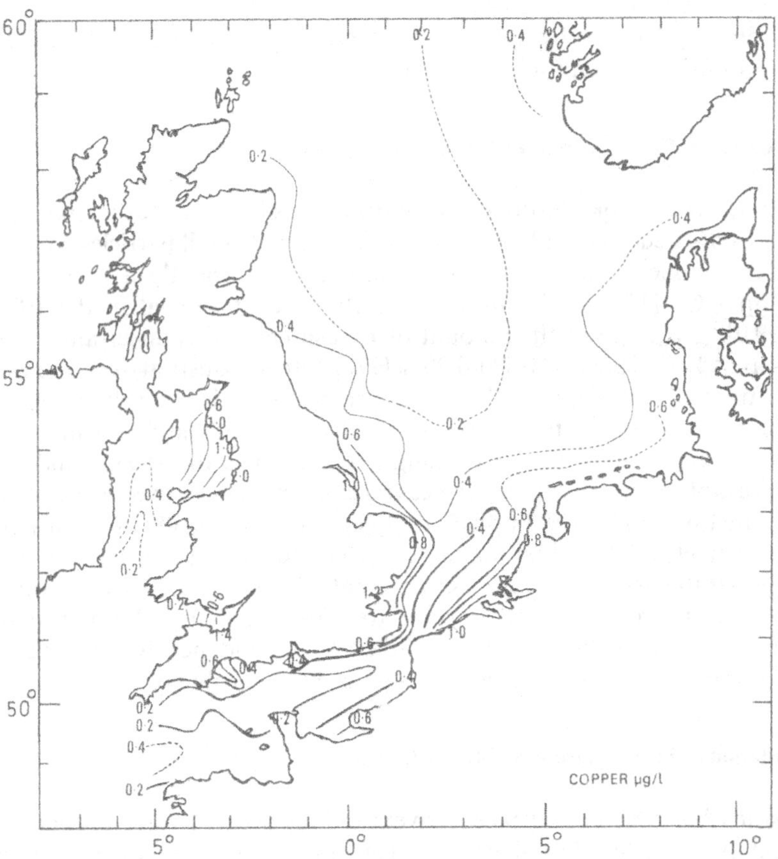

Fig. 7. Contours of copper concentration is seas around N.W. Europe [183]. High concentrations are associated with industrial pollution around the estuaries of the Rhine, Thames, Severn, Mersey, and Humber.

Contamination of rivers by copper is common in industrial areas, but biological disasters resulting from copper contamination alone are rare [181, 182]. Contamination of the sea around the margins of industrialised continents has been reported [70, 183, 184], but again biological effects of this contamination are rare and confined to a few small estuaries [185]. Figure 7 shows recent copper concentrations in the North Sea, which are notably high near the mouths of the rivers Rhine, Thames, and Humber, as well as the Severn and Mersey on the west side of Britain. Note that the additional industrial input to the air (f_{31}) contributes to the contamination of this Sea as well as to the ocean [186].

Perturbations of Flux f_{35} from Lithosphere to Soil

Human transfer of metals to the soil includes the application of fertilizers, fungicides such as Bordeaux mixture, and sewage sludges. Quantitative estimates of these transfers can be made as follows:

Fertilisers: Assume that all trace metals are added from phosphatic fertilisers, which contain 13.8% P, 100 ppm Cu, 0.05 ppm Ag, and 0.002 ppm Au [10]. If 1.12×10^{13} g P is applied to soil each year, the soils will also receive 8.1×10^9 g Cu, 4.1×10^6 g Ag, and 1.6×10^5 g Au.

Fungicides: 7×10^{10} g Cu are applied locally each year [10].

Sewage Sludge: Using the assumptions made in the last section, soils will receive 2.3×10^{10} g Cu/y, 5.8×10^8 g Ag/y, and $<3 \times 10^7$ g Au/y.

In addition a substantial proportion of the industrial emissions f_{31} will return to the soil as a perturbation of flux f_{15}. Nriagu [168] estimates that soils receive 6.7×10^{10} g Cu/y in this way.

Although these perturbations appear small compared to the natural fluxes estimated in Table 14, they can cause local pollution effects, and can sometimes cause measurable increases in metal concentrations in the surface layers of undisturbed soils. Copper toxicity has occasionally been reported in orchards and vineyards which have been repeatedly sprayed with fungicides [187]. After treatment with sewage sludge, soils can show an order of magnitude increase in total copper and gold, while total silver may increase by two orders of magnitude [129]. High soil inputs of copper have often been reported near smelters [188], and large coal-

Table 15. Enrichment of copper in the upper layers of lake sediments

Lake	Surface [Cu] (ppm)	[Cu] at depth (ppm)	Reference
L. Windermere, England	≦140 at 5 cm	22 at >40 cm	[197]
L. Lomond, Scotland	44 at 1 cm	12 at >1 cm	[198]
L. Neagh, N. Ireland	580	20 at 13–50 cm	[199]
L. Constance, Switzerland	24–92 (post–1900)	24–36 (pre–1900)	[200]
L. Erie, USA	43–109	20–46 at > 10 cm	[201]
L. Huron, USA	43–78	25–48 at > 10 cm	[202]
L. Superior, USA	113–213	30–84 at > 10 cm	[202]
L. Washington, USA	50	15	[203]
Woodhull Lake, USA	18–36 at 5 cm	5 at >25 cm	[194]
Woods Lake, N.Y., USA	15 at 1 cm	6.2 at >8 cm	[204]

fired power plants [176], which can effectively sterilise soils very close to the source. In the Watarase river basin, Japan, pollution by waste mining slimes has increased soil copper to 510–2,020 ppm in the plough layer, causing decreased rice production [189]. Livett et al. [190] have measured copper in three British peat soil profiles whose layers can be dated using carbon-14. Copper in peat from the unpolluted Glen Shieldaig in Scotland has remained constant at about 6 ppm for the last 1,100 years. At Moor House, Cumbria and Grassington, Yorkshire, the concentration of copper in soil was about 5 ppm from 1,500 to 1,750 A.D. and has since risen steadily to about 26 and 52 respectively, presumably from contamination from local industries. Urban soils can contain much more copper, e.g. up to 545 ppm on Merseyside, U.K. [191] and up to 300 ppm in Sakai City, Japan [122].

Indirect Perturbations of Flux f_{24}, from the Ocean to its Sediments

This flux has only been measurably affected near the margins of industrialised continents. For example, recent sediments in the Baltic Sea contain 70 ppm Cu compared with 35 ppm Cu in older sediments at a depth of 1.2 m [192]. Particularly high concentrations of trace metals are found where sewage sludges and wastes are dumped on the continental shelf, as off New York and in the estuaries of the Thames and Mersey, U.K. Many recent examples are given by Förstner, with some bottom sediments containing up to 12,000 ppm Cu and 190 ppm Ag [193].

A notable indicator of local contamination is the undisturbed sediment core of a freshwater lake. In industrial regions contamination is marked for copper and has occasionally been noted for silver and gold as well. Thus Galloway and Likens [194] examined the sediments of Woodhull Lake, New York. Below 25 cm depth the concentrations found were 5 ppm Cu, 0.1 ppm Ag, and 0.0001 ppm Au, while at 5 cm depth they were 18–36 ppm Cu, 0.5 ppm Ag, and 0.004 ppm Au. The same workers found surface enrichments of 20-fold for Cu and Ag, and 14-fold for Au, in sediments of Honnedaga Lake, New York. Silver has been found to be enriched up to 8-fold in the upper sediments of Lake Ontario relative to background concentrations of 1 ppm below 32 cm depth [195]. Occasionally the addition of copper salts as algicides can markedly perturb the copper content of the sediment profile, as at Lake Monona, Wisconsin [196]. More often the surface enrichment of copper must be attributed to recent increases in the aerial or fluvial input. A selection of further examples is given in Table 15. While many of these show that large local perturbations are occurring, there is insufficient data to generalise to the global situation.

References

1. Ahrens, L.H.: Geochim. Cosmochim. Acta *2*, 155 (1952)
2. Whittaker, E.J.W., Muntus, R.: Geochim. Cosmochim. Acta *34*, 945 (1970)
3. Bailar, J.C. et al.: Comprehensive Inorganic Chemistry, vol. 3, Pergamon, London 1973
4. Mason, B., Moore, C.B.: Principles of Geochemistry, J. Wiley, New York 1982

5. Wedepohl, K.H.: Handbook of Geochemistry, Springer-Verlag, Berlin 1969–1974
6. Flanagan, F.J.: Geochim. Cosmochim. Acta *37*, 1189 (1973)
7. Crocket, J.H., Macdougall, J.D., Harriss, R.C.: Geochim. Cosmochim. Acta *37*, 2547 (1973)
8. Taylor, S.R.: Geochim. Cosmochim. Acta *28*, 1273 (1964)
9. Shawe, D.R.: U.S. Geol. Surv. Prof. Paper 937 (1976)
10. Bowen, H.J.M.: Environmental Chemistry of the Elements, Academic Press, London 1979
11. Glasby, G.P.: Mar. Chem. *1*, 105 (1973)
12. Glasby, G.P.: Marine Manganese Nodules, Elsevier, Amsterdam 1979
13. Jones, D.B.: Oxford Economic Atlas of the World, 4th edit., Oxford University Press, London 1972
14. Day, F.H.: The Chemical Elements in Nature, G.G. Harrap, London 1963
15. Garrels, R.M., Christ, C.L.: Solutions, Minerals and Equilibria, Harper & Row, New York 1965
16. Cox, D.P.: The distribution of copper in common rocks and ore deposits. In: Copper in the Environment, vol. 1 (Nriagu, J.O. ed.), p. 119, J. Wiley, New York 1979
17. Sutlov, A.: Copper Porphyries, University of Utah, Salt Lake City, Utah 1974
18. Boyle, R.W.: Geol. Surv. Can. Bull. 160 (1968)
19. Linke, W.F.: Solubilities of Inorganic and Metal Organic Compounds, 4th edit., Amer. Chem. Soc., Washington 1958
20. Lakin, H.W., Curtin, G.C., Hubert, A.E.: Can. Inst. Min. Metall. Sep. *11*, 196 (1971)
21. Lakin, H.W., Curtin, G.C., Hubert, A.E., Shacklette, H.T., Doxtader, K.G.: U.S. Geol. Surv. Bull. 1330 (1974)
22. Grondev, S.: In: Biogeochemistry of Ancient and Modern Environments (Trudinger, P.A., Walter, M.R., Ralph, B.J. eds.), p. 485, Aust. Acad. Sci, Canberra 1980
23. Seward, T.M.: Geochim. Cosmochim. Acta *37*, 379 (1973)
24. Weissberg, B.C.: Econ. Geol. *65*, 551 (1970)
25. Baker, W.E.: Geochim. Cosmochim. Acta *42*, 645 (1978)
26. Boyle, E.A.: Copper in natural waters. In: Copper in the Environment, vol. 1 (Nriagu, J.O. ed.), p. 77, J. Wiley, New York 1979
27. Martin, J.M., Maybeck, M.: Mar. Chem. *7*, 173 (1979)
28. Gibbs, R.J.: Geol. Soc. Amer. Bull. *88*, 829 (1977)
29. Kharkar, D.P., Turekian, K.K., Bertine, K.K.: Geochim. Cosmochim. Acta *32*, 285 (1968)
30. Eisenreich, S.J., Hoffmann, M.R., Rastetter, D., Yost, E., Maier, W.J.: Adv. Chem. Ser. *189*, 135 (1980)
31. Reeder, S.W., Hitchon, B., Levinson, A.E.: Geochim. Cosmochim. Acta *36*, 825 (1972)
32. Figura, P., McDuffie, B.: Anal. Chem. *52*, 1433 (1980)
33. Owens, J.W., Gladney, E.S., Purtymun, W.D.: Anal. Lett. *13*, 253 (1980)
34. Wilson, A.L.: Phil. Tr. Roy. Soc. London B *288*, 25 (1979)
35. Vanderborght, B.M.R., Van Grieken, R.E.: Talanta *27*, 417 (1980)
36. Nissing, W.: Ber. Arbeitsgem. Rheinwasserwerke *32*, 83 (1975)
37. Sherief, M.K., Awadallah, R.M., Grass, F.: J. Radioanal. Chem. *60*, 267 (1980)
38. Imal, S., Muroi, M., Hamaguchi, A., Koyama, M.: Anal. Chem. *55*, 1215 (1983)
39. Viligur, K.S., Maravskii, V.I.: Gidrokhim. Mater. *71*, 21 (1978)
40. Carles, J.: Presse Therm. Clim. *116*, 117 (1979)
41. Weissberg, B.G.: Econ. Geol. *64*, 95 (1969)
42. Habib, S., Minski, M.J.: J. Radioanal. Chem. *63*, 379 (1981)
43. Hamilton, T.W., Ellis, J., Florence, T.M.: Anal. Chim. Acta *148*, 225 (1983)
44. Florence, T.M., Batley, G.E.: Talanta *24*, 151 (1977)
45. Stiff, M.J.: Water Res. *5*, 585 (1971)
46. Sterritt, R.M., Lester, J.N.: Environ. Pollut. A *27*, 37 (1982)
47. Lee, J.: Geol. Surv. Pap. Can. 79-1A, 121 (1979)
48. Sunda, W.G., Hanson, P.J.: Chemical speciation of copper in river water. In: Chemical Modelling in Aqueous Systems (Jeanne, E.A. ed.), p. 147, Amer. Chem. Soc., Washington 1979

49. Matsunaga, K., Negishi, M., Fukase, S., Hasebe, K.: Geochim. Cosmochim. Acta *44*, 1615 (1980)
50. Sholkovitz, E.R., Copland, D.: Geochim. Cosmochim. Acta *45*, 181 (1981)
51. Leckie, J.O., Davis, J.A.: Aqueous environmental chemistry of copper. In: Copper in the Environment, vol. 1 (Nriagu, J.O. ed.), p. 89, J. Wiley, New York 1979
52. Mantoura, R.F.C., Dickson, A., Riley, J.P.: Estuar. Coastal Mar. Sci *6*, 387 (1978)
53. Wagemann, R., Barica, J.: Water Res. *13*, 515 (1979)
54. Netzer, A., Beszedits, S.: Removal of copper from waste waters. In: Copper in the Environment, vol. 1 (Nriagu, J.O. ed.), p. 124, J. Wiley, New York 1979
55. Ngilmi, A., Bowen, H.J.M.: unpublished work 1983
56. Murray, C.N., Meinke, S.: J. Oceanog. Soc. Jap. *30*, 216 (1974)
57. Rohatgi, N., Chen, C.Y.: J. Water Pollut. Control Fed. 27, 2298 (1975)
58. Rohatgi, N., Chen, C.Y.: J. Environ. Eng. Div. Am. Civ. Eng. *102*, 675 (1976)
59. Thomas, D.J., Grill, E.V.: Estuar. Coastal Mar. Sci. *5*, 421 (1977)
60. Windom, H.L., Beck, K.C., Smith, R.: Southeast. Geol. *12*, 1109 (1971)
61. Duinker, J.C.: Suspended matter in estuaries. In: Chemistry and Biogeochemistry of estuaries (Olausson, E., Cato, I. eds.), p. 135, John Wiley, New York 1980
62. Galloway, J.N.: Geochim. Cosmochim. Acta *43*, 207 (1979)
63. Duinker, J.C., Nolting, R.F.: Mar. Pollut. Bull. *8*, 65 (1977)
64. Thorne, L.T., Nickless, G.: Sci. Tot. Environ. *19*, 207 (1981)
65. Bowen, H.J.M., Westgate, P.: Unpublished work, 1982
66. Boyle, E.A., Sclater, F., Edmond, J.M.: Earth Planet. Sci. Lett. *37*, 38 (1977)
67. Moore, R.M.: Earth Planet. Sci. Lett. *41*, 461 (1978)
68. Bruland, K.W.: Earth Planet. Sci. Lett. *47*, 176 (1980)
69. Schmidt, R.L.: CRC Crit Rev. Environ. Control *8*, 101 & 247 (1979)
70. Lewis, A.G., Cave, W.R.: Oceanog. Mar. Biol. Annu. Rev. *20*, 471 (1982)
71. Danielsson, L.G.: Mar. Chem. *8*, 199 (1980)
72. Masuda, N.: Hokkaido Daigaku Suisangakabu Kenkyu Iho *32*, 425 (1981)
73. Windom, H.L., Smith, R.G.: Mar. Chem. *7*, 157 (1979)
74. Buat-Ménard, P., Chesselet R.: Earth Planet. Sci. Lett. *42*, 399 (1979)
75. Chester, R., Stoner, J.H.: Mar. Chem. *2*, 17 (1974)
76. Brewer, P., Nozaki, Y., Spencer, D.W., Fleer, A.P.: J. Mar. Res. *38*, 703 (1980)
77. Jedwab, J.: Geochim. Cosmochim. Acta *43*, 101 (1979)
78. Hirata, S.: Talanta *28*, 809 (1981)
79. Brewer, P.G.: Minor elements in sea water. In: Chemical Oceanography (Riley, J.P., Skirrow, G. eds.), p. 415, Academic Press, London 1975
80. Van der Berg, C.M.G.: Mar. Chem. *11*, 323 (1982)
81. Van der Berg, C.M.G.: Anal. Proc. *20*, 458 (9183)
82. Hirose, K., Dokiya, Y., Sugimura, Y.: Mar. Chem. *11*, 343 (1982)
83. Desai, M.V.M., Ganguly, A.K.: Geology and Geochemistry of Manganese, vol. 1, p. 389, Hungarian Acad. Sci., Budapest 1980
84. Robertson, D.E.: Pacific N.W. Lab. Annu. Rep., Battelle Inst., Richland, Washington 1970
85. Krendelev, F.P., Pogrebnyak, Y.F., Melnikova, R.D.: Geokhimiya no. 3, 439 (1977)
86. Nakata, R., Okazaki, S., Hori, T., Fujinaga, T.: Anal. Chim. Acta *149*, 67 (1983)
87. Robertson, D.E.: Trans. Am. Nucl. Soc. *27*, 170 (1977)
88. May, S., Piccot, D., Pinte, G.: J. Radioanal. Chem. *27*, 353 (1975)
89. Haber, F.: Ges. Endk. Berl. *3*, 3 (1928)
90. Noddack, I., Noddack, W.: Arkiv. Zool. *32 A*, no. 4 (1940)
91. Hummel, R.W.: Analyst *82*, 483 (1957)
92. Brooks, R.R.: Analyst *85*, 745 (1960)
93. Heuss, E., Lieser, K.H.: J. Radioanal. Chem. *50*, 289 (1979)
94. Kawabuchi, K., Riley, J.P.: Anal. Chim. Acta *65*, 271 (1973)
95. Ryabinin, A.I., Lazareva, E.A.: Geokhimiya no. 12, 1870 (1980)
96. Ryabinin, A.I., Lazareva, E.A.: Geokhimiya no. 21, 249 (1981)
97. Schutz, D.F., Turekian, K.K.: Geochim. Cosmochim. Acta *29*, 259 (1965)
98. Massee, R., Maessen, F.J.M.J., de Goeij, J.J.M.: Anal. Chim. Acta *127*, 181 (1981)

99. Ahrland, S.: Phys. Chem. Sci. Res. Rep. *1*, 219 (1975)
100. Krendelev, F.P., Zhmodik, S.M., Mironov, A.G.: Geochem. Int. *15*, 156 (1978)
101. Murata, K.J.: Am. J. Sci. *258*, 769 (1960)
102. Mizutani, Y.: Jap. Geochem. J. *4*, 87 (1970)
103. Krauskopf, K.B.: Econ. Geol. *52*, 786 (1957)
104. Herron, M.M.: Glaciochemical dating techniques. In: Nuclear and Chemical Dating Techniques (Currie, L.A. ed.), p. 303, Amer. Chem. Soc., Washington 1982
105. Oana, S.: Bull. Volcanol. *24*, 48 (1962)
106. Cadle, R.D., Wartburg, A.F., Pollock, W.H., Gandrud, B.W., Shedlovsky, J.P.: Chemosphere *2*, 231 (1973)
107. Lepel, E.A., Stefansson, K.M., Zoller, W.H.: J. Geophys. Res. C *83*, 6213 (1978)
108. Vossler, T., Anderson, D.L., Aras, N.K., Phelan, J.M., Zoller, W.H.: Science *211*, 827 (1981)
109. Buat-Ménard, P., Arnold, M.: Geophys. Res. Lett. *5*, 245 (1978)
110. Dams, R., De Jonge, J.: Atmos. Environ. *10*, 1079 (1976)
111. Egorov, V.V., Zhigalovskaya, T.N., Malakhov, S.G.: J. Geophys. Res. *75*, 3650 (1970)
112. Hoffmann, G.L., Duce, R.A., Hoffmann, E.J.: J. Geophys. Res. *77*, 5322 (1972)
113. Adams, F.C., Van Craen, M., Van Espen, P., Andreuszi, D.: Atmos. Environ. *14*, 879 (1980)
114. Cunningham, W.C., Zoller, W.H.: J. Aerosol. Sci. *12*, 367 (1981)
115. Thompson, R.J.: Adv. Chem. *172*, 54 (1979)
116. Pillay, K.K.S., Thomas, C.C.: J. Radioanal. Chem. *7*, 107 (1971)
117. John, W., Kaifer, R., Rahn, K.A., Winchester, J.W.: Atmos. Environ. *7*, 107 (1973)
118. Hodge, V., Johnson, S.R., Goldberg, E.D.: Geochem. J. *12*, 7 (1978)
119. Dams, R., Robbins, J.A., Rahn, K.A., Winchester, J.W.: Anal. Chem. *42*, 861 (1970)
120. Cawse, P.A.: Rep. AERE-R 9886, H.M. Stationery Office, London 1981
121. Priest, P., Navarre, J.L., Ronneau, C.: Atmos. Envir. *15*, 1325 (1981)
122. Komai, Y.: Heavy metal pollution in urban soils. In: Heavy Metal Pollution in Soils of Japan (Kitagushi, K., Yamane, I. eds.), p. 193, Japan Sci. Soc., Tokyo 1981
123. Beavington, F., Cawse, P.A.: Sci. Total Environ. *10*, 239 (1978)
124. Rahn, K.A.: The Chemical Composition of the Marine Aerosol, Tech. Rep. Univ. Rhode Is., USA 1976
125. Cattell, F.C.R., Scott, W.D.: Science *202*, 429 (1978)
126. Boutron, C.: J. Geophys. Res. *85*, 7427 (1980)
127. Boutron, C., Martin, S.: J. Geophys. Res. *85*, 5631 (1980)
128. Ure, A.M., Berrow, M.L.: The elemental constituents of soils. In: Environ. Chem. vol. 2 (Bowen, H.J.M. ed.), p. 94, Roy. Soc. Chem., London 1982
129. Salmon, L., Cawse, P.A.: In: Soil Analysis (Smith, K.A. ed.), p. 299, M. Dekker, New York 1983
130. Iimura, K.: Heavy metals in soils. In: Heavy Metal Pollution in Soils of Japan (Kitagishi, K., Yamane, I. eds.), p. 19, Japan Sci. Soc., Tokyo 1981
131. Reaves, G.A., Berrow, M.L.: J. Soil Sci. (in press)
132. Harmsen, K.: Behaviour of Heavy metals in Soils, Agric. Publ., Wageningen, Holland 1977
133. Tyler, G.: Soil Biochem. *5*, 371 (1981)
134. Killingbeck, K.T., Wali, M.K.: Oikos *30*, 29 (1978)
135. Pellenberg, R.E., Church, T.M.: Science *203*, 1010 (1979)
136. McLaren, R.G., Crawford, D.V.: J. Soil Sci. *24*, 443 (1973)
137. Goodman, B.A., Cheshire, M.V.: Soil Sci. *27*, 337 (1976)
138. Chao, T.T., Anderson, B.J.: Chem. Geol. *14*, 159 (1974)
139. Karlin, K.D.: Inorg. Perspect. Biol. Med. *2*, 127 (1979)
140. Silverberg, B.A., Stokes, P.M., Ferstenberg, L.B.: J. Cell. Biol. *69*, 210 (1976)
141. Walker, G.: Mar. Biol. *39*, 343 (1977)
142. Ehrlich, H.L.: How Microbes cope with heavy metals, arsenic and antimony in their environment. In: Microbial Life in Extreme Environments (Kushner, D.J. ed.), p. 381, Academic Press, New York 1978
143. Christensen, E.R., Scherfig, J.: Water Res. *13*, 79 (1979)

144. North, W.J.: Calif. State Water Pollut. Board Publ. *26*, 55 (1964)
145. Simola, L.K.: In: Proc. Kuopio Meet. Plant Damages caused by Air Pollution, Sci. Pap. Sympos. (Karenlampi, L. ed.), p. 148, Kuopion Korkeakoulu, Finland 1976
146. Chino, M., Kitagushi, K.: J. Sci. Soc. Manure Japan *37*, 342 (1966)
147. Wallace, A., Alexander, G.V., Chaudhry, F.M.: Commun. Soil Sci. Plant Anal. *8*, 751 (1977)
148. Webb, M.: Devel. Toxicol. Environ. Sci. *1*, 51 (1977)
149. Hutchinson, T.C.: Copper contamination of ecosystems caused by smelter activities. In: Copper in the Environment, vol. 1 (Nriagu, J.O. ed.), p. 451, J. Wiley, New York 1979
150. Farago, M.E., Mullen, W.A., Cole, M.M., Smith, R.F.: Environ. Pollut. *A 21*, 225 (1980)
151. Byrne, A.R., Dermelj, M., Vakselj, T.: Chemosphere *10*, 815 (1979)
152. Tyler, G.: Tr. Brit. Mycol. Soc. *74*, 41 (1980)
153. Lange, O.L., Ziegler, H.: Mitt. Flor. Soziol. Arbeitsgemeinschaft N.F. *10*, 156 (1963)
154. Malaisse, F., Gregoire, J., Morrison, R.S., Brooks, R.R., Reeves, R.D.: Oikos *33*, 472 (1979)
155. Brooks, R.R., Reeves, R.D., Morrison, R.S., Malaisse, F.: Bull. Soc. Roy. Bot. Belg. *113*, 166 (1980)
156. Korobushkina, E.D., Chernyak, A.S., Mineev, G.: Mikrobiologiya *43*, 49 (1974)
157. Korobushkina, E.D., Mineev, G.G., Praded, G.P.: Mikrobiologya *45*, 535 (1976)
158. Girling, C.A., Peterson, P.J.: Trace Subst. Environ. Health *12*, 105 (1978)
159. Valente, I.M., Minski, M.J., Peterson, P.J.: CONF-800433, p. 470, Univ. Missouri, Columbia MO 1980
160. Seeliger, U., Edwards, P.: Mar. Pollut. Bull. *8*, 16 (1977)
161. Amiard, J.C.: Cah. Biol. Mar. *20*, 125 (1979)
162. Gibbs, P.E., Bryan, G.E.: J. Mar. Biol. Ass. U.K. *60*, 205 (1980)
163. Fowler, S.W.: Nature, Lond. *269*, 51 (1977)
164. Bowen, H.J.M.: The elemental contents of human diets and excreta. In: Environ. Chem., Vol. 2 (Bowen, H.J.M. ed.), p. 70, Roy. Soc. Chem., London 1982
165. Kagi, J.H.R., Nordberg, M.: Metallothionein, Birkhaüser, Basel, Swizerland 1979
166. Sabbioni, E.: Int. Wksp. Trace elements Anal. Chem. Med. Biol., Neuherberg, Germany 1980
167. Schmitz, G., Minkel, D.T., Gingrich, D., Shaw, C.F.: J. Inorg. Biochem. *12*, 293 (1980)
168. Nriagu, J.O.: The global copper cycle. In: Copper in the Environment, vol. 1 (Nriagu, J.O. ed.), p. 1, J. Wiley, New York 1979
169. Garrels, R.M., Mackenzie, F.T.: Envolution of Sedimentary Rocks, W.W. Norton, New York 1971
170. Garrels, R.M., Mackenzie, F.T., Hunt, C.: Chemical Cycles and the Global Environment, W. Kaufmann, Los Altos, CA, USA 1973
171. Govett, J.S., Govett, M.H.: World Mineral Supplies, Elsevier, New York 1976
172. Lantzy, R.J., Mackenzie, F.T.: Geochim. Cosmochim. Acta *43*, 511 (1979)
173. Mackenzie, F.T., Lantzy, R.J., Paterson, V.: Math. Geol. *11*, 99 (1979)
174. Li, Y.-H.: Geochim. Cosmochim. Acta *45*, 2073 (1981)
175. Mining Annual Review: Mining J. (1982), pp. 23–33
176. Lindberg, S.E., Andren, A.W., Randon, R.J., Fulkerson, W.: Environ. Health Perspect. *12*, 9 (1975)
177. Bertine, K.K., Goldberg, E.D.: Science *173*, 233 (1971)
178. Imhoff, K., Müller, W.J., Thistlethwayte, D.K.B.: Disposal of Sewage and other Waterborne Wastes, 2nd edit., Butterworth, London 1971
179. Berrow, M.L., Webber, J.: J. Sci. Food Agric. *23*, 93 (1972)
180. Kirk-Othmer: Encyclopedia of Chemical Technology, 3rd edit., J. Wiley, New York 1979
181. Butcher, R.W.: J. Inst. Sewage Purif. *2*, 92 (1946)
182. Shibuya, M., Kamemoto, Y.: Mizushori Gijitsu *4*, 17 (1963)
183. Lee, A.J., Ramster, J.W.: Atlas of seas around Britain, Min. of Agric. Fish & Food, London 1981
184. Waslenchuk, D.G.: Environ. Pollut. B *3*, 271 (1982)
185. Bryan, G.W.: Heavy metal contamination in the sea. In: Marine Pollution (Johnston, R. ed.), p. 185, Academic Press, London 1976

186. Cambray, R.S., Jefferies, D.F., Topping, G.: Mar Sci. Commun. *5*, 175 (1979)
187. Delas, J.: Agrochimica *7*, 258 (1963)
188. Tyler, G.: Soil Biol. Biochem. *8*, 327 (1976)
189. Morishita, T.: The Watarase river basin. In: Heavy Metal Pollution in Soils of Japan (Kitagushi, K., Yamane, I. eds), p. 165, Japan Sci. Soc., Tokyo 1981
190. Livett, E.A., Lee, J.A., Tallis, J.H.: J. Ecol. *67*, 865 (1979)
191. Parry, G.D.R., Johnson, M.S., Bell, R.M.: Environ. Pollut. B *2*, 97 (1981)
192. Erlenkeuser, H., Suess, E., Willkomm, H.: Geochim. Cosmochim. Acta *38*, 823 (1974)
193. Förstner, U.: Inorganic pollutants, particularly heavy metals in estuaries. In: Chemistry and Biogeochemistry of Estuaries (Olausson, E., Cato, I. eds.), pp. 307–328, J. Wiley, New York 1980
194. Galloway, J.N., Likens, G.E.: Limnol. Oceanog. *24*, 427 (1979)
195. Lum, K.R., Edgar, D.G.: Analyst *108*, 918 (1983)
196. Sanchez, I.: Ph. D. Thesis, Univ. Wisconsin, Madison, Wis 1971
197. Hamilton-Taylor, J.: Environ. Sci. Technol. *13*, 693 (1979)
198. Farmer, D.G., Swan, D.S., Bazter, M.S.: Sci. Total Environ. *16*, 131 (1980)
199. Rippey, B.: Environ. Pollut. B *3*, 199 (1982)
200. Müller, G.: Z. Naturforsch. *32 C*, 913 (1977)
201. Kemp, A.L.W., Thomas, R.L.: Water Air Soil Pollut. *5*, 419 (1976)
202. Kemp, A.L.W., Williams, J.D.H., Thomas, R.L., Gregory, M.L.: Water Air Soil Pollut. *10*, 381 (1978)
203. Barnes, R.S., Schell, W.R.: Physical transport of trace metals in the Lake Washington watershed. In: Cycling and Control of Metals (Curry, M.G. ed), p. 45, Environ. Prot. Agency, Cincinnati, Ohio, USA 1973
204. Heit, M., Tan, Y., Klusek, C., Burke, J.C.: Water Air Soil Pollut. *15*, 441 (1981)

Modelling the Global Carbon Cycle

G. Kratz

Mainstraße 28
D-6370 Oberursel, Federal Republic of Germany

Summary

A description of carbon cycle models, their different underlying assumptions and their results is given. To this end those transfer processes of carbon within the global environment (atmosphere, oceans, land biota) presently assumed relevant are sketched and the respective empirical material is summarised. The process of carbon cycle model constructing and its refinement is discussed with particular emphasis on the problem of the additional oceanic CO_2 uptake capacity and the question whether the land biota are a source or a sink for additional atmospheric CO_2. It turns out that the predicted doubling-time of atmospheric CO_2 concentration primarily depends on the assumed fossil fuel consumption scenario an input growth coefficient of 4%/a–5%/a leading to such a doubling in about 2,030–2,040.

Introduction

Measurements of carbon dioxide which have been made at different locations on the continents, aboard ships and aircrafts and some of which have now been conducted for more than 25 years agree that the concentration of this trace gas increases in the atmosphere. At the same time consumption of fossil fuels (coal, petroleum, natural gas, and their refining products) accelerated. Because the combustion of fossil fuels leads to CO_2 release, the increase of atmospheric carbon dioxide is assumed, at least partially, to be due to this process. Furthermore, large clearings, especially in tropical forests, and desertification may be regarded as causes for additional CO_2 emissions.

CO_2 absorbs infrared radiation with absorption being strongest in a band at 14–16 μm [1] i.e. near the intensity maximimum of tropospheric back-radiation [2]. Therefore, a warming of the troposphere is to be expected with increasing atmospheric CO_2 content (greenhouse effect) which could possibly have severe climatic and societal consequences. To assess the risk of such anthropogenically induced climatic changes the investigation of three problems appears necessary [3]:

(1) the development of future total energy consumption and the contribution of fossil fuels
(2) the dynamic behaviour of the global carbon cycle
(3) the effect of an increase in atmospheric CO_2 content on the climatic system.

Disregarding feed-back effects of temperature increases on processes within the carbon cycle and on energy consumption and assuming certain future fossil fuel combustion and land use scenarios the increase of the expected atmospheric CO_2 content can be estimated by means of adequate model simulations of the natural carbon cycle. Climate models which describe the influence of increased CO_2 concentrations on the climatic regime can then be used to approximately predict variations of climatic variables such as temperature and precipitation.

In order to provide a basis for the prediction of future CO_2 levels in the atmosphere carbon cycle models have to be validated against data concerning the variation of the dynamic behaviour of this element cycle. This means that a model which is to describe the unperturbed exchange of carbon between adjacent reservoirs should reproduce relevant empirical information when realistic perturbations are incorporated.

Since atmospheric carbon dioxide deserves special interest as a climatically relevant compound, two questions have to be dealt with: How much CO_2 is emitted into the atmosphere and which fraction of this additionally released CO_2 is taken up by other environmental reservoirs (mainly oceans and land biota). In contrast to fossil input the quantity of anthropogenically released biogenic CO_2

is not known accurately. In addition, there is almost no indication of how much CO_2 is removed from the atmosphere by regrowth of forests on cleared areas and, eventually, be growth stimulation.

From the beginning of the industrialisation until present not only the total carbon content but also the concentrations and isotopic ratios of other constituents in the individual reservoirs have been perturbed. The release of fossil carbon almost completely lacking ^{14}C leads to a dilution of atmospheric ^{14}C (Suess effect). Since fossil and (recent) biogenic materials contain less ^{13}C per mole C than atmospheric CO_2, the release of these materials leads to a similar (but weaker) effect for atmospheric ^{13}C. By atmospheric nuclear weapons explosions large quantities of ^{14}C and ^3H have been released which can be used as tracers.

The global carbon cycle models conceived so far can only be accomodated with the observed isotopic ratios, if the land biota are assumed to be at most a weak net source for atmospheric CO_2 at present. This, however, seems to be in contradiction to vegetation geographical and ecological experience (source-sink-problem, [4–6]).

To provide a closer look to the modelling procedure some methodical considerations will be presented in the next section. Subsequently empirical information about the carbon transfer processes within and between atmosphere, oceans, and land biota will form a basis for a simple model formulation which is then used to discuss some principal features of global carbon cycle models and the results obtained by them.

Methodical Considerations

The problem of constructing a carbon cycle model (like for any other tracer) is to incorporate the information about important transport and reaction processes this element undergoes in nature into a system of differential equations. The solution of these equations should satisfactorily reproduce the relevant empirical data. The process of developing such simulation model can be considered to consist of several stages:

- Defining the data set the model is to be validated against
- Defining the compartments, i.e. elements of some spatial extension assumed homogeneous with respect to the tracer concentration(s) considered. In other words, mixing within the compartments is assumed fast in relation to exchange processes.
- Defining the state variables. Each compartment i contains a certain mass N_i^k or number n_i^k of moles of the chemical species k. N_i^k, n_i^k or derived quantities such as concentrations or isotopic ratios can be regarded as state variables in the corresponding system of differential equations. (In global carbon cycle models aerodynamic, hydrodynamic, and radiation processes are usually not described by their own set of equations of motion but are rather assumed constant in time, and independent from carbon transfer. The information available about these processes may be incorporated in the design of carbon flux paths and the calculation of transfer coefficients.)

– Setting up the flux equations: formulation of the dependence of the exchange fluxes F^k_{ij} (in g C/a) or f^k_{ij} (in moles/a) from compartment i to compartment j on the state variables or other quantities related to them. Such exchange fluxes represent transport processes as well as chemical and nuclear reactions (radioactive decay). The only reaction which is not treated as a transfer from one compartment to another in carbon cycle models is radioactive decay. More detailed studies of the behaviour of certain compounds or compound classes in certain environmental reservoirs may create a need to investigate systems with several and more complicated chemical reactions occuring within some of the model compartments.
– Formulation of perturbation functions. These explicitly time-dependent functions $F^k_{e,i}$ (t) should describe the perturbation of the natural system especially by human activities such as fossil CO_2 release, modifications in land use and release of bomb ^{14}C.
– Defining the initial conditions. In many cases it seems realistic and useful to evaluate initial conditions subject to stationarity and continuity. The calculation of transfer coefficients in the oceans by initially stationary or by perturbed tracer concentration distributions is commonly referred to as calibration.
– Setting up a system of differential equations (mass balances). The time-derivatives of the N^k_i (or n^k_i, resp.) are the sum of all fluxes entering compartment i minus the sum of all fluxes leaving this compartment:

$$\frac{dN^k_i}{dt} = \sum_j (F^k_{ji} - F^k_{ij}) + \sum_e F^k_{e,i}(t) \tag{1}$$

subscript e: sources having causes external to the system
– Solving the system of differential equations. This yields the explicitely time-dependent form of the state variables and those variables derived from them. For some models this step can be carried out analytically, while particularly for models with non-linear exchange fluxes and/or complicated input functions numerical integration methods are more convenient or even essential. Computer languages for the simulation of continuous systems (as e.g. DYNAMO III or CSMP III) provide easy to handle tools for formulating and solving systems of first-order differential equations.
– Comparison of calculation results to empirical data (validation).
– Estimate of future system behaviour. In case the validation yields good agreement of empirical and calculated values the model can be used to simulate the future system behaviour under the influence of different perturbation functions (scenarios).

Of course, such a program as outlined above and its individual steps are almost never performed without restarting at previous points under different conditions. In order to systematically predict the qualitative effect(s) the variation of a certain parameter has with respect to the results to obtain a sensitivity analysis is often useful.

Carbon Dioxide in the Atmosphere

Atmospheric CO_2 Content

The atmospheric CO_2 content is normally given as mole fraction in units of parts per million on a volume basis (ppmv). Throughout this paper data on CO_2 partial pressure always refer to an atmospheric pressure of 1 atm. Adopting a mass of the total atmosphere of $m_A = 5.136 \cdot 10^{21}$ g [7] the mole fraction of CO_2 in air can be given as

$$\mu_{CO_2} = \frac{n_a}{n_A} = \frac{m_a}{m_A} \cdot \frac{M_A}{M_a} \tag{2}$$

n: mole numbers, m: masses, M: molecular weights
subscripts: a: CO_2 in the atmosphere, A: total atmosphere
$M_A = 28.9$: molecular weight of dry air [8]. This relation leads to the conversion factor (for atmospheric CO_2):

$$1 \text{ ppmv } CO_2 \triangleq 2.1346 \text{ Gt C}. \tag{3}$$

The carbon dioxide content of the atmosphere is assumed to be affected by reactions in the gas phase only to a neglegible extent; furthermore, dissolution of CO_2 in precipitation seems to be insignificant (see e.g. [9] for data). Processes being of major direct influence on the atmospheric CO_2 concentration take place at the boundary between atmosphere and oceans as well as between atmosphere and land biota.

Because of the increasing relevance the CO_2 problem has been attributed to in recent years, more and more concentration measurements have been conducted at ground stations [10–13]. The atmospheric CO_2 content has been observed continuously for more than 25 years at Mauna Loa Observatory, Hawaii, and at the South Pole [11]. Other records have been taken aboard ships [14] and aircrafts [15–17, 22].

The seasonal oscillations of atmospheric CO_2 partial pressure in the northern hemisphere of the globe exhibit an increase approximately from August to April and a decrease during the rest of the year which is mainly due to biotic assimilation and mineralisation [18]. This interpretation is in accordance with a phase shift of about half a year and a smaller amplitude of the respective South Pole record the latter resulting from the considerably smaller area on the southern hemisphere covered by vegetation with a seasonal cycle.

In more detailed studies [19] an increase in amplitude of the seasonal CO_2 cycles at Mauna Loa and Weathership P (50 °N, 145 °W) of 0.66%/a and 0.77%/a, resp., has been obtained. An analysis of the regional pattern of CO_2 partial pressures leads to the assumption of an equatorial CO_2 source which is almost independent from season. This is attributed to the fact that CO_2 is less soluble in warm than in cold water resulting in a permanently higher average CO_2 partial pressure over the equatorial oceans than over ocean regions of higher latitude. In this way, permanent outgassing of CO_2 at low latitudes is compensated for by additional dissolution in high latitudes oceans, the exchange being accomplished mainly by atmospheric circulation. The increased amplitude of the seasonal concentration variation of atmospheric CO_2 can be attributed to the following causes or a combination of these: (a) increase in fossil fuel use by which relatively more CO_2 should be released in wintertime, if the seasonal pattern is retained. (b) increase of photosynthetic activity (c) variations in the seasonal pattern of atmospheric, and (d) oceanic circulation. The latter may lead to seasonally abnormal sea-surface temperatures which cause an altered dissolution capacity for CO_2. (For a discussion see [20, 21].

Profiles of atmospheric CO_2 partial pressures with altitude which are still subject to discussion have been given by [17] and [22].

An evaluation of the Mauna Loa and South Pole data [11] shows that the average CO_2 partial pressure at these stations from January 1959 to January 1979 increased from 315.6 to 334.9 ppm the difference being 19.3 ppm $\triangleq$ 41.2 Gt C. During the same period 74.9 Gt C of fossil fuel CO_2 had been released to the atmosphere [23] such that for the period in question an airborne fraction related to fossil input α_f (1959–1979) $= 55.0\%$ results.

From the different airborne fraction for both hemispheres an interhemispheric exchange time of (1.14 ± 0.16) a is calculated [24].

If atmospheric CO_2 partial pressure records are corrected for seasonal variations, oscillations of irregular frequencies and amplitudes appear to be superimposed on the overall increase. They have been explained by the Southern Oscillation which is a large-scale multi-annual fluctuation of the atmospheric and marine circulation pattern associated with anomalies of sea-surface temperatures [25, 26].

In order to estimate the pre-industrial CO_2 partial pressure, early measurement results were compiled [189, 190] yielding baseline concentrations of about 290 and 270 ppm, resp. (The reliability of these and other data [27] has, however, been questioned [28]). Ice-core measurements [29] yielded a pre-industrial long-term average of (280 ± 20) ppm.

The record of fossil input [23, 166] can be subdivided into four stages: From 1860 to 1914 combustion of fossil fuels increased exponentially with a growth rate of 4.3 %/a. From the beginning of World War I to the end of World War II the trend is rather irregular yielding an average growth rate of about 1.6 %/a. During the period 1945–1973 combustion of fossil fuels exhibits a growth rate 4.6 %/a which decreases to 2.25 %/a in the years 1974–1980. Up to 1980, about 165 Gt C have been emitted to the atmosphere from this source.

^{13}C in the Atmosphere

The isotopic ratio $^{13}C/^{12}C$ is usually given in terms of $\delta^{13}C$:

$$\delta^{13}C\,[‰] := \left[\frac{(n^{13}/n^{12})_{sample}}{(n^{13}/n^{12})_{standard}} - 1\right] \cdot 1{,}000 \tag{4}$$

n^{13}, n^{12}: mole numbers of ^{13}C and ^{12}C, resp.

The standard n^{13}/n^{12}-ratio is 0.0112372 [42]. In the following reference is often made to the deviation of $\delta^{13}C$ from some initial value $[\Delta(\delta^{13}C)]$.

Measurements of atmospheric $^{13}C/^{12}C$ ratios for the years 1956 and 1978 [30] yielded $\delta^{13}C$ values of $-6.69‰$ and $-7.34‰$ the difference being (-0.65 ± 0.13) ‰. Additionally, seasonally resolved ^{13}C data from air samples have more recently been provided [31].

Since the atmospheric $^{13}C/^{12}C$ ratio is decreased by both fossil and biogenic release having approximately the same lower $\delta^{13}C$ value ($\approx -25‰$) a record of this quantity which should provide additional information about the total CO_2 input function is assumed to be laid down in tree-rings [32–34]. Since these data, however, exhibit considerable scatter [32–34] ambient air $\delta^{13}C$ does not seem to be the only factor determining the $\delta^{13}C$ of tree-rings. Even if isolated trees of industrially uncontaminated areas were selected and only the cellulose fractions were analysed, these data are supposedly influenced by local climatic conditions [35–37]. From 33 tree-ring records [38] only 3 fulfill all the selection criteria given by the authors, at the same time covering the entire period from 1860 to 1980. 8 tree-ring records exhibiting data for the time from 1905–1955 can be regarded as unobjectable for the same reasons yielding an overall $\delta^{13}C$ decrease of 0.16‰ with an overall standard deviation of about 0.6‰. This uncertainty range should be kept in mind, when tree-ring $\delta^{13}C$ data are to be incorporated into carbon cycle model considerations.

Since the principal fossil fuels consumed all have slightly different $\delta^{13}C$ and since their proportion varies with time average time-dependent $\delta^{13}C$ of fossil input has been calculated [39]. Since coal is isotopically heavier than oil and natural gas, this fossil fuel input $\delta^{13}C$ decreases from about $-24‰$ to $-27.5‰$. The effect on model calculated atmospheric $\delta^{13}C$ is of minor importance.

^{14}C in the Atmosphere

Analogously to definition (4), the $^{14}C/C$ ratio can be given in terms of

$$\delta^{14}C\,[‰] := \left(\frac{A_{sample}}{A_{standard}} - 1\right) \cdot 1{,}000 \qquad (5a)$$

A: specific activity (number of desintegrations per unit time and g C)
standard activity: $(1.176 \pm 0.010) \cdot 10^{-10}$ % [40].

Usually, however, quantities corrected for fractionation are given, use being made of the fact that the fractionation factors for ^{14}C are about the squares of the respective fractionation factors for ^{13}C [41, 42]:

$$\Delta^{14}C\,[‰] := \left[\left(\frac{\delta^{14}C}{1000} + 1\right) \cdot \frac{0.975^2}{\left(\frac{\delta^{13}C}{1{,}000} + 1\right)^2} - 1\right] \qquad (5b)$$

[43].

^{14}C undergoes radioactive decay with a half-life of 5,370 a and a decay constant of $\lambda^{14} = 1/(8.267\ a)$ [44].

^{14}C is naturally produced in the stratosphere by cosmic ray neutrons which react with ^{14}N to form ^{14}C. Cosmic ray irradiation is modulated by the external geomagnetic field which itself is composed of the internal geomagnetic field and those magnetic fields caused by currents of ionised particles in the upper atmosphere. Enhanced solar activity which seems to be correlated to an increase in sunspot numbers or enhanced geomagnetic moment leads to an increased screening of cosmic rays and thus to diminished ^{14}C production.

From tree-ring analysis two cycles can be regarded as occurring most frequently: (1) An oscillation of about 10,000 a with an amplitude of about 10% of the mean ^{14}C production which is attributed to geomagnetic changes. (2) The so-called De Vries oscillation or Suess wiggles of about 200 a and an amplitude of 1–2‰ of the natural atmospheric ^{14}C activity which is explained by variations of solar activity [45]. Since the decrease in ^{14}C activity in the period from 1860–1950 amounted to about 20‰ such natural variations seem necessary to be regarded when modelling the Suess effect is intended.

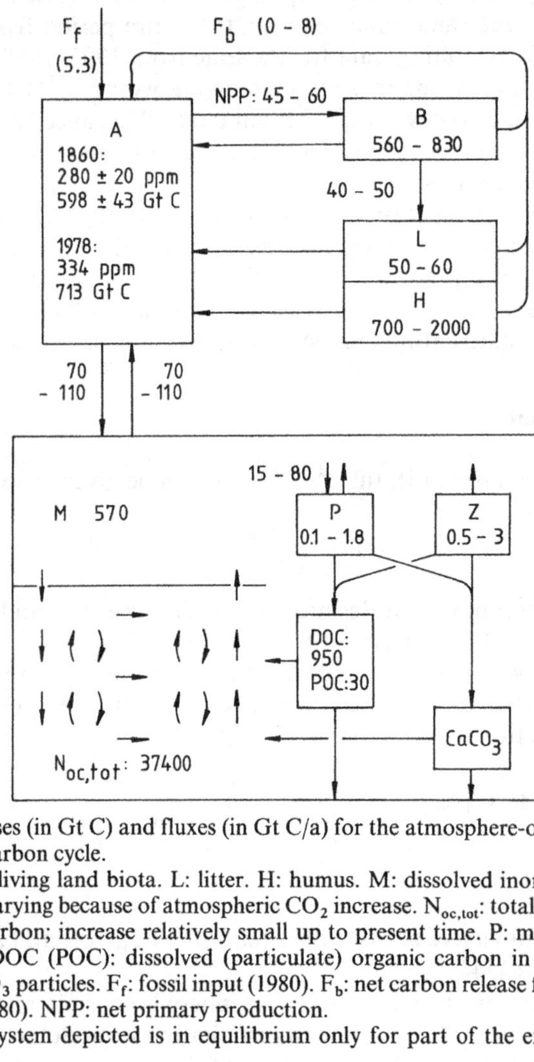

Fig. 1. Carbon masses (in Gt C) and fluxes (in Gt C/a) for the atmosphere-ocean-land biota system of the global carbon cycle.
A: atmosphere. B: living land biota. L: litter. H: humus. M: dissolved inorganic carbon in the sea-surface layer; varying because of atmospheric CO_2 increase. $N_{oc,tot}$: total oceanic mass of dissolved inorganic carbon; increase relatively small up to present time. P: marine phytomass. Z: marine zoomass. DOC (POC): dissolved (particulate) organic carbon in the ocean. $CaCO_3$: downsinking $CaCO_3$ particles. F_f: fossil input (1980). F_b: net carbon release from living and dead biogenic matter (1980). NPP: net primary production.
The biota-humus system depicted is in equilibrium only for part of the exchange flux ranges given.

The Aa-indices which give a measure for diurnal averages of magnetic activity at two antipodal stations [47] correlate with ^{14}C production rates obtained from neutron flux determinations with a correlation coefficient r = 0.67 according to:

$$\Pi^{14} = 2.130 - 0.024 \cdot \mathrm{Aa}\left[\frac{\text{atoms}}{\text{cm}_e^2 \cdot \text{sec}}\right] \tag{6}$$

cm_e^2: cm^2 earth surface [46].
Since for 1867–1967 the average Aa is 17.9 an average ^{14}C production of 1.88 atoms/cm_e^2/sec results for this period.

Investigations of the $^{14}C/C$ isotopic ratios in dendrochronologically dated tree-rings resulted in that the relative atmospheric ^{14}C content had decreased from the beginning of the industrialisation to 1950 by (-20 ± 1.2) ‰ [48].

Nuclear weapon explosions mainly in the 1960's released large quantities of ^{14}C within comparatively short time. Since the propagation of this tracer into the oceans has been observed during the GEOSECS program, it is suitable for calibration or validation of carbon cycle models. The release of bomb ^{14}C has been estimated [51]. The record of ^{14}C in the troposphere and in surface ocean water from 1962–1980 [52] which reflects this bomb effect visualises the relaxation of the $^{14}C/C$ ratios following the large perturbations in the early 1960's. Since ^{14}C exchange can be assumed to linearly depend on the $^{14}C/C$ ratios in atmospheric and marine CO_2, resp., the record of atmospheric (bomb perturbed) ^{14}C can be used as in input function to test the response of an ocean sub-model to the bomb ^{14}C perturbation, the appropriate output function being the ^{14}C record for the ocean surface layer [167].

Oceans

From the fact that the ocean is the largest sink for additional atmospheric CO_2 the relevance of appropriately modelling this environmental reservoir is evident. CO_2 is exchanged via the boundary layer between air and sea-water; the equilibrium partial pressure of dissolved CO_2 is above all determined by bicarbonate and carbonate ion concentrations together with pH. Carbon in the ocean is transferred with water mass movements and with biotic processes.

CO_2 Exchange Between Atmosphere and Oceans

Chemical equilibrium of inorganic carbon in atmosphere and ocean

The chemical equilibria

$$\begin{aligned} CO_2(\text{gas}) &\rightleftharpoons CO_2\ (\text{dissolved}) \\ CO_2 + H_2O &\rightleftharpoons (H_2CO_3) \rightleftharpoons HCO_3^- + H^+ \\ HCO_3^- &\rightleftharpoons CO_3^{2-} + H^+ \\ H_3BO_3 &\rightleftharpoons H_2BO_3^- + H^+ \\ H_2O &\rightleftharpoons OH^- + H^+ \end{aligned} \tag{7a–e}$$

exert a buffering action on the uptake of additional CO_2 by the ocean. The equilibrium constants of the above reactions are:

$$\begin{aligned}
\alpha_s &= \frac{p_{CO_2}}{[CO_2]} \\
K_1' &= [HCO_3^-][H^+]/[CO_2] \\
K_2' &= [CO_3^{2-}][H^+]/[HCO_3^-] \\
K_B' &= [H_2BO_3^-][H^+]/[H_3BO_3] \\
K_w &= [OH^-][H^+]
\end{aligned} \tag{8a–e}$$

[ ]: concentrations
$[CO_2]$: concentration of (physically) dissolved CO_2 and of H_2CO_3
α_s: solubility of CO_2 in sea-water

All equilibrium constants K' are apparent equilibrium constants, i.e. they are corrected for ion pair formation and activity coefficients. They depend on temperature and salinity. p_{CO_2} is the equilibrium partial pressure in the gaseous and in the liquid phase. Concentration of total inorganic dissolved carbon is then:

$$\Sigma C = [CO_2] + [HCO_3^-] + [CO_3^{2-}]. \tag{9}$$

The alkalinity is defined as

$$A = [HCO_3^-] + 2[CO_3^{2-}] + [H_2BO_3^-] + [OH^-] - [H^+]. \tag{10}$$

If one, in addition, defines total inorganic dissolved boron which in carbon cycle models is usually assumed constant

$$\Sigma B = [H_3BO_3] + [H_2BO_3^-] \tag{11}$$

Equs. (9) and (10) can be transformed to

$$\Sigma C = \left(1 + \frac{K_1'}{[H^+]} + \frac{K_1'K_2'}{[H^+]^2}\right)\alpha_s p_{CO_2} \tag{12}$$

and

$$A = \left(\frac{K_1'}{[H^+]} + 2\frac{K_1'K_2'}{[H^+]^2}\right)\alpha_s p_{CO_2} + \frac{K_B' \cdot \Sigma B}{K_B' + [H^+]} + \frac{K_W}{[H^+]} - [H^+]. \tag{13}$$

Substituting $\alpha_s p_{CO_2}$ in (13) by the corresponding term in (12) yields an implicit function of ΣC, A, and $[H^+]$, which is a 5th order polynomial in $[H^+]$:

$$\sum_{i=0}^{5} D_i [H^+]^i = 0 \tag{14}$$

with

$$\begin{aligned}
D_0 &= -K_1'K_2'K_B'K_W \\
D_1 &= [(A - 2\Sigma C - \Sigma B)K_2'K_B' - (K_2' + K_B')K_W]K_1' \\
D_2 &= [(A - 2\Sigma C + K_B')K_2' + (A - \Sigma C - \Sigma B)K_B']K_1' - (K_1' + K_B')K_W \\
D_3 &= (A - \Sigma B + K_1')K_B' + (A - \Sigma C + K_2')K_1' - K_W \\
D_4 &= A + K_1' + K_B' \\
D_5 &= 1
\end{aligned} \tag{15a–f}$$

[53].

Provided ΣC and A being known this polynomial can be solved iteratively (for an appropriate procedure see [54]) to give $[H^+]$ which itself yields the partial pressure of dissolved CO_2 when inserted into Eq. (12). Since the CO_2 flux from ocean to atmosphere is most often assumed proportional to the time-dependent CO_2 partial pressure in the ocean surface layer, p_s, this procedure is used to calculate this flux. The flux from atmosphere to ocean is assumed proportional to the atmospheric CO_2 partial pressure.

The above method although being correct does not bring about a fast survey concerning the response of the oceans to increased atmospheric CO_2 partial pressures. Therefore, attempts have been made to give simple relationships between the partial pressure increase Δp_s and the increase of carbon concentration in the surface layer, $\Delta\Sigma C$. The Taylor series

$$\frac{p_{CO_2}}{p_{CO_2}^0} = 1 + \left.\frac{d(p_{CO_2}/p_{CO_2}^0)}{d(\Sigma C/\Sigma C_0)}\right|_{\Sigma C=\Sigma C_0} \cdot \frac{\Delta\Sigma C}{\Sigma C_0} + \dots \tag{16}$$

subscript 0: stationary state
Δ: deviation from stationary state

yields the so-called evasion factor, if terms of second and higher orders are neglected:

$$\xi := \frac{\Delta p_{CO_2}/p_{CO_2}^0}{\Delta\Sigma C/\Sigma C_0}. \tag{17}$$

In case p_{CO_2} is explicitely identified to p_s, ξ is the evasion factor according to [53] or the Revelle factor R as defined in [4]. If p_{CO_2} is assumed equal to the CO_2 partial pressure in the atmosphere, p_a, which approximately holds in case equilibration occurs relatively fast ξ is identified to γ according to [55] or to α^{-1} according to [56]. An average value of ξ in the ocean surface layer is currently about 10, i.e. an increase of the atmospheric CO_2 partial pressure by x% causes the inorganic carbon content of the oceanic surface layer to be raised by (x/10)%.

Beside the above formulations the following expressions have been used to provide for possibilities of computing time-dependent evasion factors without determination of $[H^+]$:

$$\xi = 9 + 4.9\frac{\Delta p_s}{p_{so}} - 0.1 \cdot \left(\frac{\Delta p_s}{p_{so}}\right)^2 \tag{18a}$$

[57]

$$\xi = \xi_0 \cdot \frac{1 + 0.18\dfrac{\Delta\Sigma C}{\Sigma C_0}}{1 - 4.72\dfrac{\Delta\Sigma C}{\Sigma C_0}} \tag{18b}$$

[58].

If reference is not made to an equilibrium state, the buffer factor

$$B = \frac{\Sigma C}{p_{CO_2}} \cdot \frac{dp_{CO_2}}{d\Sigma C} \tag{19}$$

represents the buffering action of the ocean at an arbitrary point (p_{CO_2}, ΣC). Since $p_{CO_2} = p_{CO_2}(\Sigma C, A, [H^+])$, we have

$$dp_{CO_2} = \left(\frac{\partial p_{CO_2}}{\partial \Sigma C}\right)_{A,[H^+]} d\Sigma C + \left(\frac{\partial p_{CO_2}}{\partial A}\right)_{\Sigma C,[H^+]} dA + \left(\frac{\partial p_{CO_2}}{\partial [H^+]}\right)_{\Sigma C, A} d[H^+] \tag{20a}$$

Assuming changes in $[H^+]$ being negligible Eq. (20a) yields

$$dp_{CO_2} \approx \left(\frac{\delta p_{CO_2}}{\delta \Sigma C}\right)_A d\Sigma C + \left(\frac{\delta p_{CO_2}}{\delta A}\right)_{\Sigma C} dA \tag{20b}$$

which leads to

$$B := \frac{\Sigma C}{p_{CO_2}} \left[\left(\frac{\delta p_{CO_2}}{\delta \Sigma C}\right)_A + \left(\frac{\delta p_{CO_2}}{\delta A}\right)_{\Sigma C} \frac{dA}{d\Sigma C}\right]. \tag{20c}$$

If this definition of the buffer factor is used, also dissolution of $CaCO_3$ according to

$$CaCO_3 + CO_2 + H_2O \rightarrow 2HCO_3^- + Ca^{2+} \tag{21}$$

can be accounted for. Thus the dissolution of 1 mole of $CaCO_3$ increases ΣC by 1 mole while it increases the alkalinity by 2 equivalents, such that $dA/d\Sigma C$ is positive. Thus, since

$$\left(\frac{\partial p_{CO_2}}{\partial A}\right)_{\Sigma C} = -\frac{1}{\alpha_s}\frac{K_2'}{K_1'}\frac{[HCO_3^-]}{[CO_3^{2-}]}(A-W) \qquad \left(W = \frac{K_B' \cdot \Sigma B}{K_B' + [H^+]} + \frac{K_W}{[H^+]} - [H^+]\right) \tag{22}$$

is negative, in case dissolution of $CaCO_3$ should occur, the buffer factor would be reduced which corresponds to an enhanced oceanic uptake capacity of additional atmospheric CO_2. If A is assumed constant,

$$(B)_A = \frac{\Sigma C}{p_{CO_2}} \left(\frac{\delta p_{CO_2}}{\delta \Sigma C}\right)_A \tag{20d}$$

is the buffer factor according to [137] or the homogeneous buffer factor according to [64].

Constants K_1', K_2', and K_B' have been determined as functions of temperature and salinity [59–63]. By examination of these data [64, 65] it was found that the data set compiled in [62] was in best agreement with the empirical values. The corresponding temperature and salinity dependence of the solubility has been determined in [66]. Determination of the temperature dependence of the evasion factor [4] showed that ξ decreases from about 14 at 0 °C to about 8 at 30 °C (p_{CO_2} = 320 ppm).

The carbon dioxide exchange flux between atmosphere and ocean

For short-term periods and for restricted regions atmospheric and oceanic transport as well as variations of biotic production and decomposition lead to deviations from equilibrium between atmospheric and (physically) dissolved CO_2.

In most carbon cycle models CO_2 exchange between atmosphere and oceans is treated as being linearly dependent on the average CO_2 partial pressures in atmosphere and surface ocean:

$$F_{as} - F_{sa} = F_{as0}\left(\frac{p_a}{p_{a0}} - \frac{p_s}{p_{s0}}\right). \tag{23}$$

We shall now turn to the different methods employed to estimate the CO_2 atmosphere-ocean exchange-time, τ_{as}.

(1) In a stationary state, the balance of ^{14}C in the oceans is given by net uptake from the atmosphere and radioactive decay [67]:

$$\left.\frac{dn^{14}_{oc,0}}{dt}\right|_{t=t_0} = 0 = F_{as0}(\alpha^{14}_{as}Q^{14}_{a0} - \alpha^{14}_{sa}Q^{14}_{s0}) - \lambda^{14}Q^{14}_{oc,0}\cdot n_{oc,0} \tag{24a}$$

$n^{14}_{oc,0}$, $n_{oc,0}$: total mole numbers of ^{14}C and C, resp., for a stationary state (subscript 0)

$$Q^{14}_{oc,0} = n^{14}_{oc,0}/n_{oc,0}$$

α^{14}_{as}, α^{14}_{sa}: fractionation factors for ^{14}C transfer from atmosphere to ocean and vice-versa

Q^{14}_{a0}, Q^{14}_{s0}: stationary isotopic ratios in atmosphere and surface ocean.

This yields

$$F_{as0} = \frac{\lambda^{14}Q^{14}_{oc,0}n_{oc,0}}{\alpha^{14}_{as}Q^{14}_{a0} - \alpha^{14}_{sa}Q^{14}_{s0}}. \tag{24b}$$

Using values of $\lambda^{14} = 1/(8267\text{ a})$, $\Delta^{14}_{oc,0}C \doteq -160‰$, $n_{oc,0} = 37{,}400$ Gt C/12.011 = 3,114 Gmoles, $\Delta^{14}_{a0}C = 0‰$, $\Delta^{14}_{s0}C = -45‰$, $\alpha^{14}_{as} = 0.99685$ and $\alpha^{14}_{sa} = 0.98066$ leads to an average exchange flux F_{as0} of about 61.5 GtC/a or an exchange time $\tau_{as} \approx 10$ a for an atmospheric partial pressure of 280 ppm. This value is very sensitive with respect to Q^{14}_{s0} which is only inaccurately known [68].

(2) The relatively fast perturbation of $^{14}C/C$ ratios induced by nuclear weapon explosions has also been used to determine the atmosphere-ocean exchange-time which yielded $\tau_{as} = 6.8$ a [69].

(3) In [70] the ^{222}Rn deficiency in the surface ocean is used to determine the effective diffusion-to-microlayer thickness ratio $D(^{222}Rn)/\Delta z$ which can then be used to give the CO_2 exchange flux:

$$F_{as0} = O\cdot\frac{D(CO_2)}{\Delta z}\cdot\alpha_s\cdot p_{CO_2} \tag{25}$$

O: ocean surface ($362\cdot 10^{12}\,m^2$)

or $\tau_{as} = (7\text{–}13\text{a})$. Attempts have been made to associate gas exchange fluxes with wind velocities over the ocean surface. Laboratory experiments [71] indicated that the exchange flux increases with the square of the wind velocity. Shipboard Rn-measurements [72] yielded a relation

$$g = \begin{cases} 0.075\cdot\bar{v} & \text{for } 0 \leqq \bar{v} < 4\text{ m/sec} \\ 0.72\cdot(\bar{v}-4) & \text{for } \bar{v} \geqq 4\text{ m/sec} \end{cases} \tag{26}$$

g: gas transfer rate in m/d, $\bar{v}$: wind velocity in m/sec

between total gas transfer rate and wind velocity. It has been pointed out that bubbles may enhance gas exchange [73]. ^{222}Rn deficiencies according to [70] exhibit only weak correlation with wind velocities. Underlying the nonlinear relationship [26] between gas transfer and wind velocity an average wind velocity cannot be used to calculate an average gas transfer rate.

(For details of the fractionation of carbon isotopes occurring along with CO_2 exchange between atmosphere and oceans see e.g. [74, 49].)

Water Mass Transport in the Oceans

The total sea-surface amounts to $362 \cdot 10^{12}$ m^2; with a mean depth of 3,729 m an ocean volume of $1.35 \cdot 10^{18}$ m^3 results [75]. The cumulated frequency distribution of certain distances between sea-floor and surface has been compiled [75].

In a first approach the oceans may be vertically subdivided into a relatively homogeneous, well-mixed (wind stress) surface layer and the underlying deep-sea. The vertical structure of the oceans and transport form the surface to the deep-sea deserve some attention, since they set limits to the uptake capacity of the oceans for additional CO_2. An investigation of the mixed-layer depths [76] yielded an average depth of 75 m for the North Pacific. According to [77] the southward and northward extension of this mixed layer varies with season in such a way that at mid-latitudes during the summer of the respective hemisphere a mixed layer occurs which disappears in wintertime.

Below the surface layer, temperature generally decreases monotonically with depth while salinity frequently exhibits a maximum at intermediate depths indicating horizontal advection. Overall, density increases monotonically with depth such that a relatively stable stratification results above all in equatorial regions. The largest temperature and salinity gradients occur at depths between 75 and 1,000 m giving rise to refer to this depth interval as the main thermocline or more precisely as the thermohalocline or pycnocline.

Ocean currents are subdivided into surface and deep currents, the former being caused by wind stress. The direction of surface currents is determined by the direction of the prevailing winds, the rotation of the earth (Coriolis effects) and the ocean topography. Deep-ocean currents are caused by down-sinking of polar waters primarily in the Norwegian Sea (North Atlantic Deep Water, NADW) and in the antarctic region (Antarctic Bottom Water, AABW) [78, 79].

Two possibilities of the formation of cold, deep water have been suggested: (1) Cooling of polar surface water mainly in wintertime leads to suspension of density stratification. Wind stress produced convection can be built up over most parts of the water column until by horizontal surface advection of warm water or by warming of surface water a new density stratification is established [78, 79]. (2) When sea-ice is formed, less salt is incorporated into the ice than corresponds to ambient sea-water salinity giving rise to high salinity brines which can sink down once a critical density is reached [77]. Together with these causes for deep-water formation the following process may be responsible for intermediate water (600–1,200 m) formation: Cooling in winter does not lead to a complete suspension of density stratification, but causes the density horizons to be lifted strongest at the location of most severe cooling. The equal density surfaces (isopycnals)

Table 1. Estimates of annually formed deep-water (in Sv)

NADW		AABW	Reference	
8		24	79	
	15–90		85	
6			86	Authors' estimate
2.0–7.6			86	Range of values cited
30		20	87	
		3–8	83	
10–30			88	

[1] $Sv = 10^6\ m^3/sec \approx 3.125 \cdot 10^{13}\ m^3/a$)
NADW: North Atlantic Deep Water
AABW: Antarctic Bottom Water

thus exhibit a slope along which water masses may glide downwards the horizontal velocity component being by far larger than the vertical one. Such a process has been observed in the North Pacific by ^3H measurements [80].

In equatorial regions especially at the western continental slopes deep-water upwelling may occur. Since the direction of the prevailing winds points towards the equator, coastal surface water is directed towards the open sea causing the above mentioned upwelling process.

Equatorial surface water can be transported polewards by surface currents and may also sink down in temperate latitudes [81, 82].

Polar surface water sinking down as newly formed bottom water can be replaced either by upwelling intermediate and bottom water [82, 83] or at least in part by horizontally inflowing surface water [84].

Apart from large scale water mass movement (advection) tracer transport is affected by turbulent diffusion which is itself caused by the irregular motion of eddies of random size [79].

Organic Material in the Oceans

The total biotic production in the ocean is estimated as 15–80 Gt C/a with a most probable range at 23–44 Gt C/a [89, 90].

According to [89] no growth stimulation of phytoplankton by additional CO_2 could be demonstrated to date. Although phytoplankton growth can be stimulated remarkably by additional (riverine) nitrogen and phosphorus, a long-term carbon sink hardly occurs, because additional sedimentation appears too low [91].

Total marine plant biomass estimates range from 0.08 to 1.76 Gt C; total marine zoomass is assumed as 0.45–2.93 Gt C.

According to [92] the following general relationship for the sinking rate of biogenic material has been obtained

$$\bar{F}_{bd}(z) = \frac{F_{sb}}{0.0238 \cdot z + 0.212} \tag{27}$$

with $\bar{F}_{bd}$ (z) being the flux of dead biotic carbon sinking below depth z (in m) and F_{sb} being surface biotic production. Thus 50% of the assimilated carbon sink to greater depths than 75 m, or 50% are decomposed in the surface layer. According to [89] at most 15% pass the 200 m depth level (20%, if the above formula is applied).

$\delta^{13}C$ of dissolved organic carbon in the deep-sea has been found to be in the range of −21.2‰–−24.4‰ [93].

Marine $CaCO_3$

Marine $CaCO_3$ is mainly produced as structural material of microorganisms in the form of calcite and aragonite. Apart from this, inorganic precipitation occurs in limited regions (e.g. the Bahama Bank). The overwhelming part of surface sea-water is oversaturated with respect to $CaCO_3$ leading to the assumption that kinetic factors control $CaCO_3$ precipitation rather than thermodynamic ones. Also the dissolution of $CaCO_3$ according to [21] appears to be controlled by kinetic effects (e.g. organic coatings) as can be seen from the fact that in ocean layers far below the saturation depth which separates the over-saturated from the under-saturated layers solid $CaCO_3$ occurs.

Additional atmospheric CO_2 has been assumed to cause an increase in acidity leading to an enhanced uptake capacity of sea-water (see above). However, as has been pointed out [4, 94] oversaturation should first have to be suspended before dissolution can occur in surface waters. Dissolution slightly above the compensation depth (3,000–5,000 m, the depth below which carbonate free sediments accumulate [192]) appears possible only when additional CO_2 reaches the respective sediments with an appropriate deep-water current which should take at least 300 a (mean age of NADW, [4]).

In contradiction to this conception it has been assumed [95] that (1) there are regions in high latitudes undersaturated with respect to $CaCO_3$ and that (2) carbonates with high Mg-calcite films exist which are supposed to respond to enhanced p_{CO_2}. The dissolution rate is regarded too low to be important in [4].

Accelerated dissolution during sinking [96] and under enhanced p_{CO_2} from the oxidation of additionally sedimented organic matter [97] are further mechanisms suggested. Also, from total carbon concentration and alkalinity profiles in the Pacific it is concluded that some $CaCO_3$ dissolution is likely to occur above the lysocline (depth of sharp increase in $CaCO_3$ net dissolution) [178].

The ratio of dissolved $CaCO_3$ to dissolved organic matter has been estimated as 0.25 [98] and 0.23–0.56 [99], respectively. For the North Atlantic and depths below 1,000 m values of 1.4 have been found [100]. $CaCO_3$ exhibits a $\delta^{13}C$ of about 0‰.

While dissolution of $CaCO_3$ leads to an increase in both ΣC and A, dissolution of organic material only increases ΣC. According to these dissolution processes the profiles of the above quantities exhibit a positive gradient at least up to depths of about 1,000 m. Below this level admixing of bottom water with less ΣC may leads to a relatively small decrease [101, 102]. The average ΣC in low and mid-latitude surface waters is about 2.0 moles/m^3.

Table 2. Estimates of NNP, standing biomass, litterfall, litter, consumption by animals and humus

	Reference		Reference
NPP		Litterfall	
(Ct C/a)		(Gt C/a)	
77.6 (potential)	182	50	109
48,3	105	42.6	108
45.1	106	48.2–57.6	110
52.9	107		
59.9	108	Litter	
		(Gt C)	
Living phytomass		50	107
(Gt C)		59.5	108
1,081 (potential)	182	210	111
827	107		
560	108	Humus	
		(Gt C)	
Animal consumption		700	112
(Gt C/a)		1,392	113
3.3	105	3,000	115
		1,080	116
		1,635, 2,070	108
Total animal mass		1,151 (1,860)	117
(Gt C)		1,515 (at present)	117
0.5	105		

The $\delta^{13}C$ profiles of marine dissolved inorganic carbon are also influenced by the dissolution of $CaCO_3$ and of organic matter leading to decreasing $\delta^{13}C$ values with increasing depths.

The stationary $^{14}C/C$ isotopic ratio profile is mainly affected by radioactive decay, likewise causing the $\Delta^{14}C$ values to decrease with depth [103, 104].

Terrestrial Biota

Structure of the unperturbed system. The total mass of carbon assimilated by the vegetation of a certain area per unit time is referred to as gross primary production (GPP), part of which is consumed in plant (autotroph) respiration (R_a). The difference $NPP = GPP - R_a$ is then called net primary production. If, furthermore, that fraction is subtracted which is consumed by man and animals or is decomposed by microorganisms (heterotrophic respiration, R_h) the net ecosystem production $NEP = NPP - R_h$ results. For ecosystems in climax states $NEP = 0$, i.e. organic mass is not changed. All dead plants and dead parts of plants are referred to as litter the largest part of which is oxidised by microorganisms or converted to relatively stable organic compounds (humus). Animal mass in ecosystems is usually low compared to phytomass.

Some global estimates have been summarised in Table 2 yielding averages of about 52 Gt C/a for NPP and 700 Gt C for living phytomass; more detailed compilations for different types of biomes are given in [107, 108, 118].

Human Impact on the Biota

Several kinds of perturbations of the biosphere have been found to occur: inadvertent fires and clearings in extratropical forests [119, 143] as well as shifting cultivation, increase of permanent agriculture and commercial clearings in tropical forests [114, 120] and desertification [121]. Areas affected by wildfires and by shifting cultivation can be assumed to be repopulated by the respective forest ecosystems, whereas clearings in extratropical forests as well as increase of permanent agriculture and shifting cultivation on tropical forest areas lead to grassland-like ecosystems. Desertification diminishes the grassland area.

Only the above-ground living phytomass is affected by forest fires and even this portion is combusted incompletely, the burning efficiency being about one third [120]. The remaining dead above-ground portion may now be converted to relatively stable chemical compounds (e.g. charcoal). Another consequence of fires and the destruction of plant life on a forest area is immediate accelerated oxidation of humus compounds only parts of which are released to the atmosphere (about 40%) the remainder being eroded [114]. It is estimated [114] that only due to this process 2.5–7.4 Gt C/a can be released. Estimates for total anthropogenically released biogenic CO_2 range from (1.0 ± 0.6) Gt C/a [123] over 0.9–3.0 Gt C/a [6] to 4–8 Gt C/a [5].

It is discussed at present to which extent the colonisation of the USA, Canada, Australia, and New Zealand in the period from 1860–1920 led to largely cleared areas which have been abandoned or reforested after a certain time. An area cleared between 1800 and 1920 of about $1.3 \cdot 10^{12}\ m^2$ in the USA has been attributed to this pioneer effect [122] corresponding to a cumulative release of (20 ± 2) Gt C. It is further estimated that other clearings may have contributed with about 50% to this effect [6] leading to a total release of (30 ± 3) Gt C for this period.

CO_2 Storage Stimulation by Enhanced CO_2 and NO_x Concentrations

Although photosynthesis can be accelerated under increased CO_2 partial pressures it may be questioned whether this effect leads to significant changes on the ecosystem level [191]. Many investigations which indicated that such stimulation exists have been carried out under optimal supply of water and nutrients which are, however, limiting in most non-agricultural ecosystems such that additional CO_2 can be stored only to less extent [124]. Furthermore, plants may often respond with partially closing their stomata to avoid water loss rather than to enhance CO_2 uptake [125]. On the other hand, such increased water-use efficiency can itself lead to enhanced relative assimilation especially in regions with extreme water stress [126]. Then the question arises to which extent additional assimilation causes additional long-term storage. Although for the above reasons doubt is cast on the effect in question, growth stimulation is nevertheless considered possibly causing a net withdrawal of up to 1 Gt C/a [124]. It has been shown that the observed increase of the seasonal amplitude of the Mauna Loa data is in accordance with a possible CO_2 stimulation [21].

A second possibility of anthropogenically induced enhanced carbon storage has been suggested to be caused by additional NO_x which is converted to NO_3^-

on the soil surface. It is estimated that at most 1 Gt C/a may additionally be stored by this process, but at the same time attention is directed towards the noxious effects of NO_x emissions [6]. This discussion of additional carbon storage stimulation (above all on a global scale) indicates that the uncertainty concerning this problem is considerable [191].

Analytical Treatment of 3-Box Models with Exponential Inputs

As has been said above modelling the biogeochemical cycle of a tracer is essentially identical to setting up a system of first order differential equations with the compartment sizes N_i as state variables:

$$\dot{N}_i = \sum_j (F_{ji} - F_{ij}) + \sum_e F_{e,i}(t). \tag{1a}$$

Here $\dot{N}_i$ denotes the derivative of carbon masses in compartment i with respect to time, fluxes from compartment i to j are denoted by $F_{ij} \cdot F_{e,i}$ (t) are explicitely time-dependent, externally caused fluxes such as fossil and biogenic input. (They are given negative signs, when they stand for fluxes leaving compartment i.) The fluxes inherent to the system, F_{ij}, in carbon cycle models, are frequently assumed to be exclusively influenced by the sizes of the compartments i and j they connect:

$$F_{ij} = F_{ij0} \cdot \Phi_{ij}\left(\frac{N_i}{N_{i0}}, \frac{N_j}{N_{j0}}\right). \tag{28}$$

In the same way the number of moles of carbon n_i in a compartment i can be used:

$$\dot{n}_i = \sum_j (f_{ji} - f_{ij}) + \sum_e f_{e,i}(t) \tag{1b}$$

f_{ij} and $f_{e,i}$: fluxes in moles/a.

When dealing with isotopes it is common practice to give an index of the isotopic composition rather than the respective mass or mole number values:

$$R_i^k = n_i^k / n_i^{12}, \tag{29a}$$

$$Q_i^k = n_i^k / n_i, \tag{29b}$$

$$N_i = (12 + Q_i^{13}) \cdot n_i \tag{30}$$

n_i^k: number of moles of isotope k in compartment i.

Since the variation of Q_i^{13} is negligible compared to 12, N_i can be assumed proportional to n_i. The fluxes of mole numbers for ^{13}C are:

$$f_{ij}^{13} = \alpha_{ij}^{13} R_i^{13} f_{ij}^{12} \tag{31a}$$

or

$$f_{ij}^{13} = \overline{\alpha_{ij}^{13}} Q_i^{13} f_{ij} \tag{31b}$$

$\overline{\alpha_{ij}^{13}}$ and α_{ij}^{13} can be interrelated by:

$$\overline{\alpha_{ij}^{13}} = \frac{\alpha_{ij}^{13}}{1 - (1 - \alpha_{ij}^{13}) \cdot Q_{i0}^{13}} \tag{32}$$

subscript 0: stationary state value,

where use has been made of $f_{ij}=f_{ij}^{12}+f_{ij}^{13}$. ^{14}C is only a minute fraction of total carbon mole numbers such that it can be neglected when one deals with the sum of the individual carbon isotope amounts. This also leads to

$$f_{ij}^{14}=\alpha_{ij}^{14}Q_i^{14}f_{ij}\,. \tag{31c}$$

From quantum chemical considerations the relationship

$$\frac{\alpha_{ij}^{14}}{\alpha_{ji}^{14}}=\left(\frac{\alpha_{ij}^{13}}{\alpha_{ji}^{13}}\right)^2 \tag{33}$$

has been derived [41]. Using equations like (31b, c) is advantageous, because in these cases n_i^{12} and f_{ij}^{12} need not to be determined.

A 2-Box Ocean Sub-model

We shall now investigate how simple 2-box ocean sub-models respond to an exponential fossil input (no biosphere nor biogenic net transfer) i.e. under which conditions they are able to reproduce the observed airborne fraction and the Suess effect.

The state equations of carbon masses in the atmosphere, surface-ocean and deep-sea compartments are:

$$\begin{aligned}\dot{N}_a&=-k_{as}N_a+k_{sa}[\xi(N_s-N_{s0})+N_{s0}]+F_f(t)\\ \dot{N}_s&=k_{as}N_a-k_{sa}[\xi(N_s-N_{s0})+N_{s0}]-k_{sd}N_s+k_{ds}N_d\\ \dot{N}_d&=k_{sd}N_s-k_{ds}N_d\end{aligned} \tag{34a–c}$$

k_{ij}: exchange constants
$F_f(t)=F_{f0}\cdot\exp[k_f(t-t_0)]$: fossil input

$$k_f=1/(30a)\qquad k_{as}=1/(7.5a)\qquad \xi=10$$
$$k_{ds}=1/(1{,}000a)\,.$$

From steady state conditions

$$k_{as}N_{a0}=k_{sa}N_{s0} \tag{35a}$$

and

$$k_{sd}N_{s0}=k_{ds}N_{d0} \tag{35b}$$

as well as from

$$\kappa_{as}:=N_{a0}/N_{s0} \tag{36a}$$

and

$$\kappa_{oc}:=N_{a0}/(N_{s0}+N_{d0}) \tag{36b}$$

$$k_{sa}=\kappa_{as}k_{as} \tag{37a}$$

and

$$k_{sd}=\left(\frac{\kappa_{as}}{\kappa_{oc}}-1\right)k_{ds} \tag{37b}$$

follow. Transition to mass deviations from steady state masses N_{io} yields:

$$\frac{d}{dt}\Delta N_a = -k_{as}(\Delta N_a - \xi\kappa_{as}\Delta N_s) + F_f(t),$$

$$\frac{d}{dt}\Delta N_s = k_{as}(\Delta N_a - \xi\kappa_{as}\Delta N_s) - k_{ds}\left[\left(\frac{\kappa_{as}}{\kappa_{oc}} - 1\right)\Delta N_s - \Delta N_d\right], \qquad (38a\text{–}c)$$

$$\frac{d}{dt}\Delta N_d = k_{ds}\left[\left(\frac{\kappa_{as}}{\kappa_{oc}} - 1\right)\Delta N_s - \Delta N_d\right]$$

which together with mass conservation

$$N_f(t) = \int_{t_0}^{t} F_f(t')dt' = \Delta N_a + \Delta N_s + \Delta N_d \qquad (39)$$

leads to a differential equation of second order in ΔN_a:

$$\frac{d^2}{dt^2}\Delta N_a + a_1\frac{d}{dt}\Delta N_a + a_0\Delta N_a = \dot{F}_f(t) + b_1F_f(t) + b_0N_f(t) \qquad (40)$$

with

$$a_0 = b_0 + k_{as}k_{ds}\frac{\kappa_{as}}{\kappa_{oc}}$$

$$a_1 = b_1 + k_{as} \qquad (41a\text{–}d)$$

$$b_0 = \xi\kappa_{as}k_{as}k_{ds}$$

$$b_1 = \kappa_{as}(\xi k_{as} + k_{ds}/\kappa_{oc}).$$

If the onset of the perturbation can be assumed smooth such that $\frac{d}{dt}\Delta N_a(t_0) \approx 0$, and the time range considered is larger than $3/k_f$ which allows us to use the approximation $N_{f0} \approx F_{f0}/k_f$, then the solution of the differential equation (40) is given by:

$$\Delta N_a(t) = Ae^{\lambda_+(t-t_0)} + Be^{\lambda_-(t-t_0)} + Ce^{k_f(t-t_0)} \qquad (42)$$

with

$$\lambda_\pm = -\frac{a_1}{2} \pm \sqrt{\left(\frac{a_1}{2}\right)^2 - a_0}. \qquad (43)$$

The first two terms reflect the transition of the system from steady state to permanent exponential perturbation. Since both eigenvalues are purely real and negative these expressions can be neglected in long-term considerations. In this case the airborne fraction α is:

$$\frac{\Delta N_a}{N_f} =: \alpha = \cfrac{1}{1 + \cfrac{k_{as}}{k_f}\cfrac{1}{1 + \cfrac{\xi\kappa_{as}k_{as}}{k_f}\cfrac{1}{1 + \cfrac{\kappa_{as}}{\kappa_{oc}}\cfrac{1}{1 + k_f/k_{ds}}}}}. \qquad (44)$$

This relationship could be derived within the framework of the adiabatic approximation for general, not necessarily exponential input functions as well as for many other model assumptions [58, 162]. If we now assume all constants appearing in Eq. (44) except κ_{as} to be given, then we may select such a value for the surface layer depth that the observed $\alpha_f = 55\%$ can be best reproduced. Underlying the data given above together with $\kappa_{oc} = 1/60$ this can be accomplished choosing $\kappa_{as} = 1/8.8$ or a surface layer depth of $z_s = 565$ m. $k_f = 1/(22\ a)$ which is appropriate for about the last 30 years yields $\kappa_{as} = 1/9.9$ and $z_s = 636$ m. This then cannot be regarded as a representation of the oceanic well-mixed surface layer, but is entirely required by modelling conditions which is best illustrated by the fact that the surface compartment depth depends on k_f. As can be shown by the respective 1-box ocean model there is virtually no net flux to the deep-sea for the first 120 a such that a 1-box model ocean would dynamically behave similarly. Furthermore, as we shall see in what follows the reproduction of the Suess effect requires a much thinner surface layer. To overcome this difficulty it has been suggested that the land biota should be treated as a net sink for atmospheric CO_2 in such a manner that the vegetation cover is thought to respond with higher CO_2 assimilation to higher ambient CO_2 concentrations which has been expressed by biota growth factors (see below).

We shall now investigate the behaviour of the $^{14}C/C$ ratios or the Q_i^{14} under the above model assumptions. Particularly the dilution of ^{14}C in the atmosphere due to the input of fossil CO_2 almost completely lacking any ^{14}C will be of interest. To this end, we first have to set up a convenient system of differential equations which can be derived following [127] for ^{14}C as well as for ^{13}C. Starting from

$$\frac{dn_i^k}{dt} = \sum_i (f_{ji}^k - f_{ij}^k) + \sum_e f_{e,i}^k(t) \tag{1b}$$

we first neglect any fractionation such that the isotope fluxes read as

$$f_{ij}^k = Q_i^k f_{ij}. \tag{31d}$$

This implies that all steady state values Q_{i0}^k are equal. Only the isotopic ratios of fossil and biogenic input CO_2 are assumed to differ from those of the unperturbed system. Furthermore, natural production and radioactive decay of ^{14}C are neglected, since λ^{14} is about an order of magnitude less than the smallest exchange constant, k_{ds}, occurring in the system under consideration. From

$$\frac{dQ_i^k}{dt} = \frac{1}{n_i}\left[\frac{dn_i^k}{dt} - Q_i^k \frac{dn_i}{dt}\right], \tag{45a}$$

$$\frac{dQ_i^k}{dt} = \frac{1}{n_i}\left[\sum_j (Q_j^k - Q_i^k) f_{ji} + \sum_e (Q_e^k - Q_i^k) f_{e,i}(t)\right] \tag{45b}$$

is obtained. Analogously to the procedure in the case of total carbon masses, eqns. (38), we introduce deviations from steady state isotopic ratios $\Delta Q_i^k = Q_i^k - Q_{i0}^k$:

$$\frac{d}{dt}\Delta Q_i^k = \frac{1}{n_i}\left\{\sum_j (\Delta Q_j^k - \Delta Q_i^k) f_{ji} + \sum_e [Q_e^{13} - (\Delta Q_i^k + Q_{a0}^k)] f_{e,i}(t)\right\} \tag{46}$$

($Q_{i0}^k = Q_{a0}^k$ for all i).

Next use of the isotopic labels

$$l_i^k := n_i \frac{\Delta Q_i^k}{Q_{a0}^k} \tag{47}$$

yields

$$\frac{d}{dt} l_i^k = \sum_j \left(\frac{f_{ji}}{n_j} l_j^k - \frac{f_{ij}}{n_i} l_i^k \right) + \sum_e \nu_{e,i}^k \cdot f_{e,i}(t) \tag{48}$$

with

$$\nu_{e,i}^k = \frac{Q_e^k}{Q_{a0}^k} - 1. \tag{49}$$

If we assume that flux deviations from steady state values are small compared to the respective total fluxes, we may apply the linearised forms of Eq. (28):

$$f_{ij} \approx f_{ij0} \left(1 + \psi_{\underline{i}j} \frac{\Delta n_i}{n_{i0}} + \psi_{i\underline{j}} \frac{\Delta n_j}{n_{j0}} \right) \tag{50}$$

with

$$\psi_{\underline{i}j} = \left. \frac{\partial \Phi_{ij}}{\partial (n_j/n_{i0})} \right|_{t=t_0} \tag{51a}$$

and

$$\psi_{i\underline{j}} = \left. \frac{\partial \Phi_{ij}}{\partial (n_j/n_{j0})} \right|_{t=t_0}. \tag{51b}$$

In the case of the 2-box ocean sub-model

$$\psi_{\underline{a}s} = 1\,, \qquad \psi_{\underline{s}a} = \xi\,, \qquad \psi_{\underline{s}d} = \psi_{\underline{d}s} = 1$$

and all $\psi_{i\underline{j}}$ vanish since none of the fluxes f_{ij} inherent to the system is assumed to depend on carbon masses in the acceptor compartment j. This implies:

$$\frac{f_{ij}}{n_i} \approx \frac{f_{ij0} \left(1 + \psi_{\underline{i}j} \frac{\Delta n_i}{n_{i0}} + \psi_{i\underline{j}} \frac{\Delta n_j}{n_{i0}} \right)}{n_{i0} \left(1 + \frac{\Delta n_i}{n_{i0}} \right)} \tag{52a}$$

$$\frac{f_{ij}}{n_i} \approx \frac{f_{ij0}}{n_{i0}} = k_{ij}\,. \tag{52b}$$

We can now rewrite Eq. (48) to give

$$\frac{d}{dt} l_i^k = \sum_j (k_{ji} l_j^k - k_{ij} l_i^k) + \sum_e \nu_{e,i}^k f_{e,i}(t)\,. \tag{53}$$

By this procedure a system of differential equations has been obtained the only state variables of which are the isotopic labels l_i^k. [No carbon masses or mole numbers appear in Eq. (53).] If this equation is compared to the respective state equation for carbon masses [e.g. Eq. (38) for the 2-box model] then it becomes evident that isotopic ratios or their deviations from steady state depend on total carbon exchange whereas carbon mass deviations are functions of additional fluxes. More specifically, isotopic labels are not significantly influenced by the oceanic buffer factor. In addition, even if we regard the biosphere as neutral with respect to additional atmospheric CO_2, it nevertheless acts as a medium attenuating perturbations of carbon isotope ratios in the atmosphere. For, in such a case, two fluxes of equal total carbon masses enter and leave the biosphere the isotopic ratios of which differ, however, due to varying isotopic ratios in the atmosphere. Likewise, the ocean acts as such an attenuating reservoir, and we shall now apply the 2-box model to an interpretation of the Suess effect. In so doing it must be kept in mind that by neglecting the biosphere the attenuating capacity of the model ocean has to be exaggerated, if it is to reproduce the observed ^{14}C dilution in the atmosphere.

This is normally given as $\Delta_a^{14}C = -20‰$ [48] for the time from 1860 to 1950. The isotopic labels l_i^{14} are connected with the respective $\Delta_i^{14}C$ values by

$$l_i^{14} = n_i \frac{\Delta(\Delta_i^{14}C)}{1{,}000} \cdot \frac{A_{standard}}{A_{a0}} . \tag{54}$$

We can simplify this relationship by assuming $A_{standard} \approx A_{a0}$. Particularly, in the case of the atmosphere, $\Delta^{14}C \approx 0$ holds, thus giving

$$l_a^{14} = n_a \frac{\Delta_a^{14}C}{1{,}000} . \tag{54a}$$

Similarly

$$l_i^{13} \approx n_i \frac{\Delta(\delta_i^{13}C)}{1{,}000} \frac{R_{standard}^{13}}{R_{a0}^{13}} \tag{54b}$$

holds.

In the case of our 2-box ocean sub-model Eq. (53) then takes the form

$$\frac{d}{dt} l_i^{14} = \sum_j (k_{ji} l_j^{14} - k_{ij} l_i^{14}) - \sum_e \nu_{f,i}^{14} f_f(t) . \tag{53a}$$

Since for the moment we consider only one externally caused input into the system, namely fossil input into the atmosphere, the last term vanishes for the surface layer and deep-sea compartments.

$$\nu_{f,a}^{14} = \nu_f^{14} = -1 , \tag{55}$$

since $Q_f^{14} = 0$.

$$\frac{d}{dt} l_a^{14} = k_{sa} l_s^{14} - k_{as} l_a^{14} - f_f(t) . \tag{56}$$

Introducing mass conservation for ^{14}C in the form

$$-n_f(t) = -\int_{t_0}^{t} f_f(t')dt' = l_a^{14} + l_s^{14} + l_d^{14} \tag{57}$$

the system of differential equations can be transformed into one differential equation of second order in l_a^{14} in much the same way as was done above for total carbon mass deviations:

$$\frac{d^2}{dt^2} l_a^{14} + a_1^* \frac{d}{dt} l_a^{14} + a_0^* \cdot l_a^{14} = -[\dot{f}_f(t) + b_1^* f_f(t) + b_0^* n_f(t)] \tag{58}$$

with

$$\begin{aligned} a_0^* &= b_0^* + k_{as} k_{ds} \frac{\kappa_{as}}{\kappa_{oc}} \\ a_1^* &= b_1^* + k_{as} \\ b_0^* &= \kappa_{as} k_{as} k_{ds} \\ b_1^* &= \kappa_{as}(k_{as} + k_{ds}/\kappa_{oc}) \end{aligned} \tag{59a–d}$$

which are the same for l_a^{13}. Under the same conditions as given above the approximated solution of Eq. (58) is:

$$-\frac{l_a^{14}}{n_f} = -\frac{n_a}{n_f} \frac{\Delta_a^{14}C}{1{,}000} = \frac{b_0^* + b_1^* k_f + k_f^2}{a_0^* + a_1^* k_f + k_f^2}. \tag{59}$$

Assuming $\alpha_f = 55\%$, $\Delta_a^{14}C = -20‰$, and N_f (1950) = 60 Gt C, κ_{as} can then be calculated as 1/1.68 such that the respective ocean model version can reproduce the observed Suess effect. The large difference between this value and the one obtained from the airborne fraction shows that the 2-box approximation for ocean models is in poor internal consistency, but may be useful in qualitative argumentation.

3-Box Models of the Global Carbon Cycle

In this section it will be derived how and under which conditions the ratio of $\Delta(\delta^{13}C)$ and $\Delta(\Delta^{14}C)$ in the atmosphere can be employed to estimate the poorly known biogenic input from the better known fossil CO_2 release [34, 127].

We consider the isotopic labels l_i^{13} and l_i^{14} for a 3-box carbon cycle model consisting of ocean surface layer, atmosphere and land biota. The anthropogenically caused biogenic input is regarded as originating from outside the system, thus not leaving the biosphere compartment. It is assumed to exhibit about the same constant $^{13}C/C$ ratio as fossil fuel CO_2 such that $\nu_f^{13} = \nu_b^{13} = \nu_{f,b}^{13}$. The $^{14}C/C$ ratio of the biogenic input is assumed equal to Q_{a0}^{14} such that $\nu_b^{14} = 0$.

Following our above practice including ^{13}C mass conservation as

$$\nu_{f,b}^{13} \cdot [n_f(t) + n_{b,rel}(t)] = \nu_{f,b}^{13} \cdot \int_{t_0}^{t} [f_f(t') + f_{b,rel}(t')] dt' = l_a^{13} + l_s^{13} + l_b^{13} \tag{60}$$

$f_{b,rel}$ ($n_{b,rel}$): (cumulated) biogenic flux
the following differential equations of l_a^{13} and l_a^{14}, resp., are obtained:

$$\begin{aligned} \frac{d^2}{dt^2} l_a^{13} + c_1^* \frac{d}{dt} l_a^{13} + c_0^* l_a^{13} &= \nu_{f,b}^{13} \{\dot{f}_f(t) + \dot{f}_{b,rel}(t) \\ &\quad + d_1^* [f_f(t) + f_{b,rel}(t)] + d_0^* [n_f(t) + n_{b,rel}(t)]\} \\ \frac{d^2}{dt^2} l_a^{14} + c_1^* \frac{d}{dt} l_a^{14} + c_0^* l_a^{14} &= \nu_f^{14} [\dot{f}_f(t) + f_f(t) + n_f(t)] \end{aligned} \tag{61a, b}$$

with

$$\begin{aligned} c_0^* &= d_0^* + (k_{sa} + k_{as})k_{ab} \\ c_1^* &= d_1^* + k_{as} + k_{ab} \\ d_0^* &= k_{sa}k_{ba} \\ d_1^* &= k_{sa} + k_{ba}\,. \end{aligned} \tag{62a–d}$$

The approximated solutions then give

$$\frac{l_a^{13}}{l_a^{14}} = \frac{\nu_{f,b}^{13}}{\nu_f^{14}} \frac{N_f(t) + N_{b,rel}(t)}{N_f(t)} \tag{63}$$

(Stuiver's approximation).

The above relationship must, however, be used with care for the following reasons: (1) It does not hold in the transient state, i.e. during the transition from the stationary to the permanently perturbed state. Apart from this, if $\dot{F}_{b,rel}(t_0) \neq 0$, the above procedure is likewise not justified. (2) The combustion of parts of the vegetation cover not only leads to a biogenic input, but also diminishes the capacity of the biota to attenuate perturbations of the atmospheric isotopic ratios.

Moreover, the biogenic input which is treated as a net input in Eq. (61) may at least in part be taken up by the biota again which could almost not be detected on the basis of the above assumptions. Thus, what can be estimated from Stuiver's approximation is the CO_2 released by human interference into the land biota but not the net (release minus uptake) biospheric CO_2 loss. Although an error estimation is independent from the ocean sub-model conceived, it is not easily made, because it depends on the time-dependent functions for fossil and biogenic input. (Only unrealistic models in which e.g. the same exponential growth function is used to describe both inputs can be handled conveniently.)

Box Models of the Global Carbon Cycle

In this section some models are presented the compartmentalisation of which is relatively crude. Carbon reservoirs like the continental biomass or marine carbon are in many cases described by at most two compartments. (The problem associated with these models have been discussed in the preceeding section.)

Models which exclusively simulate $^{14}C/C$ ratios [128–132] are not presented here nor are those which do not contain a figure for the airborne fraction α as in important validation quantity [116, 118, 133–136].

If the model calculated airborne fraction

$$\alpha(t_2, t_1) = \frac{N_a(t_2) - N_a(t_1)}{N(t_2) - N(t_1)} \tag{64}$$

$t_1 < t_2$, $N(t)$: cumulated total input

is less than the empirical airborne fraction α_f with respect to fossil input for the time range under consideration, then the model can accomodate for a higher simulated input than has been emitted from fossil sources. In this case a model biosphere could act as a net source for atmospheric CO_2. Should, however, $\alpha(t_2, t_1)$ be greater than $\alpha_f(t_2, t_1)$, then the model biosphere must take up additional carbon dioxide, i.e. must be regarded as a sink.

The evasion factor ξ as defined in Eq. (17) is, in fact, not independent from N_s; as has been shown in [159], however, the linear approximation (ξ = const) can be used in 2-box ocean models with exponential fossil input without large error. Should a more complicated, e.g. logistic input be introduced this approximation may turn out problematic.

The biota growth factor γ is usually defined as

$$\gamma := \frac{\Delta F_{ab}/F_{ab0}}{\Delta N_a/N_{a0}} \tag{65}$$

which can also be obtained by linearisation of e.g.

$$F_{ab} = F_{ab0}\left(\frac{N_a}{N_{a0}}\right)^{\gamma}. \tag{66}$$

(For other assumptions for the assimilation kinetics are discussed in [160].)

The regrowth of vegetation on cleared areas can be modeled using the following kinetics:

$$F_{ab} - F_{ba} = F_{ab0}\left[\left(\frac{N_b}{N_{b0}}\right)^{\delta} - \left(\frac{N_b}{N_{b0}}\right)\right] \tag{67}$$

$0 \leq \delta < 1$: climax parameter.

If we start with $N_b < N_{b0}$ (N_{b0}: stationary state climax biomass), then $F_{ab} > F_{ba}$ until $N_b = N_{b0}$ i.e. a new climax state is reached.

Using a model with one compartment for atmosphere and ocean each an expression for the ^{14}C Suess effect is derived which together with the observed $^{14}C/C$ decrease is used to determine the CO_2 exchange-time from atmosphere to ocean ($\tau_{as} \approx 10$ a) [55]. Regarding the buffering action of sea-water chemistry the relative CO_2 concentration increase due to fossil input is estimated as $23\% < \alpha < 46\%$.

The system behaviour of a 3-box model consisting of atmosphere, oceanic surface layer and deep-sea is described in [137] as the solution of a differential equation of third order. The airborne fraction in response to fossil fuel input up to 1954 is calculated as depending on τ_{as} and τ_{ds}. Variation of τ_{as} in the range of 2 to 10 a has only little influence. Varying τ_{ds} from 200 a to ∞ leads to changes in the calculated airborne fraction from 67% to nearly 100% ($\xi = 12.5$). (For a critique of the method of calculating the Suess effect see [140].) For a 5-compartment model (additional living biomass and humus, $\gamma = 1$) an airborne fraction of $\alpha = 51\%$ is estimated.

A further 2-box model ocean with a surface layer of 75 m depth and a main thermocline compartment ranging from 70 to 1,000 m depth has been presented [56]. A fraction of 65–67% of a fossil puls input with the mean age of the realistic fossil CO_2 release is assumed to remain airborne. In a second model it is taken into account that the ocean stratification is less well developed in polar regions thus providing immediate contact between deep-sea and atmosphere in these parts of the oceans ($50\% \leq \alpha \leq 54\%$).

Table 3. Features of some box models

Reference		Number of boxes for			Model parameters											Airborne fraction
		Atmosphere	Inorganic dissolved C	Terrestrial biota	Ocean					Terrestrial biota				Marine biota	$CaCO_3$-dissolution	α [%]
					$1/\kappa_{as}$	τ_{as} [a]	τ_{ds} [a]	C exchange atm.-polar waters	ξ	Arrangement (1)	γ	δ	$F_{b,rel}$	Downsinking, dissolution		
55		1	1		40	7	–		10							23–46
127	(2)	1	2		0.8–2	2–10	500		12.5							75–79
	(2)	1	2	2		5	500		12.5	B/H	1	0				51
56		1	2		1	10±3			10							65–67
		1	3		1	10±3		x	10							50–54
138	$\xi_{md}=8$	1	2		1.33	8.8	1,250		10							56
	$\xi_{md}=6$	1	2		1	6.4	1,260	x	10							56
		1	2		1	6.4	1,080	x	10							56
139		2	2	2	1.83	4.5	1,600		10	L/S	0.25					65
140	(4)	1	2	2	2.0	7	1,500		ξ(t)	L/S	0.3	1				46
53	(4)	2	2	2	2.0	6.3	1,500		ξ(t)	S(L)	0.56	1		x		46
		2	2	2	4	7	1,500		ξ(t)	S(L)	0.44	1		x		46
141		1	2	1	5.3	7.5	1,209		8.9		0.2	1			x	57
127		1	2	1	1.3	7.5	1,129		8.9		0.29	1		x		54
		1	2	1		7.5	1,129		2.4		−0.11	1		x	x	54

142	1	2	2	1.74	7	1,084	(5)	S(L)	0.35	1		56
	1	2	2	6.14	7	1,084	(5)	S(L)	0	1		56
	1	2	2	6.14	7	1,084	(5)	S(L)	0.44	1	(6a)	56
	1	2	2	3.26	7	1,084	(5)	S(L)	0.44	1	(6b)	56
	1	2	2	1.44	7	1,084	(5)	S(L)	0.44	1	(6c)	56
143	1	2	2	4.5	12.5	500	(7)	L(S)	0.3	1	(8)	65

(1) B/H: living biomass and humus (arranged in sequence); L/S: long- and short-lived biomass (parallel); L(S): assimilation of long-lived biomass is controlled by amount of short-lived biomass; S(L): vice-versa

(2) α and ^{14}C Suess effect calculated for $2a \leqq \tau_{as} \leqq 10a$ and for $200a \leqq \tau_{ds} < \infty$

(3) α and ^{14}C Suess effect estimated

(4) In addition to (2) α and ^{14}C Suess effect calculated for $1/4 \leqq \kappa_{as} \leqq 1$; $\xi(t)$: calculation of the evasion factor according to the chemical reactions involved

(5) $\xi(p_s/p_{s0}) = -0.122\,(p_s/p_{s0})^2 + 5.36\,(p_s/p_{s0}) + 3.6$

(6) $F_{b,rel} = d \cdot F_f$; $F_{b,rel}$: CO_2 released by vegetation destruction; a) $d=0.32$, b) $d=0.18$, c) $d=0.08$

(7) Depending linearly on ΔN_a $\quad \xi(\Delta N_a) = \xi_0 + d \cdot (\Delta N_a/N_{a0}) \quad \xi_0 = 9 \quad d=4$

(8) $F_{b,rel} = \dot{M}_b = 0.03 \cdot M_b \left(1 - \frac{M_b}{315\,\text{Gt C}}\right)$

$M_b(1860) = 3.5\,\text{Gt C}$

A similar result has been obtained [138] using models which likewise admit the reflection of direct carbon exchange between atmosphere and ocean. An attempt is made to account for the distinct buffering action in warm and cold surface waters or for a steady state net circulation of CO_2 from warm surface waters via the atmosphere to polar surface waters ($\alpha \approx 56\%$).

A 4-box model consisting of atmosphere, land biota, surface layer of the oceans and deep-sea with exclusively linear, donor-controlled fluxes results in a small simulated increase because the buffer mechanisms is neglected and a biota growth factor of $\gamma = 1$ is assumed [147]. This situation is not changed significantly, if the assimilation of CO_2 by land plants is described by a Michaelis-Menten kinetics instead of a term proportional to atmospheric CO_2 concentrations [144, 145]. A further model with exclusively linear, donor-controlled exchange fluxes yields an airborne fraction of about 8–12% [157].

In a 2-box ocean model [139] the atmosphere-ocean exchange-time ($\tau_{as} = 4.5$ a) is derived from the decrease of the atmospheric ^3H concentration after the bomb input. With a surface layer depth of about 110 m and a biota growth factor of $\gamma = 0.25$ an airborne fraction of 65% is obtained.

The box model studied most intensively which is to describe the exchange behaviour not only of total carbon but also of $^{13}C/C$ and $^{14}C/C$ has been presented in several variations [53, 127, 140, 141, 146]. Apart from an atmosphere subdivided into troposphere and stratosphere, short- and long-lived biota compartments are presented the growth of the short-lived biota being assumed to be controlled by the respective amount of carbon in the long-lived biota compartment. Like in the preceeding models the ocean is described by 2 boxes. For the first time, the evasion factor is explicitely calculated from chemical equilibrium constants and is not prescribed as being constant. $\tau_{as} \approx 6.3$ a and $\tau_{ds} = 1{,}500$ a given, the Suess effect of $-19.9‰$ and an airborne fraction of $\alpha = 46\%$ are best reproduced, if $\kappa_{as} = 1/2$ and $\gamma = 0.56$ are assumed. A smaller biota growth factor of 0.44 is only compatible with κ_{as} being diminished to 1/4 leading to an exaggerated oceanic attenuation capacity for $^{14}C/C$ perturbations in the atmosphere thus the calculated Suess effect being changed to $-18.0‰$.

From this model the difficulties associatied with describing the response behaviour of the global carbon cycle to external perturbations by models with a few boxes become transparent. In all these cases it is inevitable to assume either large biota growth factors or large ocean surface layers. It is calculated [148] that in order to reproduce $\alpha = 56\%$ with $\gamma = 0$ a surface layer depth of 812 m has to be introduced (565 m or 636 m, respectively, with the parameter sets used in the preceeding section). In some of the modifications of this model $CaCO_3$ dissolution near the ocean surface as well as in the deep-sea has been incorporated. Following these calculations an effect of $CaCO_3$ dissolution in the deep-sea would become apparent after about 300 a, while the assumed complete $CaCO_3$ dissolution (surface layer plus deep-sea) would have an effect already in the first half of the next century.

This model conception has then been applied to the simulation of an additional biogenic input which can be accomplished for by extending the surface layer depth ($\alpha = 56\%$) [142].

In reference [143] the carbon cycle is described by a 5-box model which apart from exhibiting the indicated properties of box models (large γ, small κ_{as}) contains a biosphere the carbon mass of which is controlled by the amount of short-lived biotic material. In this way, one can describe how in a relatively short period of time additional CO_2 is extracted from the atmosphere and fixed in the land biota. Apart from this, a logistic biogenic input has been defined. κ_{as} is determined according to the decrease of the bomb produced atmospheric ^3H peak. From this model simulation an airborne fraction of 65% is obtained.

An 8-box model has been constructed [148–155] which in addition to an atmospheric CO_2 compartment contains boxes for inorganic dissolved carbon in the oceanic surface layer and the deep-sea, land biota, humus, phytoplankton, zooplankton, dead organic material in the oceans, and ocean sediments. Apart from a land biota growth factor of $\gamma=0.81$ the uptake of carbon by phytoplankton is proportional to the CO_2 partial pressure in the surface layer. This means that with respect to total dissolved inorganic carbon a biota growth factor of $\gamma=\xi\approx 9$ or of $\gamma\approx 1$ with respect to atmospheric CO_2 results for this flux. Under these conditions an airborne fraction of $\alpha=17\%$ for purely fossil input is obtained [150].

In reference [156] it is attempted to more precisely describe the carbon transfer by the marine biota using compartments for dead organic matter in shallow waters and in the deep-sea, resp. Land biota have not been accounted for. Similarly to the model above, phytoplankton assimilation is proportional to atmospheric CO_2 concentrations as well as phytoplankton reservoir size. Such bilinear mechanisms are also introduced for zooplankton grazing rate and the sinking rate of dead organic matter. The evasion factor is $\xi=1$ and deep-sea carbon dioxide is not in equilibrium with HCO_3^- and CO_3^{2-}, but is exchanged with surface layer CO_2. For a total input of 300 Gt C until the year 2000 $\alpha=39\%$ is obtained.

Hoffert [158] starts with a latitude dependent distribution of fossil input but the increase in atmospheric CO_2 partial pressure is calculated after a spatial average has been computed. The net flux from atmosphere to ocean is assumed proportional to the difference between an equilibrium partial pressure $p_{a,\infty}$ (t) and the respective partial pressure of CO_2 in the atmosphere, p_a (t) ($\xi=$const):

$$\frac{dp_a}{dt}=k_{as}[p_{a,\infty}(t)-p_a(t)]+g_p\cdot F_f(t) \tag{68a}$$

[$g_p=0.4685$ ppm/Gt C, see Eq. (3)].

Under conservation of carbon mass this can be transformed to

$$\frac{d}{dt}\Delta N_a=-k_{as}\left[\Delta N_a-\frac{1}{1+\dfrac{1}{\xi\kappa_{oc}}}N_f(t)\right]+F_f(t) \tag{68b}$$

which is equivalent to an ocean sub-model consisting of one box of the size of the total ocean reservoir ($\alpha\approx 42\%$). In addition to this so-called finite ocean model an infinite model ocean is presented in which $p_{a,\infty}(t)=p_{a0}$ implying

$$\frac{d}{dt}\Delta N_a=-k_{as}\Delta N_a+F_f(t) \tag{68c}$$

which is a special case of Eq. (34a) with $\xi=0$ ($\alpha\approx 31\%$).

General Considerations Concerning More Detailed Ocean Models

As can be seen from the preceeding sections the need for more detailed ocean models arose mainly from two reasons: It could be expected that such efforts were to reproduce the Suess effect and the atmospheric CO_2 increase within the framework of one model conception which implied that the source-sink problem would be at least mitigated.

The concentration of an oceanic tracer at a certain depth can be influenced by turbulent diffusion, advection, biotic uptake, and release and, in the case of unstable isotopes, by radioactive decay. This is expressed in the following equation for a water column of depth-independent cross-section:

$$\frac{\partial c^k(z,t)}{\partial t} = \frac{\partial}{\partial z}\left[K(z)\frac{\partial c^k}{\partial z}\right] - \frac{\partial}{\partial z}[w(z)\cdot c^k] + \varphi(z)\ \{-\lambda^k c^k\} \tag{69}$$

z: depth
c^k: concentration of isotope k
K(z): diffusion constant
w(z): upwelling velocity
$\varphi(z)$: dissolution of biotic material per unit volume and per unit time.

[69] is a partial differential equation for all carbon isotopes all quantities of which appear as independent of geographic position. (In the next section we shall, however, deal with models in which the horizontally inhomogeneous current pattern of the oceans has been taken into account.) As boundary conditions to the above equation the surface layer concentrations must at any time be equal to the concentrations in the upper boundary of the deep-sea and exchange with the sea-floor is normally assumed to vanish.

It is now possible to generate functions K(z), w(z), and $\varphi(z)$ from the stationary state assumption for at least three significantly different concentration or isotopic ratio profiles. This is one way of calibrating an ocean model. In the other, the above generally depth-dependent transfer coefficients are varied until reasonable agreement between empirical tracer profiles and those calculated under the condition of a simulated external perturbation is obtained.

The numerical solution of differential Eqs. (69) can be achieved by finite difference methods; in most cases the deep-sea is subdivided into a number of compartments which are suitably chosen thinnest where the greatest concentration gradients occur, i.e. between surface layer and 1,000 m.

We shall now turn to a special case of (69), namely the box diffusion model [161]. The only transport considered is depth-independent turbulent diffusion

$$\frac{\partial c_d}{\partial t} = K\frac{\partial^2 c_d}{\partial z^2} \tag{70}$$

with the above boundary conditions in the form of

$$c_d(z_s, t) = c_s(t) \tag{71a}$$

and

$$-K\frac{\partial c_d(z_d, t)}{\partial z}=0 \tag{71b}$$

$c_d(z,t)$: concentration in the deep-sea
$c_s(t)$: concentration in the surface layer
z_s, z_d: surface layer and deep-ocean depth.

This implies that the stationary state total carbon concentration has to be assumed depth-independent whereas $\Delta^{14}_{d0}C$ varies with depth because of radioactive decay. For an exponential input the profile of the concentration deviation from stationary state and after the system has passed the transient state is then:

$$\Delta c_d(z,t)=\Delta c_s(t)\cdot\exp\left[-\sqrt{\frac{K}{k_f}}\cdot(z-z_s)\right] \tag{72}$$

([161, for a derivation see e.g. [163]). From (72) the mean penetration depth

$$\bar{z}:=\frac{1}{\Delta c_s(t)}\int_{z_s}^{z_d}\Delta c_d(z,t)\,dz \tag{73a}$$

is obtained as

$$\bar{z}=\sqrt{\frac{K}{k_f}} \tag{73b}$$

if the perturbed profile can be assumed not to be influenced by the total depth of the deep-sea z_d, i.e.

$$\sqrt{\frac{K}{k_f}}\Big/ z_d \ll 1$$

Thus, for $K=4{,}000\,m^2/a$, as has been obtained by calibration using stationary state radiocarbon distribution, $\bar{z}=296\,m$ results. Calculation of the airborne fraction under exponential input gives [58, 161, 162]

$$\alpha=\cfrac{1}{1+\cfrac{k_{as}}{k_f}\cfrac{1}{1+\cfrac{\xi\kappa_{as}k_{as}}{k_f}\cfrac{1}{1+\cfrac{z_d}{z_s}\cfrac{\tanh(z_d/\bar{z})}{z_d/\bar{z}}}}}. \tag{74}$$

To reproduce the observed airborne fraction $\alpha_f=55\%$ with an exponential input rate coefficient of $k_f=1/(22\ a)$ as valid for 1950–1973 a turbulent diffusivity of about 18,000 m^2/a is necessary, if the calculation is to be based on the same parameter values as above. The mean penetration depth into the deep-sea for such a turbulent diffusivity ($z=629$ m) plus the thickness of the surface layer ($z=75$ m) is in good agreement to the depth of a 2-compartment model ocean which is also to reproduce the empirical airborne fraction (see above). Indeed, it could be shown that the behaviour of carbon masses as represented by a box diffusion model can be simulated using a 2-box model with an appropriate surface layer

Table 4. Features of some detailed ocean sub-models of the global carbon cycle

Reference		Number of compartments for						Model parameters											Airborne fraction
								Ocean						Terrestrial biota				Marine-biota	
		Atmosphere	Surface layer	Intermediate water	Deep-sea	Terrestrial biota	Marine biota	Distribution of cross-section area	w [m/a]	K [m²/a]	τ_{as} [a]	C exchange atm.-polar waters	ξ (2)	Arrangement (3)	γ	δ	$F_{b,rel}$	Down-sinking, dissolution	α [%]
161		1	1	37	5	1		0_0		3,987	7.7		10		0.2	1			54.5
164		1	1	37	5	1		0_0		4,000–6,000	7.7–5.1				0.0–0.2	1			62.5 53.2
171		1	2	2	8	4		0(z)	(4)		5	x	ξ(t)	B/H	Variable		(5)	x	70–75
172	(12) 2B	1	812 m					0_0		τ_{ds} = 1,386a			ξ(t)						53
	(12) 2BD	1	595 m					0_0		3,255			ξ(t)						53
	(12) 2AD	1	700 m					0_0	1	1,385			ξ(t)						53
	(12) MB	1	2					0_0	(4a)				ξ(t)						53
162		1	1	continuous					0–6	500–6,000	7.7		ξ(t)						59
58		1	1							(6)	7.7		ξ(t)						56
176		1	2		20	1		0_0	3	6,000 +4·z(m)	13.7	x	(9)						67
51		1	1	5	7	2		0(z)	(8)		10.2		ξ(t)	L/S	0			(9)	66
170	Model 2	1	1	10	8			0(z)			6.3		ξ(t)	Source:	9 Gt C				
	Model 3a	1	1	10	8			0(z)			6.3		ξ(t)	sink:	17 Gt C				
	Model 3b	1	1	10	8			0(z)			6.3		ξ(t)	sink:	10 Gt C				

174	1	2	5	7	2	3	0(z)		4,000	6.3		$\xi(t)$	L/S	0–0.1			x	65 (10)
175	1	4	220				0_0	5	1,500 23,000 100,000	2.5						120 Gt C (11)		55
169	1	2	continuous				0_0		2,200– 12,600	8.9– 11.7	x	9–12						57–68
173	1	2	3	7	42		0_0		4,000		x	$\xi(t)$		0.5		6.9 Gt C1a	x	64
49, 50	2	2	74	20	48		0(z)	(12)	(12)	6.3	x	$\xi(t)$			0.3	0.8 Gt C (1959– 1978)	x	
179	1	4	2	6				(12)	(12)	10–12	x	$\xi(t)$					x	74–75
180			15,000				general circulation model				x	$\xi(t)$						65

(1) 0_0: depth-independent ocean cross-section area; 0(z): depth-dependent ocean cross-section area
(2) Evasion factor for low latitudes, if more than one ocean region has been modelled
$\xi(t)$: calculation of the evasion factor according to the chemical reactions involved
(3) See note (1), table 3
(4) Depth-dependent according to an assumed stationary state $\Delta^{14}C$ distribution; a) with a turnover time of 8 a
(5) $F_{b,rel}(t) = 0.5 \cdot F_f(t)$
(6) $K(z) = 26{,}192 \cdot \exp[-z/(745\ m)]\ [m^2/a]$; $\bar{K} = 5{,}270\ m^2/a$

(7) $$p_s = \frac{1}{\alpha_s \frac{K_1'}{K_2'}} \cdot \frac{(2\Sigma C - A)^2}{A - \Sigma C}$$

(8) Calculated from realistic depth distribution of ΣC and $\Delta^{14}C$ in the ocean
(9) For explanation see text
(10) Read from Fig. 3 [174] for AD 2,000
(11) According to [34]
(12) 2B: 2-box ocean sub-model; 2BD: ocean sub-model consisting of surface-layer compartment and diffusive deep-sea; 2AD: same as 2BD including upwelling; MB: multi-box ocean sub-model [171]

depth [141]. Although in the assumptions of the box diffusion model in its original version [161] a biota growth factor of $\gamma=0.2$ is still needed to reproduce the empirical α_f, it is the achievement of this concept that it can at the same time reproduce the Suess effect without changing the underlying physical model parameters.

It has then been pointed out that a neutral biosphere can be described, if the ocean uptake capacity is assumed an upper limit, which is possible, if the entire uncertainty ranges of K and k_{as} are taken into account [164]. Several attempts have been made to describe an enhanced oceanic uptake capacity for additional CO_2 mainly by introducing fast transfer mechanisms in high latitudes model regions which will be discussed in the next section.

The box diffusion model concept including a flux to describe polar outcrops has been applied to $\delta^{13}C$ values to estimate the biogenic input by a so-called deconvolution procedure [38]. According to these calculations the biosphere must be assumed a stronger source of additional CO_2 than fossil input for most of the time from 1860–1980, whereas in the period of the Mauna Loa record biogenic inputs turn out to be still 30% of fossil CO_2 release. The reservations noted in context with the representativity of tree-ring $^{13}C/C$ data for global atmospheric $^{13}C/C$ values and those with respect to the range of applicability of Stuiver's approximation apply to this argumentation.

Detailed Ocean Sub-models

Ocean Sub-models with Detailed Vertical Structure

The box diffusion model (see preceeding section) has been extended to describe the enhancing action of increased continental nitrogen and phosphorus run-off onto the marine biotic production including the temperature dependence of CO_2 solubility in the oceanic surface layer [165]. The relevance of the results obtained is, however, questioned by the fact that on the basis of an input of 101 Gt C in the period from 1900 to 1980 (150 Gt C according to [23] and [166]) $\alpha\approx 1$ is obtained if no temperature variation is allowed for.

In order to provide a description of deep-sea outcrops in polar surface waters a direct exchange contact between deep-sea and atmosphere on 10% of the total ocean surface area has been allowed for [167]. Calibration with the stationary state profile of ^{14}C in the ocean yields lower values for the turbulent diffusion constant (2,224 m^2/a) and for the global stationary atmosphere-ocean exchange flux (13.3 moles/m^2/a vs. 18.0 moles/m^2/a in the version without outcrops). The reproduction of the ^{14}C record of the surface ocean layer after the bomb input required higher values for both K and k_{as} in either case; this seems at least in part to be due to the neglected downsinking of biogenic material which even if not influenced by total inorganic CO_2 provides a possibility for the attenuation of $^{14}C/C$ perturbations in the surface layer. If this process is not taken into account in an appropriate manner, turbulent diffusion has to be assumed enhanced such that the total attenuation capacity of the oceans can be represented. The enhanced diffusivity applies to CO_2 perturbations as well, such that in the outcrop case an airborne fraction of $\alpha=0.528$ [$k_f=1/(22.5\ a)$] is reached.

After a box diffusion model with immediate downsinking of polar surface water to the ocean floor had been suggested and used for stationary state calculations [168] this concept was treated analytically by means of the adiabatic approximation for various forms of inputs as well as for depth-dependent diffusion constants [58, 162]. By comparing calculated and empirical airborne fractions an effective biogenic input flux could be derived for each combination of upwelling velocity and turbulent diffusivity. Following these calculations a neutral biosphere results if e.g. $10{,}000\ m^2/a \leqq K \leqq 12{,}000\ m^2/a$ and $4\ m/a \leqq w \leqq 6\ m/a$.

A model in which the airborne fraction is calculated by the same method exhibits a model atmosphere exchanging carbon dioxide with a mixed surface layer and a homogeneous compartment of polar water. The mixed layer is connected with two not directly coupled diffusive oceans which are to describe the different mixing behaviour of Atlantic ocean on the one hand and Pacific plus Indian ocean on the other hand [169]. The empirical α_f can be reproduced if the surface of the homogeneous polar water column extends to 20% of the total ocean surface.

The uptake capacity of additional CO_2 as described by five ocean models two of which (model 1a and b) are 2-box models with different surface layer depths has been compared [170]. Three of these models exhibited realistic depth dependence of oceanic cross sections; 10 compartments have been provided for the intermediate waters in 75 to 1,000 m depth. In model 2 a flux of water is modeled to sink down from the surface and horizontally passing the different layers of the bulk of the ocean; then these water masses are described as immediately upwelling. In model 3a upwelling is described as an upwards directed water mass motion, whereas in model 3b also downwelling is assumed as passing each deep-sea layer from the overlying one such that global diffusion is described in this case. The models are calibrated against stationary total dissolved inorganic carbon and ^{14}C profiles which implies different water fluxes from and to the surface layer. This has the same effect as a marine biota growth factor would have. Models 3a and 3b lead to a possible total biogenic release of 17 and 10 Gt C, respectively, for the time period from 1860 to 1975.

A further class of box diffusion models was provided with a separate high latitude surface water compartment which was described as exchanging water masses with the low latitudes surface waters and with the deep-sea which itself was not differentiated with respect to latitude at this stage of model development.

In the model concept [171] such a polar surface water box covering one third of the ocean surface is in direct exchange with the atmosphere, the low latitudes surface waters and the intermediate waters (2 boxes each about 500 m deep). Additionally, deep-sea water formation is described as originating from this compartment. Unidirectional upwelling (no diffusion) is provided for the bulk of the ocean as is dissolution of biotic material and biotic production in the surface layers. An airborne fraction with respect to fossil input of 70–75% is obtained in the standard version. With a turnover time of 8 a (which is estimated a factor of 2.5 too low) in the intermediate water compartments the model can be forced to reproduce $\alpha = 53\%$ [172].

Within this modelling concept the polar water surface compartment has been enlarged to occupy one half of the ocean surface and to extend to 400 m depth

[173]. In addition, an intermediate plus deep-water formation rate of 75 Sv is introduced. Intermediate water is represented by one compartment of 325 m and two compartments of 378 m depth. Although the turnover time of the high latitude surface compartment is only 6 a an airborne fraction of 68% is obtained for the case of purely fossil input.

The model described in [174] is a version of [170] extended by a compartment for polar waters being introduced into the intermediate water compartments. For the year 2000 an airborne fraction of 65% is obtained.

It can be concluded from the models referenced to here that neglecting diffusion and/or extending the thickness of the intermediate water compartments beyond a certain level reduces the described oceanic uptake capacity despite the introduction of rapidly downsinking polar water masses. So in a next stage of development both diffusion and advection were attempted to be modelled under consideration of their latitudinally varying pattern.

An ocean model with three regions (0–20°, 20–40°, 40–90°) of different depth-independent diffusivities (1,500, 23,000, and 100,000 m^2/a), and rapid downsinking of water masses in high latitudes which balance upwelling in the diffusive oceans has been succeedingly applied to ^3H, ^{14}C, and total carbon perturbations [175]. From the calculations with ^3H an atmosphere-ocean exchange-time of $\kappa_{as} = 2.5$ a has been derived. For an initial value of 280 ppm for the atmospheric CO_2 partial pressure and a cumulated net input of 120 Gt C from biotic sources (mainly before 1940) in addition to fossil input the Mauna Loa data can be reproduced. Best fit to the data is achieved with an upwelling velocity of 5 m/a which, however, does not influence the model results significantly.

In a preliminary study [176] an ocean model has been presented exhibiting two regions of high and low latitudes, respectively, these being separated at 30 °N and S. Rapid downsinking of water in high latitudes and depth-independent upwelling (3 m/a) in the bulks of the diffusive ocean is provided for. Turbulent diffusion is assumed equal in both regions and depending on depth according to

$$K(z) = 6{,}000 + 4 \cdot z \; [m^2/a] \tag{75}$$

z: depth in m

In the calculation of the CO_2 partial pressure in the surface layers borate concentration is neglected causing an error of 35%. Calculations yield an airborne fraction of $\alpha \approx 67\%$.

The two regions of the model ocean as developed in [49, 50, 177] which are separated at 40 °N and S exhibit ocean cross sections, turbulent diffusivities, upwelling velocities, horizontal fluxes from the high to the low latitudes deep-sea as well as dissolution of dead organic matter all depending on depth. Calibration has been made using stationary ΣC, $\delta^{13}C$, and $\Delta^{14}C$ profiles. The turbulent diffusivities immediately underneath the surface layers are 8,000 m^2/a in the low latitudes and 100,000 m^2/a (exponentially decreasing) in the high latitudes. Water mass downsinking from the high latitudes surface compartment amounts to 35.8 Sv leading to an upwelling velocity of 5 m/a in the low latitudes. This model provides an upper limit of the oceans' response capacity to external perturbations. While the Suess effect (−18.6‰ vs. −20.0‰ according to [48]) and the bomb ^{14}C input into the low latitudes deep-sea could be reproduced well the uptake capacity of the high latitudes deep-sea is overemphasised. With this model the Mauna Loa and South Pole CO_2 data could be reproduced under the assumption that the biosphere is neutral with respect to additional CO_2.

Ocean Sub-models with Detailed Horizontal Structure

To obtain steady state solutions for transport coefficients a 12-box model of the ocean has been devised [179] with surface, intermediate, deep and bottom water compartments for the Atlantic and the combined Pacific and Indian ocean. The Arctic and Antarctic oceans are represented by surface and deep-water compartments. The depth of the surface water compartments ranges from 135 to 235 m. Steady state profiles of total dissolved inorganic carbon, the alkalinity, phosphorus, oxygen, and radiocarbon have been used for calibration. Underlying the processes represented by Eq. (69) several sets of steady state fluxes have been derived yielding a total downwelling in the antarctic ocean of 30 Sv and a turbulent diffusion constant of about 3,150 m^2/a for exchange between surface and intermediate layers in the main oceans. It was found that better spatial resolution is needed to describe the oceanic behaviour in response to fossil CO_2 and bomb ^{14}C perturbations ($\alpha = 74\%–75\%$).

The general circulation of water masses in the oceans has been used to simulate carbon uptake and transport behaviour [180]. A realistically shaped ocean is resolved into water columns of $5° \times 5°$ cross section area which themselves consist of five layers the surface layer being 75 m in depth. The only process described is advection as caused by the stationary temperature and salinity field. For the first time carbon cycling in the ocean has been simulated using a 3-dimensional model reflecting at least the qualitative geographical pattern of surface distribution of ΣC and p_s. The relatively thick layers of intermediate waters seem to prevent the model from displaying an appropriate capacity to withdraw additional CO_2 from the atmosphere ($\alpha = 65\%$, presumably for 1860–1980).

Detailed Sub-models of the Terrestrial Biota

Because of the larger complexity of the terrestrial biosphere and the human impact exerted on it, models to display transfer processes in this important part of the global carbon cycle in a detailed manner have appeared only recently. Not only is adequate modelling hampered by large differences in biome behaviour and in the type of human influence but, furthermore, a considerable lack of information exists on the amount of the area of a certain biome subject to a special kind of transformation with respect to time as well as on the question of growth stimulation by additional CO_2.

In reference [181] a model of 10 regions and 12 biome types is presented which potentially exhibits 120 ecosystems. In practice they turned out to be much less, because every biome is restricted to a few regions at most. For each biome type a curve representing biomass and soil carbon density after the time of a certain impact is elaborated. According to a prescribed history of annually transformed areas, these are exposed to carbon loss and regain by regrowth (no CO_2 stimulation effect is considered) as determined by the impact curves. In the case of forest harvest the cleared biomass is introduced into four decay pools with decay-times of 1, 10, 100, and 1,000 a, respectively. The principal result of these calculations is that the biosphere turned out to be a source for additional atmospheric CO_2 from 1860 to 1970 (148 Gt C vs. 116 Gt C from fossil sources in the same time range) with an annual net input in 1970 of 2.2–4.7 Gt C/a. The calculation net CO_2 release has been larger from the assumed clearings for agricultural purposes than on cleared forest areas which have been allowed to be regrown by model forest ecosystems again.

In the model presented in [173] the biosphere is subdivided into six biome types (tropical forests, temperate forests, grasslands, agricultural area, human area, tundra, and deserts) each of which exhibits leaf, branch, stem, root, litter, humus, and charcoal compartments. (The charcoal compartments are to describe the formation of stable carbon compounds by incomplete combustion of biomass.) All kinetics are linearly donor-controlled except in two cases: the assimilation flux is described as being stimulated by additional atmospheric CO_2 and other effects ($\gamma=0.5$). After a certain impact the internal response of the system is suspended and the empty compartments are refilled with constant steady state NPP without losing carbon to the atmosphere. Transformation of biome area is represented in an exchange matrix with a history which is in context to population growth. For the year 1980 this modelling concept shows that the biosphere is a small source for atmospheric CO_2 (0.34 Gt C) which, if compared to the model described above [181], seems to be caused by the biota growth factor and by relatively fast regrowth. From the data given e.g. for the composed tropical forest biome an approximate regrowth time of (biomass/NPP)$=369.7/27.8$ a≈ 13–14 a results.

In a five-biome model (tropical and extra-tropical forests and grassland, deserts) consisting of living phytomass and soil carbon for each biome it has been attempted to describe several forms of impact (inadvertent fires and intended clearings for agricultural purposes in the extra-tropical forests as well as shifting cultivation, increase of permanent agriculture and commercial clearings in the tropical forests, desertification) and the response of the biomes concerned [49, 50, 183]. To this end the living biomass compartments of forest biomes were subdivided into 20 age classes each being allowed to regrow according to Eq. (67) with a climax parameter $\delta=0.3$ implying a time period of $t_{95}\approx 100$ a needed for the accumulation of 95% of the climax biomass after the impact had occurred. Charcoal formation and erosion after forest clearings have been accounted for. The model is connected to an ocean sub-model referred to in the preceeding section [49, 50, 177]. Essentially, two scenarios have been simulated:

(1) no pioneer effect. The biosphere is described as a net source (1.3 Gt C/a in 1980) and the atmospheric CO_2 data cannot be reproduced well. Since for the time after 1980 the different forms of impact into the tropical forests have been assumed to compete with each other for the remaining area this model biome has almost vanished in the year 2060.

(2) pioneer effect included. By modelling large clearings in the extra-tropical forests centered around 1890 the model biosphere is provided with considerable re-

growth capacity such that for the time range of the Mauna Loa data it is neither a source nor a sink. The atmospheric CO_2 measurements are well reproduced in this case.

A scenario with less impact into the tropical forests has also been run resulting in the destruction of about half the tropical forests in 2060 and reproducing the Mauna Loa data with a smaller pioneer effect.

Detailed Atmosphere Sub-models

All data underlying the land transformation rates are highly uncertain, because a detailed census seems difficult to establish. Therefore an independent source of information of the source strength of tropical forest clearings has been looked for. To this end it has been attempted to explain the partial pressure and $\delta^{13}C$ of atmospheric CO_2 as recorded at different latitudinal positions using detailed models of atmospheric circulation and of the latitude-dependence of carbon input and output from the unperturbed biosphere, the oceans and the fossil input. Such models should enable the decision on whether a substantial tropical forest source of additional atmospheric CO has to be accounted for.

One of these models [184] treats the atmosphere as consisting of 20 equal mass columns themselves consisting of 8 layers each (4 in the troposphere and 4 in the stratosphere). Circulation is represented mainly by advection terms which are tested against the spatial distribution of CCl_3F. As boundary conditions the mean annual CO_2 source or sink activities of ocean and fossil input depending on latitude are prescribed as to reproduce the atmospheric meridional CO_2 concentration profile in the early 1960's. Then, an additional source of carbon located in the tropics of 1,2 or 5 Gt C/a has been introduced the effect of which on simulated CO_2 concentrations for 1980 has been compared to the respective CO_2 data. It is concluded that tropical deforestation may at the most have amounted to 2 Gt C/a. Here, the question arises what the results would have been, if the calibration would have comprised a tropical forest clearing source.

A one dimensional meridional diffusion model for the atmosphere is used to describe the differences between the latitude-dependent CO_2 concentration profiles for the years 1962 and 1980. This can be accomplished by fossil input together with stationary sources of which the equatorial oceans appear the most distinct. A possible source from forest clearing is regarded as being at most within the uncertainty range (=2 Gt C/a).

Predicted Future CO_2 Increase

In view of the possible climatic variations due to increased atmospheric CO_2 concentrations it is of interest to assess the general trend. Both model assumptions and fossil input should influence the resulting calculated atmospheric CO_2 contents. (For appropriate conversion factors of energy consumption per unit time in TW to fossil CO_2 release in Gt C/a for the different types of fossil fuels see [194].)

Since fossil fuels are non-renewable resources, their consumption is often described by a logistic function:

$$\frac{dN_f}{dt} := F_f = k_f N_f \left(1 - \frac{N_f}{N_{f\infty}}\right) \tag{75}$$

F_f: fossil input (Gt C/a)
$N_f(t)$: fossil fuels consumed until time t (Gt C)
k_f: growth coefficient (1/a)
$N_{f\infty}$: final cumulated input.

Estimating $N_{f\infty}$ one has to distinguish between the reserves, i.e. those fossil deposits which can be mined under present technical and economic conditions, and the resources which comprise all assumed supplies.

The doubling of atmospheric CO_2 concentration with respect to its pre-industrial level is considered climatically significant. This quantity is therefore used to compare the response of different models to about the same external input, and for a brief discussion of the consequences of different scenarios of future fossil fuel consumption. As is shown in Table 6 the box model [141], the box-diffusion model [57] and an advection-diffusion model with latitude-dependent oceanic CO_2 uptake capacity [49, 50] give a doubling-time period ranging from 2020 to 2030 in response to about the same fossil input assumptions.

Table 5. Estimates of the final cumulated fossil input ($N_{f\infty}$) in Gt C [193]

a) Reserves			
	1972 [194]	1975 [195]	1976 [196]
Coal	614	388	1,291
Petroleum	72	80	275
Natural gas	29	39	182
Oil shales		27 (1)	21
Oil sands	41	32	35
$N_{f\infty}$ (2)	882	707	1,950
b) Resources			
	1972 [194]	1975 [195]	1976 [197]
Coal	4,301	1,314	
Petroleum	246	237	
Natural gas	161	278	
Oil shales		200 (1)	
Oil sands	323	222	
$N_{f\infty}$ (2)	5,157	2,392	about 7,000

(1) These figures refer to shales of at least 1 ton shale oil per 40 tons of rock.
(2) These figures are not only the sum of the individual amounts but also contain the cumulated inputs until the reference year (126 Gt C for 1972, 141 Gt C for 1975, and 146 Gt C for 1976)

Table 6. Model responses to logistic input functions (Eq. (75)) as expressed in doubling-times $t(2 \cdot p_{a0})$

	Box model [141]	Box diffusion model [57]	Advection diffusion model [49, 50]
$N_{f\infty}$ (Gt C)	5,000	7,000	7,000
k_f (%/a)	4.53	5	4.4
$t(2 \cdot p_{a0})$	2,030	2,020	2,025

The latter model [49, 50] has been run with different external inputs among which some are of the following type:

$$F_f = k_f N_f \left(1 - \frac{N_f}{N_{f\infty}}\right)^n; \qquad n=2 \tag{76}$$

(see Fig. 2a)

Logistic functions modified this way may be used to describe a more slowly growing fossil input rate in the starting period of fossil energy consumption. In addition, a low energy scenario with almost constant fossil input is considered.

The model results have been compiled in Fig. 2b which exhibits a doubling-time range from 2,025 to 2,090. Even in the case of the low energy scenario a doubling of the atmospheric CO_2 content is only deferred but is not avoided. This is the case only in scenarios of very low fossil energy consumption which are assumed possible by serious energy saving efforts (end use adapted forms of energy supply) and the extended use of renewable energy resources [200].

For the models the estimated CO_2 doubling-time is above all determined by the time-behaviour of the fossil input and only to a minor degree by the different model assumptions. (As we have seen in preceeding sections, this does not hold for other quantities such as isotope ratios, however.)

Conclusion

From the ocean models presented it becomes clear that if model oceans were to absorb all non-airborne fossil fuel CO_2 (neutral biosphere) their simulated uptake capacity would have to be an upper limit of what is compatible with isotope data. The question thus arises whether marine transfer processes not reflected in the models appear likely to withdraw atmospheric CO_2. At the same time, such processes should leave the isotopic ratios as calculated as the response to perturbations in the atmosphere (Suess effect, bomb effect) relatively unaffected which implies that such processes would essentially have to be non-linear.

Dissolution of $CaCO_3$ and additional sedimentation of biotic material the growth of which is assumed to be stimulated by additional nitrogen and phosphorus in river run-off have been suggested the additional carbon storage efficiency of which is still subject to discussion.

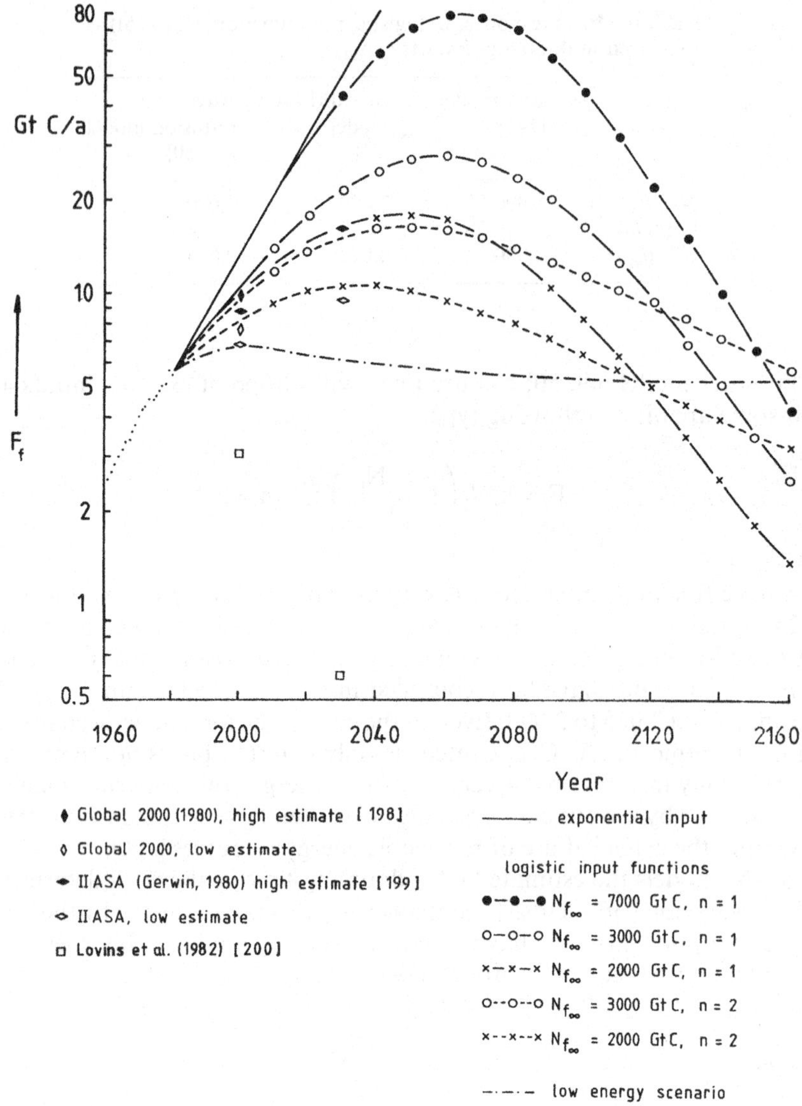

Fig. 2a. Scenarios for future fossil fuel consumption.
IIASA: International Institute for Applied System Analysis

Furthermore, anomalies of sea-surface temperature (such as occurring in connection with the Southern Oscillation) can cause enhanced CO_2 solubility. An abnormal sea-surface cooling leads to increased dissolution of atmospheric CO_2 in the surface layer. A long-term oceanic uptake of at least part of this additionally dissolved CO_2 is possible, if surface water concerned is transferred to deeper layers before normal surface temperatures are re-established. In case no such transfer takes place until the surface layer returns to normal equilibrium CO_2 partial pressures, however, the additionally dissolved CO_2 is lost completely to the atmosphere again.

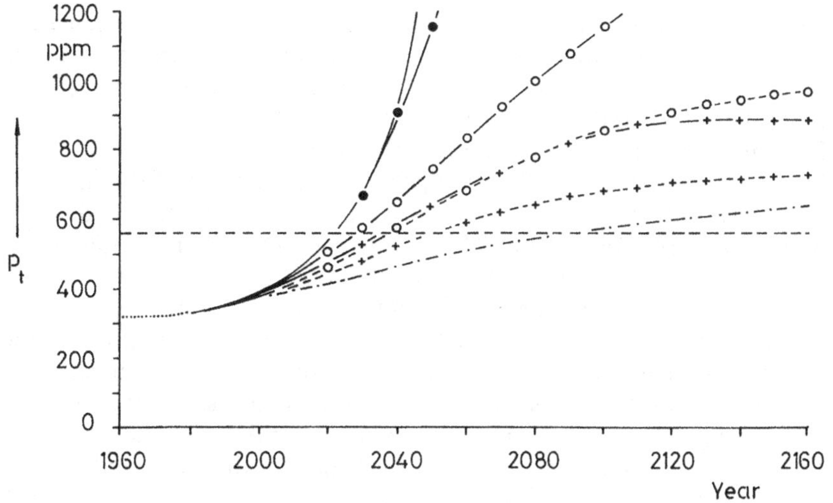

Fig. 2 b. Atmospheric CO_2 partial pressures as calculated in [49, 50] as responses to the fossil input scenarios given in Fig. 2a.
Dashed straight line: 560 ppm ($2 \cdot p_{a0}$)

Concerning the biosphere two mechanisms have been suggested to effectively remove CO_2 from the atmosphere: stimulation of assimilation and storage of CO_2 as biotic material as well as regrowth. Both these mechanisms will be difficult to establish empirically on the biome level. Model calculations show that growth of temperate and boreal zone forests may well be assumed as withdrawing as much carbon dioxide form the atmosphere as is released by global forest destruction and desertification.

Areas covered by regrowing ecosystems must, of course, have undergone an impact which may either have been anthropogenic or may have had natural causes (e.g. forestfires). Such events are likely to have recurred through all history [91] although the rates of areas concerned can well be assumed exceptionally high at present. This means that the assumption of a stationary initial state in model calculations may introduce some uncertainty [187, 188]. Furthermore, even the assumed stationary distribution of deep-sea $^{14}C/C$ may in part be a result of an ocean circulation pattern which is different from the modern one [188].

All these uncertainties have to be kept in mind when projections for future atmospheric CO_2 concentrations are made, which differ, of course, according to the time-behaviour of the underlying inputs. On the basis of a variety of input functions an advection-diffusion model [49, 50] yields a time range of doubling of the atmospheric CO content from 2,025 to 2,090 (Fig. 2b). For such an increase of the CO_2 concentration in the atmosphere climate models predict a most probable global temperature increase of 2 K [201].

Perspectives in Carbon Cycle Modelling

Apart from the elucidation of unknown or poorly known carbon transfer mechanisms and their functional dependence on certain environmental parameters such as CO_2, other nutrients, temperature, and precipitation, two general trends of further research appear to be important: methodical studies on the interpretation of incomplete or contradictory data, respectively, as well as further disaggregation of the spatial and time dimensions of the system.

The simultaneous use of tracers in the ocean [179] raises many problems among which are the criteria for an appropriate calculation of the stationary exchange coefficients in the cases of incompatible and indeterminate data, respectively, and the application of suitable statistical tests to determine that volume region which is represented by a certain set of data.

To describe atmospheric and marine transfer processes on a regional and seasonal scale, general circulation models have been applied [180, 202]. In the case of the atmosphere more detailed modelling can result in a more precise localisation of sources and sinks of biospheric CO_2 and an estimation of their total, seasonally varying magnitude, if correct boundary conditions for the oceans (including the marine biota) are provided for. Also the fact that the combustion gases of large forest fires due to their high temperatures exhibit a relatively large buoyancy and thus probably circumvent the normal atmospheric circulation at least partially may turn out to be important in this context.

Models of the general circulation of the ocean (e.g. [180]) which are able to reproduce marine tracer concentration distributions may be used to estimate the oceanic storage capacity for additional CO_2 under different temperature conditions, as may e.g. be expected as a result of the CO_2 greenhouse effect. This then would provide a more realistic look at the global carbon cycle, since the important CO_2 induced climate feedback onto the behaviour of the oceans could be accounted for.

Acknowledgement

I should like to thank H. Bröhl, Dr. U. Fischbach, Prof. G. H. Kohlmaier, and Dr. E.-O. Siré (Institut für Physikalische und Theoretische Chemie, Universität Frankfurt/M.) for extremely valuable discussions.

Financial support by the Deutsche Forschungsgemeinschaft and the Stiftung Volkswagenwerk is gratefully acknowledged.

References

1. Martin, P.M., Barker, E.F.: Phys. Rev. *41,* 219 (1932)
2. Robinson, G.D.: Advances in Geophysics, H.E. Landsberg and J. Van Miegham (eds.). London, Academic Press 1970, cited from: Schneider, S.H., Kellogg, W.W.: Chemistry of the Lower Atmosphere, S.I. Rasool (ed.), p. 203, New Nork, Plenum 1973
3. Bach, W.: Carbon Dioxide, Climate, and Society, J. Williams (ed.), p. 141, Pergamon, Oxford 1978
4. Broecker, W.S., Takahashi, T., Simpson, H.J., Peng, T.-H.: Science *206,* 409 (1979)
5. Woodwell, G.M., Whittaker, R.H., Reiners, W.A., Likens, G.E., Delwiche, C.C., Botkin, D.B.: Science *199,* 141 (1978)
6. Hampicke, U., Bach, W.: Münstersche Geographische Arbeiten *6,* 37 (1980)
7. Verniani, F.: J. Geophys. Res. *71,* 385 (1966)

8. Handbook of Chemistry and Physics, R.C. Weast, M.J. Astle (eds.), Boca Raton, Fla., CRC Press 1980
9. Kempe, S.: The Global Carbon Cycle, B. Bolin, E.T. Degens, S. Kempe, P. Ketner (eds.), p. 317, New York, Wiley 1979
10. Kelley, J.J.: An analysis of carbon dioxide in the arctic atmosphere at Point Barrow, Alaska, University of Washington, Dept. of Sciences Technical Report, Office of Naval Research Contract 447(24) (NR 307-252), 1964
11. Bacastow, R.B., Keeling, C.D.: Carbon Cycle Modelling, B. Bolin (ed.), p. 103, New York, Wiley 1981
12. Lowe, D.C., Guenther, P.R., Keeling, C.D.: Tellus *31,* 58 (1979)
13. Beardsmore, D.J.: Carbon Dioxide and Climate: Australian Research, G.I. Pearman (ed.), p. 41, Canberra, Australian Academy of Sciences 1980
14. Bacastow, R.B., Keeling, C.D., Whorf, T.P., Wong, C.S.: Proc. Conf. on analysis and interpretation of atmospheric CO data, Bern, September 14–18, 1981
15. Bolin, B., Keeling, C.D.: J. Geophys. Res. *68,* 3899 (1963)
16. Bischof, W., Bolin, B.: Tellus *18,* 155 (1966)
17. Bolin, B., Bischof, W.: Tellus *22,* 431 (1970)
18. Junge, C.E., Czeplak, G.: Tellus *20,* 422 (1968)
19. Keeling, C.D.: Proc. Carbon Dioxide Research Conf.: Carbon Dioxide, Science and Consensus, United States Department of Energy CONF-820970, p. II.3, Washington, D.C. 1983
20. Cleveland, W.S., Freeney, A.E., Graedel, T.E.: J. Geophys. Res. *88,* 10934 (1983)
21. Kohlmaier, G.H., Revelle, R., Keeling, C.D.: In preparation, 1984
22. Bischof, W., Fabian, P., Borchers, R.: Nature *288,* 347 (1980)
23. Rotty, R.M.: Distribution of and Changes in Industrial Carbon Dioxide Production, Oak Ridge, Tenn., Institute for Energy Analysis ORAU/IEA-82-2 (1982)
24. Bacastow, R.B., Keeling, C.D.: Carbon Cycle Modelling, B. Bolin (ed.), p. 241, New York, Wiley 1981
25. Bacastow, R.B.: Nature *261,* 116 (1976)
26. Bacastow, R.: J. Geophys. Res. *84,* 3108 (1979)
27. Stanhill, G.: Climatic Change *4,* 221 (1982)
28. Keeling, C.D.: Science *202,* 1109 (1978)
29. Neftel, A., Oeschger, H., Schwander, J., Stauffer, B., Zumbrunn, R.: Nature *295,* 220 (1982)
30. Keeling, C.D., Bacastow, R.B., Tans, P.P.: Geophys. Res. Lett. *7,* 505 (1980)
31. Mook, W.G., Koopmans, M., Carter, A.F., Keeling, C.D.: J. Geophys. Res. *88,* 10915 (1983)
32. Freyer, H.D., Belacy, N.: J. Geophys. Res. *88,* 6844 (1983)
33. Francey, R.J.: Nature *290,* 232 (1981)
34. Stuiver, M.: Science *199,* 253 (1978)
35. Tans, P.P., Mook, W.G.: Tellus *32,* 268 (1980)
36. Farquhar, G.: Carbon Dioxide and Climate: Australian Research, G.I. Pearman (ed.), p. 105, Canberra, Australian Academy of Sciences 1980
37. Leavitt, S.W., Long, A.: Tellus *358,* 92 (1983)
38. Peng, T.-H., Broecker, W.S., Freyer, H.D., Trumbore, S.: J. Geophys. Res. *88,* 3609 (1983)
39. Tans, P.: Carbon Cycle Modelling, B. Bolin (ed.), p. 127, New York, Wiley 1981
40. Karlén, I., Olsson, I.U., Kallenberg, P., Killici, S.: Arkiv Geofysik *4,* 465 (1964)
41. Bigeleisen, B., Mayer-Goeppert, M.: J. Chem. Phys. *15,* 261 (1947)
42. Craig, H.: Geochim. Cosmochim. Acta *3,* 53 (1953)
43. Stuiver, M., Robinson, S.W.: Earth Planet. Sci. Lett. *23,* 87 (1974)
44. Stuiver, M., Pollach, H.A.: Radiocarbon *19,* 355 (1977)
45. Castagnoli, G., Lal, D.: Radiocarbon *22,* 133 (1980)
46. Stuiver, M., Quay, P.D.: Science *207,* 11 (1980)
47. Mayaud, P.-N.: J. Geophys. Res. *77,* 6870 (1971)
48. Stuiver, M., Quay, P.D.: Earth Planet. Sci. Lett. *53,* 349 (1981)
49. Kratz, G.: Dissertation, Frankfurt/M., 315 pp., 1983

50. Kratz, G., Kohlmaier, G.H., Bröhl, H., Fischbach, U., Siré, E.-O.: unpublished, 1984
51. Killough, G.G.: Health Phys. *38*, 269 (1980)
52. Nydal, R., Lövseth, K.: J. Geophys. Res. *88*, 3621 (1983)
53. Bacastow, R., Keeling, C.D.: Carbon and the Biosphere, G.M. Woodwell and E.V. Pecan (eds.), p. 86, Springfield, Va., U.S. Atomic Energy Commission CONF-720510 (1973)
54. Bacastow, R.: Carbon Cycle Modelling, B. Bolin (ed.), p. 95, New York, Wiley 1981
55. Revelle, R., Suess, H.E.: Tellus *9*, 18 (1957)
56. Broecker, W.S., Li, Y.-H., Peng, T.-H.: Impingement of man on the oceans, D.W. Hood (ed.), p. 287, New York, Wiley 1971
57. Siegenthaler, U., Oeschger, H.: Science *199*, 388 (1978)
58. Siré, E.-O., Kohlmaier, G.H., Kratz, G., Fischbach, U., Bröhl, H.: Z. Naturforsch. *36 a*, 233 (1981)
59. Buch, K.: Havsforskning Inst. Skr. *151*, 1951, cited from [65]
60. Edmond, J.M., Gieskes, J.M.T.M.: Geochim. Cosmoschim. Acta *34*, 1261 (1970)
61. Hansson, I.: Deep-Sea Res. *20*, 4641 (1973)
62. Lyman, J.: Ph.D. Thesis, Los Angeles, 1976, cited from [65]
63. Mehrbach, C., Culberson, C.H., Hawlwey, J.E., Pytkowicz, R.M.: Limnol. Oceanogr. *18*, 897 (1973)
64. Sundquist, E.T., Plumer, L.N., Wigley, T.M.L.: Science *204*, 1203 (1979)
65. Takahashi, T., Kaiteris, P., Broecker, W.S., Bainbridge, A.E.: Earth Planet. Sci. Lett. *32*, 458 (1976)
66. Weiss, R.F.: Mar. Chem. *2*, 203 (1974)
67. Broecker, W.S.: Chemical Oceanography, New York, Harcourt Brace Jovanovitch 1974
68. Bolin, B., Keeling, C.D., Bacastow, R.B., Björkström, A., Siegenthaler, U.: Carbon Cycle Modelling, B. Bolin (ed.), p. 1, New York, Wiley 1981
69. Stuiver, M.: J. Geophys. Res. *85*, 2711 (1980)
70. Peng, T.-H., Broecker, W.S., Mathieu, G.G., Li, Y.-H.: J. Geophys. Res. *84*, 2471 (1979)
71. Kanwisher, J.: J. Geophys. Res. *68*, 3921 (1963)
72. Kromer, B.: Dissertation, Heidelberg, 1979
73. Hasse, L., Liss, P.S.: Tellus *32*, 470 (1980)
74. Siegenthaler, U., Münnich, K.O.: Carbon Cycle Modelling, B. Bolin (ed.), p. 249, New York, Wiley 1981
75. Menard, H.W., Smith, S.M.: J. Geophys. Res. *71*, 4305 (1966)
76. Bathén, K.H.: J. Geophys. Res. *77*, 7138 (1972)
77. Drake, C.L., Imbrie, J., Knauss, J.A., Turekian, K.K.: Oceanography, New York, Holt, Rinehardt and Winston 1978
78. Sverdrup, H.U., Johnson, M.W., Fleming, R.H.: The Oceans, Englewood Cliffs, Prentice Hall 1942
79. Dietrich, G., Kalle, K., Krauss, W., Siedler, G.: Allgemeine Meereskunde, Berlin, Borntraeger 1975
80. Fine, R.A., Östlund, H.G.: Earth Planet. Sci. Lett. *49*, 447 (1980)
81. Quay, P.D., Stuiver, M.: Radiocarbon *22*, 607 (1980)
82. Broecker, W.S., Peng, T.-H., Engh, R.: Radiocarbon *22*, 565 (1980)
83. Weiss, R.F., Östlund, H.G., Craig, H.: Deep-Sea Res. *26 A*, 1093 (1979)
84. Peterson, W.H., Rooth, C.G.H.: Deep-Sea Res. *23*, 273 (1976)
85. Stommel, H., Aarons, A.B.: Deep-Sea Res. *6*, 217 (1960)
86. Aarons, A.B., Stommel, H.: Deep-Sea Res. *14*, 441 (1967)
87. Broecker, W.S.: J. Geophys. Res. *84*, 3218 (1979)
88. Broecker, W.S., Peng, T.-H., Stuiver, M.: J. Geophys. Res. *83*, 6179 (1978)
89. de Vooys, C.G.N.: The Global Carbon Cycle, B. Bolin, E.T. Degens, S. Kempe and P. Ketner (eds.), p. 295, New York, Wiley 1979
90. Degens, E.T., Kempe, S., Spitzy, A.: Handbook of Environmental Chemistry, vol. 1 C, O. Hutzinger (ed.), Berlin, Springer 1984
91. Kohlmaier, G.H., Bröhl, H., Fischbach, U., Kratz, G., Siré, E.-O.: Carbon Dioxide: A Text on Current Views and Developments in Energy/Climate Research, W. Bach, A. Crane, A. Berger and A. Longhetto (eds.), p. 93, Dordrecht, Reidel 1983
92. Suess, E.: Nature *288*, 260 (1980)

93. Williams, P.M., Gordon, L.I.: Deep-Sea Res. *17,* 19 (1970)
94. Skirrow, G.: Chemical Oceanography, vol. 2, J.P. Riley and G. Skirrow (eds.), p. 1, London, Academic Press 1975
95. Wollast, R., Garrels, R.M., Mackenzie, F.T.: Am. J. Sci. *280,* 831 (1980)
96. Berner, R.A., Honjo, S.: Science *211,* 940 (1981)
97. Smith, S.V.: Science *211,* 838 (1981)
98. Li, Y.-H., Takahashi, T., Broecker, W.S.: J. Geophys. Res. *74,* 5507 (1969)
99. Edmond, J.M.: Deep-Sea Res. *21,* 455 (1974)
100. Kroopnick, P.: Earth Planet. Sci. Lett. *49,* 469 (1980)
101. Takahashi, T., Broecker, W.S., Bainbridge, A.E.: Carbon Cycle Modelling, B. Bolin (ed.), p. 271, New York, Wiley 1981
102. Takahashi, T., Broecker, W.S., Bainbridge, A.E.: Carbon Cycle Modelling, B. Bolin (ed.), p. 159, New York, Wiley 1981
103. Stuiver, M., Östlund, H.G.: Radiocarbon *22,* 1 (1980)
104. Östlund, H.G., Stuiver, M.: Radiocarbon *22,* 25 (1980)
105. Whittaker, R.H., Likens, G.E.: Carbon and the Biosphere, G.M. Woodwell and E.V. Pecan (eds.), p. 281, Springfield, Va., U.S. Atomic Energy Commission, CONF-720510, 1973
106. Lieth, H.: Primary Productivity of the Biosphere, H. Lieth and R.H. Whittaker (eds.), p. 237, Berlin, Springer 1975
107. Whittaker, R.H., Likens, G.E.: Primary Productivity of the Biosphere, H. Lieth and R.H. Whittaker (eds.), p. 303, Berlin, Springer 1975
108. Ajtay, G.L., Ketner, P., Duvigneaud, P.: The Global Carbon Cycle, B. Bolin, E.T., Degens, S., Kempe and P. Ketner (eds.), p. 129, New York, Wiley 1979
109. Reiners, W.A.: Carbon and the Biosphere, G.M. Woodwell and E.V. Pecan (eds.), p. 303, Springfield, Va., U.S. Atomic Energy Commission, CONF-720510, 1973
110. Meentemeyer, V., Box, E.O., Thompson, R.: BioScience *32,* 125 (1982)
111. Esser, G., Asselmann, I., Lieth, H.: Mitt. Geol. Paläonthol. Inst. Univ. Hamburg, SCOPE/UNEP special issue *52,* 39 (1982)
112. Bolin, B.: The Biosphere, p. 47, San Francisco, Freeman 1970
113. Kovda, V.A.: 10th Intern. Congress Soil Sci., Moscow 1974, cited from [114]
114. Buringh, P.: SCOPE-Conf. on the Role of the Terrestrial Vegetation in the Global Carbon Cycle, Woods Hole, May 6–11, 1979
115. Bohn, H.L.: Soil Sci. Soc. Am. J. *40,* 468 (1976)
116. Baes, C.F., Goeller, H.E., Olson, J.S., Rotty, R.M.: The Global Carbon Dioxide Problem, Oak Ridge National Laboratory ORNL-5195, Oak Ridge, Tenn. 1976
117. Schlesinger, W.H.: SCOPE-Conf. on the Role of the Terrestrial Vegetation in the Global Carbon Cycle, Woods Hole, May 6–11, 1979
118. Olsen, J.S., Pfuderer, H.A., Chan, Y.-H.: Changes in the Global Carbon Cycle and the Biosphere, Oak Ridge National Laboratory ORNL/EIS-109, Oak Ridge, Tenn. 1978
119. Wong, C.S.: Science *200,* 197 (1978)
120. Seiler, W., Crutzen, P.J.: Climatic Change *2,* 207 (1980)
121. Mensching, H.: Geogr. Rdsch. *31,* 350 (1979)
122. Clawson, M.: Science *204,* 1168 (1979)
123. Bolin, B.: Science *196,* 613 (1977)
124. Hampicke, U.: Experientia *36,* 776 (1980)
125. Lemon, E.: The Fate of Fossil Fuel Carbon Dioxide in the Oceans, A. Malahoff and N.R. Anderson (eds.), p. 97, New York, Plenum 1976
126. Rosenberg, N.J.: Climatic Change *3,* 265 (1981)
127. Keeling,C.D.: Environment International *2,* 229 (1979)
128. Craig, H.: Tellus *9,* 1 (1957)
129. Houtermans, J.C., Suess, H.E., Oeschger, H.: J. Geophys. Res. *78,* 1897 (1973)
130. Plesset, M.S., Dugas, D.J.: J. Geophys. Res. *72,* 4131 (1967)
131. Rafter, T.A., O'Brien, B.J.: Radiocarbon Variations and Absolute Chronology, I.U. Olsson (ed.), p. 355, New York, Wiley, Stockholm, Almqvist & Wiksell 1970
132. Nydal, R.: J. Geophys. Res. *73,* 3617 (1968)
133. Eriksson, E., Welander, P.: Tellus *8,* 155 (1956)
134. Eriksson, E.: Tellus *30,* 350 (1978)

135. Gardener, R.H.: Mankin, J.B., Emanuel, W.R.: Ecol. Modelling *8*, 313 (1980)
136. Smil, V., Milton, D.: Atm. Env. *8*, 1213 (1974)
137. Bolin, B., Eriksson, E.: The Atmosphere and the Sea in Motion, B. Bolin (ed.), p. 130, New York 1959
138. Kohlmaier, G.H.: Carbon Cycling Modelling, B. Bolin (ed.), p. 307, New York, Wiley 1981
139. Machta, L.: Carbon and the Biosphere, G.M. Woodwell and E.V. Pecan (eds.), p. 21, Springfield, Va., U.S. Atomic Energy Commission, CONF-720510, 1973
140. Keeling, C.D.: Chemistry of the Lower Atmosphere, S.I. Rasool (ed.), p. 251, New York, Plenum 1973
141. Keeling, C.D., Bacastow, R.B.: Energy and Climate, National Academy of Sciences (ed.), p. 72, Washington 1977
142. Zimen, K.E., Offermann, P., Hartmann, G.: Z. Naturforsch. *32a*, 1544 (1977)
143. Revelle, R., Munk, W.: Energy and Climate, National Academy of Sciences (ed.), p. 140, Washington 1977
144. Kohlmaier, G.H., Fischbach, U., Kratz, G., Siré, E.-O.: Carbon Dioxide, Climate, and Society, J. Williams (ed.), p. 111, Oxford, Pergamon 1978
145. Kohlmaier, G.H., Siré, E.-O., Kratz, G., Fischbach, U.: Ber. Bunsenges. *82*, 1218 (1978)
146. Keeling, C.D.: Interaction of Energy and Climate, W. Bach, J. Pankrath and J. Williams (eds.), p. 129, Dordrecht, Reidel 1980
147. Avenhaus, R., Fenyi, S., Frick, H.: Environment International *2*, 379 (1980)
148. Bacastow, R.B., Björkström, A.: Carbon Cycle Modelling, B. Bolin (ed.), p. 29, New York, Wiley 1981
149. Niehaus, F.: A non-linear eight level tandem model to calculate the future CO_2 and C-14-burden to the atmosphere, IIASA research memorandum RM-76-35, Laxenburg 1976
150. Niehaus, F.: Computersimulation langfristiger Umweltbelastung durch Energieerzeugung: Kohlendioxyd, Tritium und Radio-Kohlenstoff, Basel, Birkhäuser 1977
151. Niehaus, F.: I.A.E.A. Bull. *21*, 2 (1979)
152. Niehaus, F.: Man's Impact on Climate, W. Bach, J. Pankrath and W.W. Kellogg (eds.), p. 285, Amsterdam, Elsevier 1979
153. Niehaus, F., Williams, W.: J. Geophys. Res. *84*, 3123 (1979)
154. Niehaus, F.: Experientia *36*, 792 (1980)
155. Niehaus, F.: Climatic Variations and Variability: Facts and Theories, A. Berger (ed.), p. 641, Dordrecht, Reidel 1981
156. Cramers, J., Myers, A.L.: Atm. Env. *6*, 563 (1972)
157. Gowdy, C.M., Mulholland, R.J., Emanuel, W.R.: Int. J. Syst. Sci. *6*, 965 (1975)
158. Hoffert, M.I.: Atm. Env. *8*, 1225 (1974)
159. Laurmann, J.A., Spreiter, J.R.: Climatic Change *5*, 145 (1983)
160. Kohlmaier, G.H., Fischbach, U., Kratz, G., Siré, E.-O., Hirschberger, J., Schunck, W.: Man's Impact on Climate, W. Bach, J. Pankrath and W.W. Kellogg (eds.), p. 161, Amsterdam, Elsevier 1979
161. Oeschger, H., Siegenthaler, U., Schotterer, U., Gugelmann, A.: Tellus *27*, 168 (1975)
162. Kohlmaier, G.H., Siré, E.-O., Kratz, G., Fischbach, U., Bröhl, H.: Ber. Bunsenges. *84*, 1076 (1980)
163. Doetsch, G.: Einführung in Theorie und Anwendung der Laplace Transformation, Basel, Birkhäuser 1958
164. Oeschger, H., Siegenthaler, U., Heimann, M.: Interactions of Energy and Climate, W. Bach, J. Pankrath and J. Williams (eds.), p. 107, Dordrecht, Reidel 1980
165. Jørgensen, S.E., Mejer, H.: Ecol. Modelling *2*, 19 (1976)
166. Keeling, C.D.: Tellus *25*, 174 (1973)
167. Siegenthaler, U.: J. Geophys. Res. *88*, 3599 (1983)
168. Hoffert, M.I., Callegari, A.J., Hsieh, C.-T.: Carbon Cycle Modelling, B. Bolin (ed.), p. 287, New York, Wiley 1981
169. Crane, A.J.: Tellus *34*, 398 (1982)
170. Killough, G.G., Emanuel, W.R.: Tellus *33*, 274 (1981)
171. Björkström, A.: The Global Carbon Cycle, B. Bolin, E.T. Degens, S. Kempe and P. Ketner (eds.), p. 403, New York, Wiley 1979

172. Bacastow, R.B., Björkström, A.: Carbon Cycle Modelling, B. Bolin (ed.), p. 29, New York, Wiley 1981
173. Goudriaan, J., Ketner, P.: Climatic Change (submitted, 1983)
174. Munro, J.K., Olson, J.S.: Energy and Ecological Modelling, W.J. Mitsch, R.W. Bosserman and J.M. Klopatek (eds.), p. 45, Amsterdam, Elsevier 1981
175. Viecelli, J.A., Ellsaesser, H.W., Burt, J.E.: Climatic Change *4,* 281 (1981)
176. Pearman, G.I.: Carbon Dioxide and Climate: Australian Research, Australian Academy of Sciences (ed.), p. 79, Canberra 1980
177. Kratz, G., Kohlmaier, G.H., Siré, E.-O., Fischbach, U., Bröhl, H.: Radiocarbon *25,* 459 (1983)
178. Fiadeiro, M.: Earth Planet. Sci. Lett. *49,* 499 (1980)
179. Bolin, B., Björkström, A., Holmén, K., Moore, B.: Tellus *35 B,* 206 (1983)
180. Maier-Reimer, E.: A Carbon Cycle Model, Part I: Concept and CO_2 Storage (unpublished, 1983)
181. Moore, B., Boone, R.D., Hobbie, J.E., Houghton, R.A., Melillo, J.M., Peterson, B.J., Shaver, G.R., Vörösmarty, C.J., Woodwell, G.M.: Carbon Cycle Modelling, B. Bolin (ed.), p. 365, New York, Wiley 1981
182. Bazilevich, N.I., Rodin, L.Ye.: Soviet Geography. Review and Translation *12,* 24 (1971)
183. Kohlmaier, G.H., Kratz, G., Bröhl, H., Siré, E.-O.: Energy and Ecological Modelling, W.J. Mitsch, R.W. Bosserman and J.M. Klopatek (eds.), p. 57, Amsterdam, Elsevier 1981
184. Pearman, G.I., Hyson, P., Fraser, P.J.: J. Geophys. Res. *88,* 3581 (1983)
185. Heimann, M., Keeling, C.D.: J. Geophys. Res. (in press, 1984)
186. Keeling, C.D., Heimann, M.: J. Geophys. Res. (in press, 1984)
187. Oeschger, H., Heimann, M.: Proc. Conf. on the Analysis and the Interpretation of Atmospheric CO_2 Data, Bern, September 14–19, 1981
188. Bolin, B.: Carbon Cycle Modelling, B. Bolin (ed.), p. 314, New York, Wiley 1981
189. Callendar, G.S.: Tellus *10,* 243 (1958)
190. Wigley, T.M.L.: Climatic Change *5,* 315 (1983)
191. Gates, D.M., Strain, B.R., Weber, J.A.: Physiological Plant Ecology IV, O.L. Lange, P.S. Nobel, C.B. Osmond and H. Ziegler (eds.), p. 503, Berlin, Springer 1983
192. Kennett, J.P.: Marine Geology, Prentice Hall, Englewood Cliffs 1982
193. Kohlmaier, G.H., Fischbach, U., Kratz, G., Bröhl, H., Schunck, W.: Experientia *36,* 769 (1980)
194. Perry, H., Landsberg, H.H.: Energy and Climate, p. 35, National Academy of Sciences, Washington, D.C. 1975
195. Bundesanstalt für Geowissenschaften und Rohstoffe: Die künftige Entwicklung der Energienachfrage und deren Deckung, Hannover 1976
196. Hubbert, M.K.: Energy and the Environment: Cost-Benefit Analysis, R.A. Karam and K.Z. Morgan (eds.), p. 3, Pergamon, Oxford 1976
197. Rotty, R.M.: Carbon Dioxide, Climate, and Society, J. Williams (ed.), p. 263, Pergamon, Oxford 1978
198. Global 2000 – Der Bericht an den Präsidenten, Verlag Zweitausendeins, Frankfurt/M. 1980
199. Gerwin, R.: Die Welt-Energieperspektive, DVA, Stuttgart 1980
200. Lovins, A.B., Lovins, L.H., Krause, F., Bach, W.: Wirtschaftlichster Energieeinsatz: Lösung des CO_2-Problems, Verlag C.F. Müller, Karlsruhe 1982
201. Bach, W., Breuer, G.: Umschau *80,* 520 (1980)
202. Fung, I., Prentice, K., Matthews, E., Lerner, J., Russel, G.: A 3-D Tracer Model Study of Atmospheric CO_2: Response to Seasonal Exchanges with the Terrestrial Biosphere, Goddard Space Flight Center, New York 1983

List of Abbreviations

General

superscript k ($k=12, 13, 14$):	Quantity related to carbon isotope k
no superscript:	Quantity related to total carbon
subscript i:	Quantity related to compartment i
subscript ij:	Quantity related to transition from compartment i to compartment j
subscript 0:	steady-state value
Δ:	deviation from stationary state value
N_i:	carbon mass
n_i:	number of moles
F_{ij}:	flux of carbon mass per unit time
f_{ij}:	flux of moles per unit time
$F_{e,i}$, $f_{e,i}$:	externally caused fluxes
k_{ij}:	exchange constant
$\tau_{ij}=1/k_{ij}$:	exchange time

Isotopes

$R_i^k=n_i^k/n_i^{12}$, $Q_i^k=n_i^k/n_i$	isotopic ratios
$l_i^k=n_i\cdot\Delta Q_i^k/Q_{aO}^k$	isotopic label
α_{ij}^k:	fractionation factor

$$\delta_i^{13}C[‰]=(R_i^{13}/R_{standard}^{13}-1)\cdot 1{,}000$$

$$\delta_i^{14}C[‰]=(Q_i^{14}/Q_{standard}^{14}-1)\cdot 1{,}000$$

$$\Delta_i^{14}C[‰]=\left[\frac{Q_i^{14}/Q_{standard}^{14}}{(R_i^{13}/R_{standard}^{13})^2}\cdot 0.975^2-1\right]\cdot 1{,}000$$

$\lambda^{14}=1/(8267\,a)$:	decay constant of ^{14}C
Π^{14} [atoms/cm/sec]:	^{14}C production in the atmosphere

Significance of Special Subscripts

a:	atmosphere
s:	ocean surface layer
d:	deep-sea
oc:	total ocean
b:	living biota

Atmosphere

p_a:	CO_2 partial pressure
$\alpha(\alpha_f)$:	airborne fraction (with respect to purely fossil input)

Oceans

[ ]:	concentrations
α_s:	solubility of CO_2 in sea-water
K_1', K_2', K_B', K_W:	equilibrium constants
A:	alkalinity
c(ΣC):	(total carbon) concentrations
ΣB:	total boron concentration
ξ:	evasion factor
B:	buffer factor
z:	depth
z_s:	depth of surface layer
z_d:	depth of deep-sea
$\bar{z}$:	mean penetration depth
$O(362 \cdot 10^{12}\,m^2)$:	ocean surface area
$\kappa_{as} = N_{aO}/N_{sO}$	
$\kappa_{oc} = N_{aO}/(N_{sO} + N_{dO})$	
K[m^2/a or cm^2/sec]:	coefficient of turbulent diffusion
w[m/a]:	upwelling velocity
φ[moles/m^3/a]:	rate of dissolution of biotic material per unit volume and per unit time

Biota

γ:	biota growth factor
δ:	climax factor

Externally Caused Inputs

$N_f(F_f)$:	cumulated (annual) fossil input
k_f:	growth coefficient of exponentially assumed fossil input rate
$N_{b,rel}(F_{b,rel})$:	cumulated (annually) released biogenic CO_2

Chemical Limnology

T. Frevert

Chair of Hydrology, University of Bayreuth
D-8580 Bayreuth, Federal Republic of Germany

Summary

Limnology as the science of lakes and open freshwaters is presented from the point of view of some physical and chemical aspects. Although lakes represent only a minute proportion of the global freshwater reservoir their indicative value for polluting or eutrophying anthropogenic impacts and their morphological and chemical (subsequently also biological) individuality make them worth being considered as a subject of special research. Besides basic morphometric features of lakes some physico-chemical factors (being considered as "master variables") which control the chemical composition of lake waters are discussed. The (bio)chemical pertinence of single chemical elements (e.g. phosphorus) is mentioned in view of eutrophication processes. Two case studies [L. Constance (Gnadensee) and L. Kinneret] exemplify the annual development of some basic hydrochemical conditions in the open waters (pelagial) of lakes at different latitudes but similar geochemistry of their catchment areas.

Introduction

Science can be divided into two fields:

- the *conceptual* sciences investigating nature due to mental schemes where the objectives of research are arbitrarily chosen in order to find general categories or networks of interrelationships,

- the *subject-related* sciences investigating certain parts of nature in trying to apply generalizing concepts to the individual conditions of real systems in space and time.

Limnology, as the science of lakes and freshwaters is surely a subject-related discipline. From the very beginning of modern limnology [usually dated back to Forel's early investigations on Lake Geneva (Léman) in the 1890's] limnologists knew that, virtually, they had to be scientific universalists but, in fact, they soon had to learn that this goal was hardly to be reached.

Since then the discussion whether limnology should be understood as "holistic" or "analytic" never ended. August Thienemann, father of the "biocenotic equilibrium", was among the first limnologists who stressed the idea of lakes as organized (i.e. feedback controlled) networks of physical, chemical, and biological interrelationships. Emphasizing the dynamic (i.e. time dependent) character of such interactions he understood limnology as the science concerned with all aspects of biotic-abiotic relationships in the lake as a whole [1].

It was mostly the fate of holistic views to remain superficial. Another disadvantage is that such views are "non-conceptual;" they stress the complexity of the system, i.e. the need for the researcher to apply many different aspects simultaneously to find appropriate resolutions of the problems.

In contrast analytic or conceptual interpretations of limnology, having been predominant since the 1930's, focus on single concepts in view of certain details which are deduced from conceptual sciences and considered as preponderant for the whole lake; the corresponding notions are consequently "master" variables (e.g. pH, redox potential, D.O. concentration etc.).

Comparing holistic and conceptual interpretations it turns out that the first ones are usually realistic but wrong (for detailed informations) whereas the second ones are correct (for detailed informations) but unrealistic.

There is no doubt that it is one of the fascinating goals of recent limnology to reconcile both understandings by means of computerisation or lake modelling. In a comprehensive approach using extensively computer techniques and all sources of informations it should be possible to create a "holistic" interpretation of lakes which, however, will exceed by far what was ever be known or concluded by a single scientist.

Nonetheless, it should never be forgotten that the reliability of a computer model depends strictly on the correctness *and* the validity (or appropriateness) of its conceptual implications which are, first of all, a result of detailed observation, analytic simplification, precise measurements and intelligent imagination. Thus, more than ever it is the task of modern limnologists to be specialists and not universalists, i.e. to elaborate computerisable notions from the physical, chemical or biological point of view in order to make sure that the "holistic" conclusions of the computer are reliable, i.e. realistic as well as correct in the straightforward sense of physico-chemical or biological conceptions.

The time of limnologists "knowing everything" about lakes is over; the time of interdisciplinary teams of specialists (including mathematicians) being limnologists due to their mutual interest in lakes, not in the same discipline, has been reached. Thienemann's idea is more vivid than ever but due to a cooperating team of scientists not to a single person and brain.

In the forthgoing text limnology is reduced mainly to some physical and chemical aspects. Thus, the reader should be aware that this is just *one* point of view of limnology but not necessarily the only one. A comprehensive description of limnology was given by Wetzel [2].

The Global Freshwater Reservoir and Man

About 71% of the surface of the earth is covered with water. The water is filling in surface depressions of the upper earth crust with a mean depth of 3,800 m. However, the majority of the resulting global reservoir of condensed water of 1.4×10^9 km^3 is more or less salty (ocean water and water of salt lakes). Only the neglectable order of magnitude of about 2% is freshwater where all lakes contain 125,000 km^3 or 3% of this freshwater reservoir. In Table 1 the different compartments of the global water body are compared. The global water budget is characterized by the higher evapotranspiration compared to precipitations on the oceans and vice versa on the continents. Thus, the continuous transport of evaporated water from sea to land is balanced by a corresponding transport of precipitated water from land to sea (hydrological cycle). Mean data are given in Table 2. Considering the fact that all terrestrial life depends fundamentally upon the purity and typically low salinity of the freshwater body it becomes clear that all kinds of impact of man on nature has eventually to be judged by its consequences on the global freshwater reservoirs.

Man's use of freshwater is immense. A total need of 3,000 km^3 $\times$ year^{-1} was estimated [3], that means 800 m^3 per person and year. However, not each continental freshwater reservoir can be used in the same way; only precipitated water bodies (freshwater lakes, groundwater) and the running waters are accessible for nutritional and industrial purposes.

The total freshwater runoff of the continents is estimated to be 38.6×10^3 km^3 per year. Assuming a human population of 4×10^9 individuals the world population could thus increase to 13 times more people than today (i.e. to 52 billion individuals); however, under such circumstances each droplet of the continental

Table 1. Global water reservoirs and their renewal rates

Reservoir	$10^3 \cdot$ km^3	% Total	Renewal rate (years)
World ocean	1,370,000	97.6	37,000
Polar ice, glaciers	29,000	2.1	16,000
Groundwaters	4,000	0.3 (95)*	300
Lakes	125	0.009 (3)*	1–1,000
Salt lakes	104	0.008	1–1,000
Soil humidity	67	0.005 (1.6)*	0.77
Running waters	1.2	$9 \cdot 10^{-5}$ (0.03)*	~0.04
Atmospheric water content	14	$9 \cdot 10^{-4}$ (0.33)*	0.02

* As refered to the total freshwater reservoir ($4.2 \cdot 10^6$ km^3)

Table 2. The global hydrological budget

Global compartment	Earth	Ocean	Continents
Surface ($10^6\ km^2$)	510	361	149
Precipitation ($10^3\ km^3\ year^{-1}$)	513	411*	100
Mean precipitation (litre) per m^2 and year	1,002	1,139	671
Evapotranspiration ($10^3\ km^3\ year^{-1}$)	513	448	65
Mean evapotranspiration (litre) per m^2 and year	1,002	1,241	436

* Further inputs from tributaries 35,000 $km^3 \cdot year^{-1}$ and polar ice waters 2,000 $km^3 \cdot year^{-1}$

Table 3. Estimated data on global human water use until the end of the century* (from: M. Holy: Water and the environment. Irrigation and Drainage paper no. 8, Food and Agriculture Organization, Rom 1971)

	1967	2000	% Annual increase
Agriculture			
Irrigation	1,400,000	2,800,000	2.1
Livestock	58,800	102,200	1.7
House keeping	19,800	38,300	2.0
Other users			
House keeping	73,000	278,900	4.1
Mining and industries	437,700	2,231,000	5.0
Total	1,989,300	5,450,400	3.1

* in $10^6\ m^3$

runoff would have to be used by man. The latter will occur in about 100 years (supposed that mankind doubles all 30 years).

Not all runoff water is potable or good enough for industrial purposes. Increasing pollution with organic wastes and, particularly, electrolytes deteriorates the restricted freshwater resources in accumulating components of low oxidation numbers, i.e. high chemical energy content which, however, can hardly be used because of its widespread distribution within the water body. The same is due to the increase of salinity in freshwaters. Dissolving salts into water is a strict irreversible process bound to a considerable entropy increase. The reverse process is thus improbable and needs an input of free energy which will be much higher than the energy gain (if any) during the dissolution process.

Man's need for freshwater still increases. In Table 3 the FAO estimate on human water use until the end of the century is summarized. The data indicate that the municipal and industrial need for water will exceed by far the need in agriculture.

Considering the tiny role of lakes in the global freshwater cycle limnology seems not worthwhile to be mentioned.

However, lakes are sensitive indicators for detecting man's dealing with his environment. This is mainly due to their "conservative" character. The indicative value R_i of a freshwater body depends upon the product between input rate of nutrients or pollutants, its physical, chemical or biological response, and the exchange rate of the water,

$$R_i = \frac{\text{input} \cdot \text{response}}{\text{exchange}}$$

Assuming stationary states, running waters are usually characterized by high water exchange rates compared to input and response whereas lakes may perform extreme low water exchange rates so that, generally, $R_i^{river} < R_i^{lake}$. If the water exchange rate drops below the response (reaction) rates a lake becomes "individual," i.e. it performs characteristics which cannot simply be extrapolated from its tributaries. If the response rate is higher than water exchange and input rate (which is usually the case with deep lakes) the lake can be considered independently from its tributaries;

exchange > input > response → river, brook

input > exchange > response → lake (controlled by its tributaries)

response > input > exchange → lake (independent of its tributaries).

Further, the sensitive and individual responses to all kinds of impacts or perturbations make lakes not only excellent indicators but create also their characteristic beauty and recreational value. Man's drive to study lakes or to learn limnology is probably much more due to these impressions than to the role of lakes within the hydrological cycle.

Typology of Lakes

As was already mentioned, lakes can be investigated due to physical, chemical or biological parameters of predominant significance. Corresponding "master" variables may be water temperature and stratification pattern, D.O./H_2S concentration pattern, pH and/or redox potential (pε), phosphorus and/or nitrogen concentration, assimilation/dissimilation ratio as well as morphometric notions. In a first approach lakes can also be differentiated by their geomorphological origin [2, 4].

Geomorphological Origin

Lakes arise from water covered tectonic basins (the majority of the "old" lakes like L. Tanganyika, L. Baikal, L. Tahoe, L. Kinneret, Dead Sea etc. dating back to more than 1 million years ago), from former volcanoes (e.g. caldera lakes) or even earthquake activities and landslides, or from glacial activity (the majority of North American and European lakes which belong to the relatively "young" lakes dating back to 20,000–30,000 years ago).

Lakes may also arise from very local conditions like dolines filled with water, or subterrestrial impoundments in caves, lakes formed by river activities including oxbow lakes, kettle lakes from melted glacial ice blocks, dune lakes of aeolian origin, shore lakes arising from current-built bars on a shore line (which however are not always freshwaters), and ombrogenic lakes like bog lakes or lenswater lakes on coral islands. Lakes can be permanent and temporary, while the latters are usually very shallow and thus strongly dependent on changing groundwater tables (e.g. L. Neusiedel, L. Chilwa).

Morphology

The main morphometric parameters are [2]

- maximum length, i.e. longest distance between opposite shorelines
- maximum breadth (b_m), i.e. longest distance to the shoreline perpendicularly drawn from the maximum length
- volume (V), i.e. integral of the areas of each stratum of the basin at successive depth, or approximately

$$V = \sum_n \left(\frac{n}{3} [a_1 + a_2 + \sqrt{a_1 a_2}] \right)_n$$

where

a_1, a_2 = upper and lower surface of the nth stratum

n = vertical depth of the stratum

- maximum depth (Z_m), i.e. the greatest depth of a lake
- mean depth ($\bar{Z}$), i.e. the ratio between volume (V) and water surface (a_0)
- relative depth (Z_r), i.e. maximum depth as a percentage of the mean diameter of the lake surface, or

$$Z_r = \frac{50 Z_m \sqrt{\pi}}{\sqrt{a_0}}$$

- shoreline development (D_L), i.e. the ratio between the length of the shoreline (L) and the length of a circle line with the mean surface diameter of the lake;

$$D_L = \frac{L}{2\sqrt{\pi A_0}}$$

D_L is unity if the shoreline is circular.

In Table 4 shoreline development data of some lakes are given; the higher D_L the more significant will be the impact from the shoreline area or the groundwater to the lake; in general the deep and volcanic or tectonic lake basins perform lower shoreline developments than glacial or very shallow lakes.

The main volumetric parameter is the relative volume development as defined by the ratio between mean depth and maximum depth, i.e. $\bar{Z}/Z_m$. For $\bar{Z}/Z_m$ = 0.33 the lake basin equals a right elliptic cone; for $\bar{Z}/Z_m$ = 0.46 it equals an elliptic sinusoid. As a rough rule, it can be stated that the higher $\bar{Z}/Z_m$ the lower will be the relative impact of the sediments of the lake on chemical or biological processes within the water body.

Table 4. Shoreline development (D_L) of some lakes of different origin

Lake	Origin	D_L	$\bar{Z}/Z_m$
Crater	volcanic	~1.0	0.82
Biwa	tectonic	1.3	0.39
Kinneret	tectonic	1.2	0.60
Erie	glacial	2.1	0.33
Constance (Upper lake)	glacial/tectonic	2.3	0.39
Constance (Lower lake)	glacial	3.2	0.45

The value will be highest in deep tectonic lakes with a low shoreline development and lowest in the shallow lakes of glacial or aeolian origin (cf. Table 4).

The hypsographic curve is the depth-area curve, i.e. is a percentage of the relationship between lake areas of certain depths.

Water Temperature and Thermal Stratification

Sunlight impinges on the lake surface and is partly reflected and partly it penetrates into the water. The reflection of direct and unpolarized sunlight (r) is described by Fresnel's law

$$r = \frac{1}{2}\left[\frac{\sin^2(i-r)}{\sin^2(i+r)} + \frac{\tan^2(i-r)}{\tan^2(i+r)}\right]$$

with i = angle of incidence of light and r = angle of refraction. Thus r is mainly dependent on the angle of the sun relative to the perpendicular on the water surface. The penetrating light is scattered or absorbed by $[H_2O]_n$ molecules or molecule clusters ($n \geqq 9$) where ~53% of the light energy is transformed into heat in the first meter of water depth. The absorption as a function of depth (Z) is described as

$$I_Z = I_o \cdot e^{-\varepsilon_{tot} Z}$$

Z = water depth (m) and $I_{o,Z}$ = light intensity at the surface and in Z meter depth, respectively. ε_{tot} refers to the *in situ* conditions and may be differentiated at least into the three terms ε_w (absorption coefficient in destilled water) ε_p (absorption coefficient as due to dissolved or particulate matter), ε_c (colour of the water); thus $\varepsilon_{tot} = \varepsilon_w + \varepsilon_p + \varepsilon_c$. Absorption coefficients range from 0.2 (some caldera lakes) to 4.0 (highly eutrophicated lakes during algal blooms) [2].

The heating of the upper water layer leads to a lower density (ϱ_T) and thus to a lower viscosity of the water (η_T). In Table 5 some values of η_T as a function of temperature (K) relevant for most lakes are listed. The kinematic viscosity η_R is given as η_T/ϱ_T. From 5 to 35° it increases by a factor of 2.16 which is sufficient to create a thermal discontinuity within the water body which is stable and divides lakes into thermal strata of far reaching physical, biological, and chemical independence [5, 6]. The transport of heated surface water into deeper water layers is mainly due to turbulence or eddy diffusion. In a simplifying approach the criti-

Table 5. Viscosity (η_T) and kinematic viscosity (η_k) of pure water as dependent on temperature (5–34 °C or 273–307 K)

T	°C	η_T 10^{-3} Pa · sec^{-1}	η_k 10^{-6} · m^2 · sec^{-1}	
278	5	1.561	1.561	ratio = 1.55 (temperate lakes)
283	10	1.306	1.304	
288	15	1.138	1.139	
293	20	1.002	1.004	ratio = 1.63 (subtropical lakes)
298	25	0.890	0.892	
303	30	0.798	0.801	
308	35	0.719	0.723	
		$\Delta = 0.84 \cdot 10^{-3}$	$\Delta = 0.84 \cdot 10^{-6}$	

cal shear stress between two water layers disrupting their density stratification may be considered as due to a laminar flow at the interface of two fluids.

The so-called relative thermal resistance (RTR) is consequently due to the changing density or the increasing kinematic viscosity with depth (Z). Viscosity or density of the water and the critical (i.e. still) laminar flow velocity (v) are constant and proportional to the Richardson number (N_R) [7],

$$N_R = \frac{g(d\varrho/dZ)}{\varrho_T(dv/dZ)^2},$$

where g = acceleration of gravity.

For a given initial density at temperature T of the surface water we may integrate and simplify for certain depths (Zn)

$$N_R^{Zn} = \text{const.} \cdot \frac{\Delta\varrho}{\Delta v^2}.$$

As long as $N_R^{Zn} > 0.25$ there is laminar flow and no turbulent mixing across the density interface. If $N_R^{\Delta Z}$ drops below 0.25 the heated overlying water will be mixed with the deeper and colder water layer. As v depends on the wind attack at the water surface it will decrease to the deeper and colder water layers while ϱ_T or η_k will increase (cf. Table 5). If eventually $\frac{\Delta\varrho}{\Delta v^2} \cdot \text{const.} \geqq 0.25$ the current velocity is too low to effect turbulence or eddy diffusion and the progressing of heated water to the depth is stopped resulting in a thermal discontinuity or *thermocline* which separates the heated and mixed upper-layer water (the *epilimnion*) abruptly from the cold and more or less stagnant bottom water (*hypolimnion*). Their thermal gradient (the *metalimnion*) is the more stable the greater η_k (epilimnion)/η_k (hypolimnion) which will be the case in most subtropical lakes as compared to lakes in temperate or cold latitudes (cf. Table 5).

The reverse destratification process happens in a different way. Cooled surface water sinks without wind action into deeper water layers as long as its density exceeds the density of the surrounding epilimnetic waters. The process continues while the whole epilimnetic water body gradually and simultaneously cools down.

The thermal gradient correspondingly progresses into deeper water and becomes less abrupt. Finally, any temperature gradient to deeper waters disappears or the thermocline reaches the bottom of the lake. The now homothermal lake water can fully be overturned and mixed by wind action since the mixing barriers disappeared.

The different annual pattern of stratified and overturned periods of a lake and the thermal stability of the thermocline are still the most basic features of lake typology [8]. Lakes are *monomictic* or *dimictic* if they overturn once or twice a year; they are *oligomictic* or *polymictic* if they overturn never or daily (as many tropical and shallow lakes do). If the thermocline ranges within 0–20 °C a lake is *temperate* mono- or dimictic; if it ranges within 10–30 °C the lake is warm-monomictic (Table 5). Lakes whose water temperatures are always $\geqq 25$ °C are tropical and usually polymictic; lakes whose maximum temperatures always remain at ~ 5 °C are arctic and cold-monomictic, i.e. the short ice-free summer time leads to an overturn but not to a lasting thermocline. During the initial non-stationary thermal conditions in springtime lakes may perform intermediate stratifications with several thermoclines (if relative cold and warm periods rapidly alternate in springtime) or partial overturn if (mostly from hydrochemical reasons) the lake water is not fully overturned and mixed although homothermal conditions may exist. The latter is extreme with *meromictic* lakes which either by submerse mineral springs (crenogenic) or by biogenic accumulation of dissolved and particulate components [e.g. Fe(II), Mn(II), Cl^-, PO_4^{3-}, organic matter) retain continuously an unmixed bottom water.

Although the heatflow in lakes can be described one-dimensionally from atmosphere to the water through the air/water interface (other heat sources or sinks are usually neglectable) it is difficult to calculate the heat budget as due to heat conduction and evaporation. Both notions are dependent upon wind action, relative humidity of the air and air and water temperature at the air-water interface. As an empirical approach the Bowen ratio (B) is defined as [9]

$$B = \frac{Q_S}{Q_L},$$

where Q_S means heat convection and diffusion through the interface and Q_L the evaporation enthalpy of the water. B is a function of the air and water temperature (t_a, t_w), the corresponding saturation vapour pressures (P_a, P_w), and of the atmospheric pressure (P)

$$B = 0.66^{P/1000} \cdot \frac{t_w - t_a}{P_w - P_a},$$

where P is the atmospheric pressure in mbar.

The accumulation rate of heat energy (Q_T) in a lake is generally described as

$$Q_T = Q_R + Q_L + Q_S$$

dim $J \cdot cm^{-2}$, or

$$Q_T = Q_R + (1 + B)\, Q_L,$$

where Q_R means the net gain of heat energy of the water due to the radiation input (dim $J \cdot cm^{-2}$), i.e.

$$Q_R = \text{radiation input} - \text{reflected radiation} - 10^{-22.87} \cdot T^4.$$

Assuming a stationary state (i.e. no temperature change detectable in the considered water body) Q_T can be approximated [for $Cp(H_2O) = 4.1855$ J]

$Q_T \simeq \bar{Z} \cdot t_{ao} \cdot 4.1855$ dim $J \cdot cm^{-2}$ (shallow lakes)

$Q_T \simeq \frac{1}{2} Z_{th} \cdot t_{ao} \cdot 4.1855$ (lakes with $\bar{Z}/Z_m = 0.33$)

$Q_T \simeq Z_{th} \cdot t_{ao} \cdot 4.1855$ (deep lakes, i.e. lakes, where hypolimnion $\gg$ epilimnion)

Z_{th} = depth of the thermocline (cm)
t_{ao} = temperature (°C) of the surface water.

In a similar manner the evaporation enthalpy can be estimated (for $\varrho H_2O = 1$)

$$Q_L \simeq 2469.5 \cdot Z_{evap} \quad \text{dim} J \cdot cm^{-2}$$

Z_{evap} = depth of water loss by evaporation (cm).

Example:

Given L. Kinneret with $\bar{Z} = 26$ m (or 2600 cm), $t_{ao} = 29$ °C, $t_a = 35$ °C and a mean atmospheric pressure of 1013 mbar; what is the net input of heat energy Q_R until the epilimnetic temperature is t_{ao} if the initial water temperature after winter overturn was $t_w = 14$ °C, $Z_{th} = 15$ m and $Z_{evap} = 5$ cm?

(1) Calculation of the Bowen ratio

$$B = 0.66^{1.013} \cdot \frac{14-35}{17-56},$$

$$B = 0.35.$$

(2) Calculation of Q_T (as $\bar{Z}/Z_m = 0.6$ use the approximation for deep lakes)

$$Q_T \simeq 1500 \cdot 29 \cdot 4.1855,$$

$$Q_T \simeq 1.821 \cdot 10^5 \, J \cdot cm^{-2}.$$

(3) Calculation of Q_L

$$Q_L \simeq 2469.5 \cdot 5,$$

$$Q_L \simeq 1.235 \cdot 10^4 \, J \cdot cm^{-2}.$$

(4) Calculation of Q_R

$$Q_R \simeq 1.821 \cdot 10^5 - 1.35 \cdot 1.235 \cdot 10^4,$$

$$Q_R \simeq 165.4 \, kJ \cdot cm^{-2}.$$

Tributaries

The water regime of the majority of lakes is controlled by their tributaries much less by ground water intrusion or submerse springs [2]. Tributaries also interconnect a lake geochemically as well as biologically with its catchment area. The impact on the lake conditions by single tributaries depends first of all upon their inflow rate relative to the water budget of the lake and to the degree of mixing of the incoming waters with the lake water.

In general, the incoming water of a certain temperature and dissolved or particulate matter load will flow into the lake water layer of equal density [10]. Thus, three different types of progressing into the lake are possible: *overflow* if the incoming water density is lower than the density of the lake water, *interflow* if it corresponds to the density of metalimnetic water, and *underflow* if it exceeds the density of the lake water. From a statistical viewpoint it is the normal case that natural tributaries to lakes in temperate latitudes enter into the metalimnion as it is improbable that their water temperature will be still the same as at the spring (~4 °C, i.e. the temperature of the hypolimnion) or will already be heated like epilimnetic waters. Additionally, incoming running waters contain very often reasonable concentrations of suspensoids increasing their densities. If artificially heated or cold and loaded with particulate matter of very small size (e.g. clayey particles or colloids) the water will show overflow and underflow, respectively.

Incoming waters of a tributary may have different flow velocities. The point of entrance into the lake and the point where the water is densimetrically equilibrated will, therefore, be changing. The distance between the two points is defined by the *critical section depth* (Z_F) which is a function of the flow rate (v) and density (ϱ_T) of the incoming water, the cross sectional area (A_o) at the critical section depth and the Froude number (F_o) [11],

$$F_0 = \frac{(v/A_0)}{\sqrt{g \cdot \frac{\Delta\varrho}{\varrho_T} \cdot Z_F}},$$

where $\Delta\varrho$ means the difference of the densities of the incoming and the lake water. For given v/A_o and $\Delta\varrho$ the incoming water removes the lake water until a lake depth of Z_F meters prior to penetrating into the lake water layer of equal density. Until then its velocity will be significantly lowered which enhances the estuarine sedimentation of the majority of detritus of greater particle size. In larger lakes the pathway of progressing river waters may also be influenced by geostrophic forces [12].

The mixing between the incoming and the lake water is a difficult process which can hardly be discussed in a general way [13]. Nevertheless, in many lakes the mixing rate is low enough that larger tributaries, deviating markedly for certain variables from the lake conditions, can be detected throughout the lake basin.

Waves

Lakes are in no ways static systems. By wind action or changing atmospheric pressures their water bodies are in permanent motion leading to various kinds of progressive waves with different but relative small wave lengths, to different kinds of currents and gyres and to standing waves or the so-called "seiches" (originating from the French word sèche = dry) with relative long wave lengths. Seiches may be at the water surface or internal, the latter mainly at boundary layers of thermal or density discontinuities. In bigger lakes they are able to flood the one side of a shallow lake shore while drying the opposite side. Internal seiches may be hydrochemically or biologically relevant as they can change the pertinent chemical conditions in redox stratified lakes within shorter periods of time. Seiches may be *uni-* or *polynodal*, i.e. there are one or several nodal points (nodes) with no vertical movement of the water.

In view of the morphology of lakes the uninodal surface seiche can be used for empirical calculations of the mean depth ($\bar{Z}$ in meters) [14]. Given $\bar{Z}$ < width of the lake < length of the lake (rectangular basin) the period of the seiche (t) is given by Merian's formula,

$$t = 0.033\, l/n \sqrt{g\,\bar{Z}} \quad \text{dim min},$$

where l = length of the lake (m), g = acceleration of gravity ($9.8\ \text{m} \cdot \text{sec}^{-2}$), n = number of nodes. For example, the period of the primary (uninodal) surface seiche in L. Kinneret (n = 1) was found to be 24.8 min. The seiche stands in west-east direction (l = 12 km). The mean depth $\bar{Z}$ is therefore,

$$\bar{Z} = \frac{1.089 \cdot 10^{-3} \cdot l^2}{9.8 \cdot t^2}$$

or

$$\bar{Z} = \frac{156{,}816.0}{6{,}027.39} = 26.02\ \text{m}.$$

Internal seiches usually have greater amplitudes than surface seiches. As was mentioned they occur mainly at thermal discontinuities and can be uni- or polynodal analogous to the surface seiches. In the most simple approach of a rectangular basin separated into two water bodies of different densities, epi- and hypolimnion, the period (t) can be estimated by Watson's formula

$$t = \frac{2l}{\sqrt{\dfrac{(\varrho_1 - \varrho_2)}{\left(\dfrac{\varrho_1}{h_1} + \dfrac{\varrho_2}{h_2}\right)} \cdot g}},$$

where h_1 = height of the lower stratum with higher density, h_2 = height of the upper stratum with lower density, ϱ_1, ϱ_2 = corresponding densities ($\text{g} \cdot \text{cm}^{-3}$). In view of Watson's formula the surface seiche is a special case of an internal seiche where the density of the stratum with lower density (i.e. air) is neglected.

The discussion of progressive waves or currents which comprise surface as well as internal waves is much more difficult and a subject of extensive research. From a hydrochemical or biological point of view, particularly internal progressive waves or currents are pertinent as they may provide accelerated interchange between single strata of a stratified water body ("upwelling") and thus enhancing dispersion of nutrients or pollutants. A discussion on the present understanding and concepts of progressive waves and current phenomena was presented by Csanady [15].

Factors Controlling the Chemical Composition of Lake Waters

Geochemistry versus Hydrochemistry

Particularly in view of inorganic components the composition of lacustrine waters is a result of the average of atmospheric precipitations, the geochemistry and the geomorphological relief of the catchment area. Natural rainwater is poorly concentrated in ions but highly "aggressive" since its content of acids (mainly carbonic acid) is usually unbuffered and depresses the pH value of rainwater to about 5.6. (Only recently the acidity of rainwater increased in many areas of the world as a result of an increased emission of SO_2 and NO and NO_2 gases into the atmosphere [16]).

If such slightly acid rainwater comes into touch to solid groundrock, metal oxides or hydroxides go into solution leading to considerable disequilibria between the dissolved carbonic acid and the dissolved metal oxides tending to be equilibrated by acid-base reactions.

Besides CO_2 rainwater contains also the oxidant dissolved oxygen (D.O.). Further disequilibria arise, therefore, if D.O. containing rainwater comes into touch with groundrock compounds of low or minimum oxidation numbers (high electron load, reduced state). Their equilibration by electron transfer or redox reactions is, however, much more controlled by the reaction kinetics than analogous acid-base reactions. In the most simple case, we may assume similar reaction rates for both types of reaction, and prevailing minimum oxidation numbers of the groundrock compounds; the reactions will be

$$MeX_{solid} + \tfrac{1}{2}O_2 + 2\,H_2O \rightarrow MeOOH_{solid} + H_3O^+ + X^- \tag{1}$$

$$\overset{(-1)}{HX} + O_2 + H_2O \rightarrow \overset{(3+)}{XO_2^-} + H_3O^+ \tag{2}$$

$$MeOOH_{diss.} + H_3O^+ \rightarrow MeO^+_{diss.} + 2\,H_2O \tag{3}$$

(small numbers indicate the oxidation numbers of X).

Reactions (1) to (2) indicate that oxidation reactions increase significantly the acidity of the solution [17], the resulting acids will only partly be neutralized by reactions with basic hydroxides or oxides. They also indicate that the dissolution process of ions is mainly an acid-base rather than a redox reaction.

Since *dissolved* acids and bases react extremely rapid, their reaction kinetics will surely not control the water composition. Redox reactions (reactions 1, 2) and the dissolution process are much slower; therefore, we suppose for the overall kinetic coefficients (k):

$$k_{redox} \simeq k_{diss} \ll k_{acid/base}.$$

With dissolution and redox processes acids or bases (H_3O^+ or OH^-) are involved in an antithetic manner. In order to understand the geochemical reactions occurring in the catchment area of a lake and leading to the characteristic chemical composition of the water of its tributaries it is, therefore, necessary to consider the acidity of the incoming rainwater and the possible geochemical redox and dissolution kinetics. High acidity (i.e. low pH's) enhances the dissolution process but depresses redox processes. On the other hand, if the incoming rainwater has a low acidity which is rapidly neutralised by geochemical reactions, the pH in solution will be higher and redox processes are enhanced. From the point of view of groundrock geochemistry, this kind of negative feedback control between oxidation (acidifying) and dissolution (neutralising) processes will balance the acidity of the solution and the amount of dissolved compounds from the groundrock.

A conclusion which can be drawn from this consideration is that:

- the oxidative increase of acidity may lead to runoff water pH's which are below the values of the incoming rainwater
- pH minima (acidity maxima) as due to the oxidation of the groundrock components will hardly be reached; pH values $\ll 3$ are thus exceptional in natural waters [18].

Following the general theory of pH the natural runoff waters (from a groundrock catchment area) should kepp acidities corresponding to $\sim 3 < \text{pH} < 5.6$, even after having entered a lake.

That the latter is not the case is the result of photosynthesis, the spatial separation of the redox products of photosynthesis by gravity and lack of mixing action. These processes occur predominantly in the waters of the tributaries and mainly the lake itself. They can again be reduced to some chemical reactions analogous to reaction 1–3,

$$HCO_3^- \underset{\text{enzyme}}{\rightarrow} \langle CO_2 \rangle + OH^- \tag{4}$$

$$\langle CO_2 \rangle + H_2O \underset{\text{enzyme}}{\overset{h\nu}{\rightarrow}} \langle CH_2O \rangle + \underline{\underline{O_2}}\uparrow \tag{5}$$

$$\langle CH_2O \rangle_{\text{sed.}} + 2\,Me^{3+} + 6\,OH^- \underset{\text{enzyme}}{\rightarrow} 2\,Me^+ + HCO_3^- + 4\,H_2O \tag{6}$$

$$XO_2^- + \langle CH_2O \rangle \underset{\text{enzyme}}{\rightarrow} X^- + \underline{\underline{CO_2}} + H_2O \tag{7}$$

$$Me^+_{\text{diss}} + X^-_{\text{diss}} \rightleftarrows MeX_{\text{solid}} \tag{8}$$

Equations 4 and 5 characterize the process of photosynthesis leading to an increase in pH (release of OH^-) and the reduction of inorganic (oxidised) carbon to organic (reduced) carbon. Since the whole process is bound to an uptake of free energy [19] the opposite direction of the process should be probable to occur in the absence of light ($h\nu$) and photosynthetic organisms. Its low kinetics, however, prevent a rapid reverse reaction, and spatial separation of $\langle CH_2O \rangle$ and O_2 can become effective by escaping of gaseous O_2 into the atmosphere and sinking of organic carbon into the hypolimnetic lake water or the sediment [20].

At this point it becomes clear why interfaces like the thermal discontinuity at the metalimnion or the sediment water interface are pertinent for the (bio)hydrochemistry of lakes. The separation of the photosynthetic end products $\langle CH_2O \rangle$ and O_2 from each other will necessarily result in a lack of dissolved oxygen as an electron acceptor for oxidation processes in the deeper lake waters (due to the mixing pattern of the lake) and the deeper sediment layers (if the sediment is not resuspended). Oxidation should even come to an end when dissolved oxygen (the oxidant) is disappearing in the water or the sediment. Under the extreme circumstance where organic carbon is fully separated from its inorganic surroundings this may indeed be the case.

Often the sedimentation process through the dissolved oxygen containing epi- and hypolimnion is, however, long enough to oxidise the major part of the reduced organic carbon. Even if dissolved oxygen is depleted other terminal electron acceptors will be used for the microbially mediated oxidation process [21–23]. Equations 6 and 7 describe this "anoxic" oxidation implying that oxic and anoxic oxidation of organic carbon is releasing the weak acids HCO_3^- and $H_2CO_3^*$ into the solution [24] but the anoxic one is also reducing the strong acids HXO_2 and Me^{3+}. The reduced electron acceptors Me^+ and X^- (Eqs. 6 and 7) can again precipitate according to equation 8; eventually, the electron cycle is closed interconnecting the acidifying oxidation processes in the catchment area with the de-acidifying anoxic reduction processes in the lake, mainly in its sediments. We see that the occurrence of a *redoxocline* (spatial discontinuity of redox intensity) in the catchment area is bound to the groundrock/water interface whereas in the lake the redoxocline may be in the sediment, at the sediment-water interface or (e.g. in highly productive lakes) at the thermal or chemical density discontinuities in the water itself (thermally stratified and/or meromictic lakes).

pε Value, Redox Potential Eh, pe Value

We saw that the main chemical processes occurring in catchment area and lake (incl. its tributaries) can be summarized as dissolution-precipitation, proton exchange (acid-base), and electron exchange (redox) reactions. Conceptually valid parameters indicating the equilibria of these reactions are therefore most pertinent ("master variables" [25]) like pε (redox potential Eh), pH, pK_{sp} (the neg. decadic logarithm of the solubility product). In order to exemplify how pε values can be used we consider the oxidation of a common compound in catchment areas of lakes:

$$\underbrace{FeS_2}_{\text{groundrock}} + \underbrace{3.5\,H_2O + 3.75\,O_2}_{\text{rain}} \rightleftarrows \underbrace{FeOH^{2+} + 2\,SO_4^{2-} + 2\,H_3O^+}_{\text{precipitate or runoff}} . \tag{9}$$

The free enthalpy (ΔG_R) can be calculated from the free formation enthalpies (ΔG_f),

$$\Delta G_R^0 = \sum_n \Delta G_f^0\,(\text{products}) - \sum_n \Delta G_f^0\,(\text{educts}), \tag{10}$$

with n = the stoichiometric coefficients and 25 °C. Introduced into Eq. (9) we obtain

$$\Delta G_R^0 = (-229.4 - 2 \cdot 744.6) - (-160.2 - 3.5 \cdot 237.18 - 3.75 \cdot 16.32),$$
$$\Delta G_R^0 = -1718.6 + 1051.5 = -667.1\ \text{kJ} \cdot \text{mol}^{-1} .$$

The significant negative free reaction enthalpy indicates that – after equilibration – the proportion of the reactants will fully be shifted to the right side of the equation, i.e. to the side of the oxidized Fe and S species.

The free reaction enthalpy is connected to the law of mass action,

$$\Delta G_R = -2.303 \cdot RT\ \lg K \tag{11}$$

or

$$\Delta G_R^0 = -5.71\ \lg K^\circ\ \text{for}\ 25\ ^\circ\text{C}.$$

We thus may calculate the equilibrium constant K° as $K^\circ = 10^{116.8}$ indicating that, as long as dissolved oxygen is unlimitedly available, the oxidation of FeS_2 should occur quantitatively.

Basically bound to all oxidation (or reduction) reactions is the electron transfer between an electron donor (Fe^{2+}, S_2^{2-}) and an electron acceptor (O_2). Thus, the electron is considered as a reactant; we write

$$FeS_2 + 9\,H_2O \leftrightarrows FeOH^{2+} + 2\,SO_4^{2-} + 17\,H^+ + 15\,e^- \tag{12}$$

$$O_2 + 4\,e^- + 2H_2O \leftrightarrows 4OH^-. \tag{13}$$

Thus,

$$K_1 = \frac{[FeOH^{2+}][SO_4^{2-}]^2[H^+]^{17}[e^-]^{15}}{[FeS_2]},$$

$$K_2 = \frac{[OH^-]^4}{P_{O_2}[e^-]^4}$$

or

$$[e^-]_1^{15} = K_1 \cdot [FeS_2]/([FeOH^{2+}] \cdot [SO_4^{2-}]^2 \cdot [H^+]^{17})$$

and

$$[e^-]_2^4 = [OH^-]^4/(P_{O_2} \cdot K_2).$$

In p-notation ($p\varepsilon \equiv -\lg[e^-]$) we obtain

$$p\varepsilon_1 = \tfrac{1}{15}pK_1 + \tfrac{1}{15}pFeS_2 - \tfrac{1}{15}pFeOH^{2+} - \tfrac{2}{15}pSO_4^{2-} - 1.13\,pH$$

and

$$p\varepsilon_2 = pOH^- + \tfrac{1}{4}\lg P_{O_2} - \tfrac{1}{4}pK_2.$$

Hence, $p\varepsilon$ is a conceptual notion independent of certain reactants but characterized by a certain fictive "electron concentration (activity)" which defines the thermodynamic (energetic) level at which the electron exchange equilibrates. The resulting $p\varepsilon$ value for the whole reaction will, therefore, correspond to $[e^-]_2/[e^-]_1$ or $p\varepsilon_2 - p\varepsilon_1 = p\varepsilon$; the latter is related to the equilibrium constant,

$$p\varepsilon = -\tfrac{1}{n}\,pK.$$

Thus, as long as $p\varepsilon_2 > p\varepsilon_1$ [i.e. $(e^-)_1 > (e^-)_2$] K becomes positive or (in our case) FeS_2 will being oxidized after thermodynamic equilibrium is established.

The $p\varepsilon$ value is also called the "redox potential" where the latter (given as Eh in mV) is related to as $p\varepsilon = Eh/RT \cdot F^{-1}$.

As shown before oxidation processes are often combined proton and electron transfer reactions. The different influence of pH and $p\varepsilon$ (as master variables of acid-base and redox reactions) can be demonstrated in $p\varepsilon$- and pH stability diagrams (Fig. 1). In such diagrams the negative decadic logarithm of the concentrations (activities) of the different molecular species is plotted against $p\varepsilon$ (for pH = const.) and pH (for $p\varepsilon$ = const.) with the same units.

Even if we do not go into much detail how to read such diagrams it is obvious that proportions between the oxidized and reduced species may change per $p\varepsilon$ unit much more than the acid and the conjugate base does per pH unit.

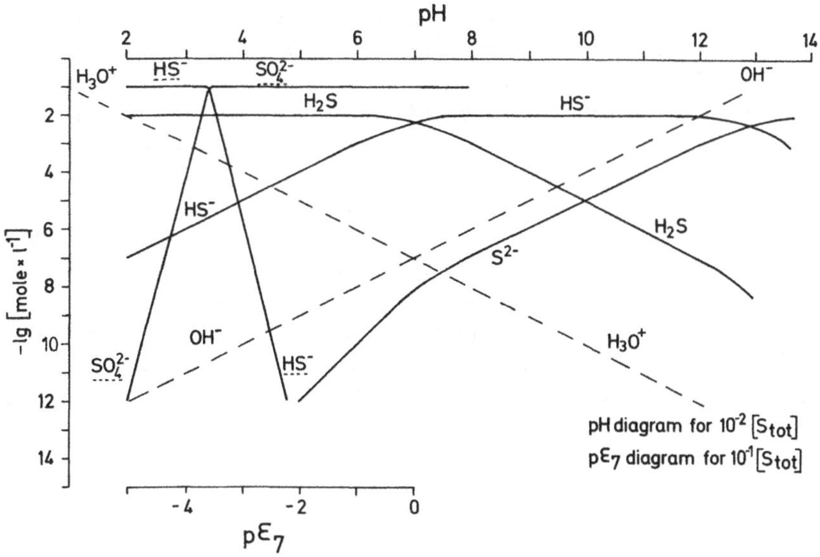

Fig. 1. pH and pε stability diagrams for the systems $H_2S/HS^-/S^{2-}$ and SO_4^{2-}/HS^-, respectively. pH and pε_7 in same units; pε_7 means pε as referred to pH = 7

For instance, if 90% of one mole of total acid is transfered into 90% of one mole of the conjugate base a minimum pH increase of two units is necessary. The process is accompanied by the transfer of exactly 90% of one mole of H^+. In the analogous case of redox reactions even more than 99% of the oxidized (reduced) species can be transfered into the reduced (oxidized) one. This stoichiometric difference between acid-base and redox reactions is the main reason why redoxoclines in lakes are usually extreme sharp with changes of the predominant oxidized or reduced species within few millimeters (in sediments) or centimeters (in water).

The pε value is a redox intensity parameter; it defines the proportions between relatively oxidized to relatively reduced compounds under equilibrium. Their absolute concentration level (e.g. whether they are present at 10^{-3} or 10^{-6} M) is, however, determined by the redox buffer intensity $\beta_{p\varepsilon} = \frac{d[e^-]}{dp\varepsilon}$ or $\beta_{p\varepsilon} = \frac{1}{V_0} \cdot \frac{d[e^-]}{dp\varepsilon}$, where V_0 is a volumetric unit (e.g. liter). As $[e^-]$ represents the concentrations of electron donors or acceptors $\beta_{p\varepsilon}$ means the stability of a certain pε against the addition of reductants (electron donors) and oxidants (electron acceptors), respectively. Unfortunately, unique $\beta_{p\varepsilon}$'s can usually not be determined in a straightforward way (lack of redox equilibria in natural solutions, unknown stoichiometry of real redox reactions, pH dependence of pε values) [26–29]. However, it is possible to define a redox buffer capacity $\beta_{\Delta pe} = \frac{1}{V_0} \cdot \frac{\Delta[e^-]}{\Delta pe}$ using pe as the redox variable [30]. The pe value, defined as $-\lg a_e$-(sensor) is a notion corresponding to the detectable response of an (usually) electrochemical sensor (e.g. Pt metal), a_e-(sensor), when immersed into a solution of different redox intensity, i.e. different energetic levels of electron transfer between its dissolved components. Thus, as we can measure $\Delta[e^-]$ *and* pe, β_{pe} is the operationally defined ratio between the addition of oxidants or reductants per volume solution and the corresponding change of the sensor detected pe value. For instance, if the pe values during oxic and anoxic periods of the hypolimnetic waters of a lake have been detected, it is possible to calculate the redox buffer capacity β_{pe} of the D.O. depletion in the lake.

The interrelationship between pe and β_{pe} (and β_{pH}, cf. next paragraph) with oxic and anoxic redox reactions in lakes can be made visible by a closed system hydraulic model (Fig. 2): By photosynthetic action electrons on a maximum energetic level enter the system. The outlets of the system correspond to possible

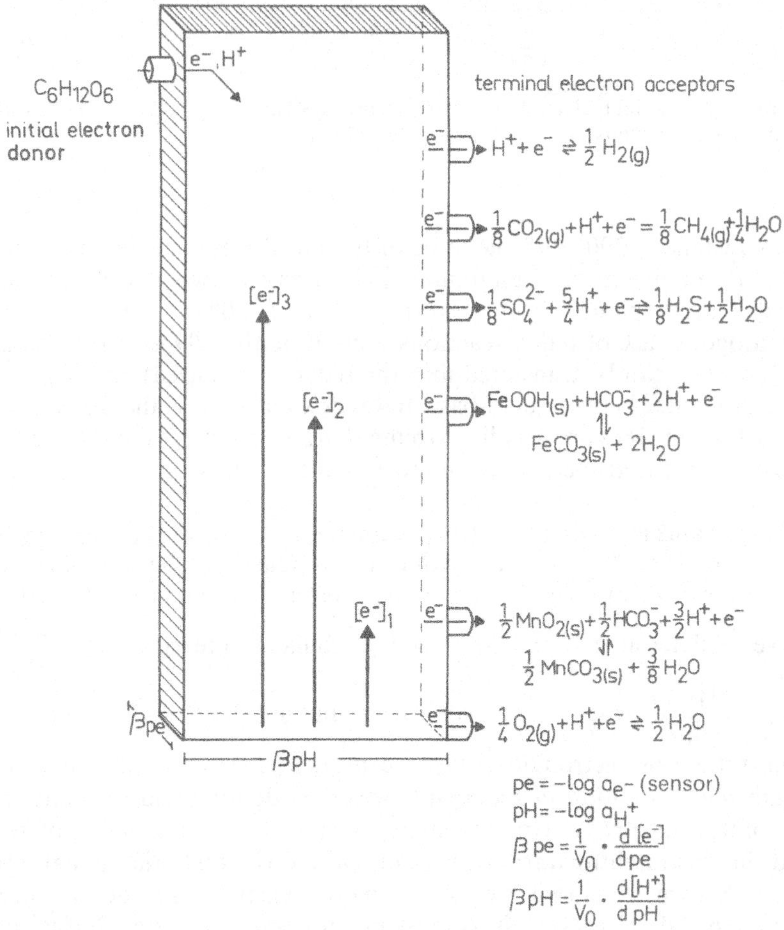

Fig. 2. Relative hydraulic model of major redox (electron transfer) reactions in lacustrine environments due to unit redox and acid-base buffer capacities in the medium

electron acceptors used by mainly chemoorganotrophic microorganisms to gain metabolic energy [31, 32]. D.O. (O_2) as electron acceptor involves the maximum energy gain. If D.O. is depleted the corresponding "outlet" from the system is closed; the (fictive) electron concentration (corresponding to the hydrostatic level in the model) will increase until reaching the next-possible outlet of electron transfer. Due to the presence of appropriate electron acceptors and metabolizing microorganisms [33] this process continues until water itself (i.e. H_3O^+ ions) is reduced defining the strongest possible reductant or weakest electron acceptor in aqueous solutions.

The length and depth of the model represents the redox and acid-base buffer capacities β_{pe} and β_{pH}, respectively. These imply the following conditions:

(a) $\beta_{pe} = \beta_{pH} \rightarrow$ the ratio between reduced and oxidised species is equally sensitive to additions of strong acids (bases) or strong reductants (oxidants) both given in equival. $\cdot l^{-1}$ (e.g. O_2/H_2O);

(b) $\beta_{pe} < \beta_{pH} \rightarrow$ the ratio between reduced and oxidised species is more influenced by acid (base) than by reductant (oxidant) additions to the system (e.g. SO_4^{2-}/H_2S);

(c) $\beta_{pe} > \beta_{pH} \rightarrow$ the ratio between reduced and oxidised species is widely acid-base independent (e.g. Fe^{3+}/Fe^{2+} at $pH < 3$).

Our model in Fig. 2 represents the (idealised) condition (a) over the whole range of different electron transfer levels (corresponding to different pe values). In natural systems like lake waters or, particularly, their sediments this may be different.

The redox-chemical available equivalents of O_2 in a shallow stratified lake with iron rich sediments can be much lower than the corresponding Fe(III) equivalents of the sediment upper layer [34, 35]. Lakes in calcareous gypsum enriched catchment areas will have much higher concentrations of redox relevant SO_4^{2-} equivalents than D.O. equivalents. Thus, β_{pe} can differ markedly due to the basic geochemistry of a lake catchment [36] as well as due to different pe levels. The higher β_{pe} will be the more resistive will be the solution against changes of the pe value. Following the sequence of different electron acceptors in natural systems by pe measurements we encounter along the pe scale values which are more frequent or characteristic for the system than adjacent ones. This is nicely be demonstrated when a sediment containing water body after being saturated with D.O. is anoxically sealed and the pe values being monitored in the sediment-water interface (Fig. 3 a). Starting with positive values in the overlying water, the sediment surface and the sediment upper layer the first redox buffer range (refered to pH 7.5) is between $pe_{7.5} = 6.5$–4.5 (D.O. buffering); the next range is between 3.5 (in the overlying water) and -3.1 in the sediment refering mainly to the Fe(III) $\rightarrow$ Fe(II) reduction [37–40] creating a sharp pe gradient between the overlying water and the sediment corresponding to the Fe_{tot} concentration gradient. Eventually, the $pe_{7.5}$ buffers between -3.5 (overlying water) and -5.5 refering to the $SO_4^{2-} \rightarrow H_2S$ reduction [41–43] which again will be more significant in the sediment than in the overlying water. Figure 3 b shows the opposite development during re-aeration of the system. Again certain pe values prove to be more stable although the fully aerated system values at the sediment surface and the overlying water remain at a lower level indicating an irreversible redoxchemical change in the overlying

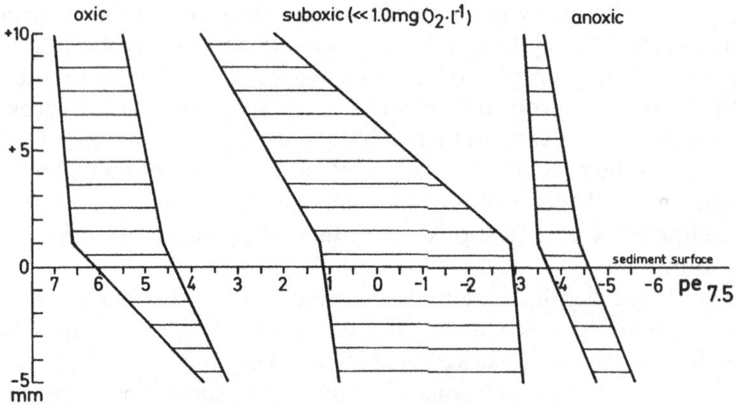

Fig. 3a. $pe_{7.5}$ values at the sediment-water interface of L. Constance (upper lake, 253 m depth) in a closed system developing from oxic to anoxic conditions

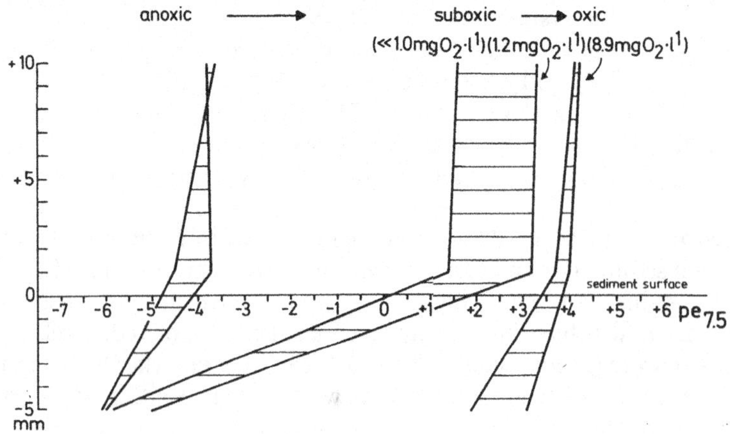

Fig. 3b. As 3a but in an open system developing from anoxic to oxic conditions by reaeration ($pe_{7.5}$ means pe as referred to pH = 7.5)

water (probably by prior release and only partly reprecipitating of Fe(II) from the sediment) [44]. As D.O./H_2O, Fe(III)/Fe(II) and $SO_4^{2-}/H_2S(HS^-)$ are the main redox couples in many lakes, their pattern of oxic/anoxic pe values will be similar.

pH Value

We mentioned the pertinence of the pH in solutions where electron transfer, i.e. redox reactions occur. Low pH values indicating high acidic solutions decrease the oxidation intensity whereas high values indicating alkaline (base) conditions decrease the reducing intensity.

In the lacustrine environment acid-base reactions occur as a direct proton transfer on H_2O

$$HSO_4^- + H_2O \rightleftarrows H_3O^+ + SO_4^{2-}, \tag{1}$$

as proton transfer due to the existence of oxides or hydroxides of multivalent cations

$$Fe^{3+} + 4\,H_2O \rightleftarrows Fe(OH)_2^{+} + 2\,H_3O^{+}, \tag{2}$$

or as a proton transfer from H_2O on dissolved anions

$$CO_3^{2-} + H_2O \rightleftarrows HCO_3^{-} + OH^{-}. \tag{3}$$

In diluted aqueous solutions (the normal case in lakes and rivers) H_3O^+ and OH^- are thermodynamically determined by the autolysis constant K_w of water. The intensity of acid-base, i.e. proton transfer reactions is, therefore, comprehensively defined by the concentration of oxonium ions (H_3O^+) in solution, i.e. $pH = -lg [H_3O^+]$.

Previously, we exemplified natural acid-base processes by their significance with oxidations of groundrock components during chemical weathering in the catchment area of a lake. In fact many lakes are not embedded into a groundrock but into a sedimentary rock catchment area, where the sedimentary rock itself is already the result of a former lacustrine or marine diagenetic development [45]. Due to the historical photosynthetic activity with subsequent carbonate precipitation (see below) those sedimentary rocks release reasonable amounts of base equivalents buffering the acid content of atmospheric precipitations (mainly CO_2), and from redox reactions.

$$MeCO_3 + CO_2 + H_2O \rightleftarrows Me^{2+} + 2\,HCO_3^{-}\cdot \tag{4}$$

Unlike to groundrock, the pH values of sedimentary rock runoff waters (particularly from calcareous sedimentary rock) should be determined by the $CO_2/HCO_3^-/CO_3^{2-}$ acid-base chemistry. Another acid buffering reaction is due to the weathering of silicates,

$$SiO_3^{2-} + 2\,H_3O^{+} \rightleftarrows H_2SiO_3 + 2\,H_2O. \tag{5}$$

After equilibration H_2SiO_3 will be predominant. However, compared to carbonates most other inorganic pH buffering components (like silicates, phosphates, aluminates, metal hydroxides) are much less dissolvable, and usually much less concentrated in the runoff waters, i.e. in the tributaries of a lake.

In introducing, additional to the pH value, the β_{pH} value $\left(\beta_{pH} = \frac{1}{V_0} \cdot \frac{d[H_3O^+]}{dpH}\right)$ as the acid-base buffer capacity it becomes clear that, analogous to the redox buffering, the acid-base buffering capacity at different pH values will change. In Fig. 4a a pH buffering diagram typical for oxic tributary or lake water from a carbonate rich catchment area is shown. The chosen composition and concentrations are typically found in many lakes. If β_{pH} increases the corresponding pH value is increasingly stable. Similar to what was the effect of different redox buffering capacities there are "traps of pH values" which hardly can be crossed because of intensive buffering (arrows in Fig. 4a). Necessarily, these pH values correspond to the acid-base constants of the $CO_2/HCO_3^-/CO_3^{2-}$ system (pK_a values) fixing them at about 8.4 if the acid-base components are equal and between 6.4 and 8.4 if the acid input prevails (catchment area) and between 8.4 and 10.4 if the acid output prevails (during photosynthesis in epilimnetic (euphotic) waters). It is obvious that the influence of silicates is secondary, where the influence of other

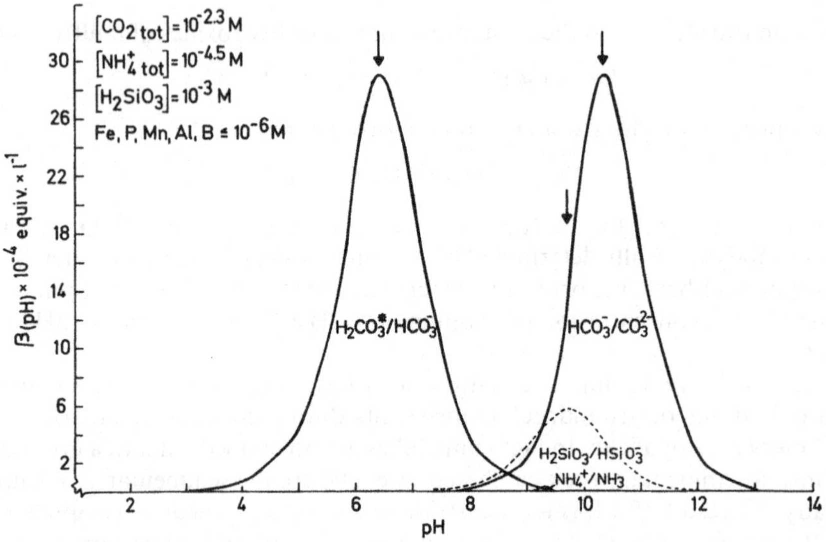

Fig. 4a. pH buffer capacities (β_{pH}) at different pH values due to common oxic lacustrine conditions (autolysis of H_2O neglected)

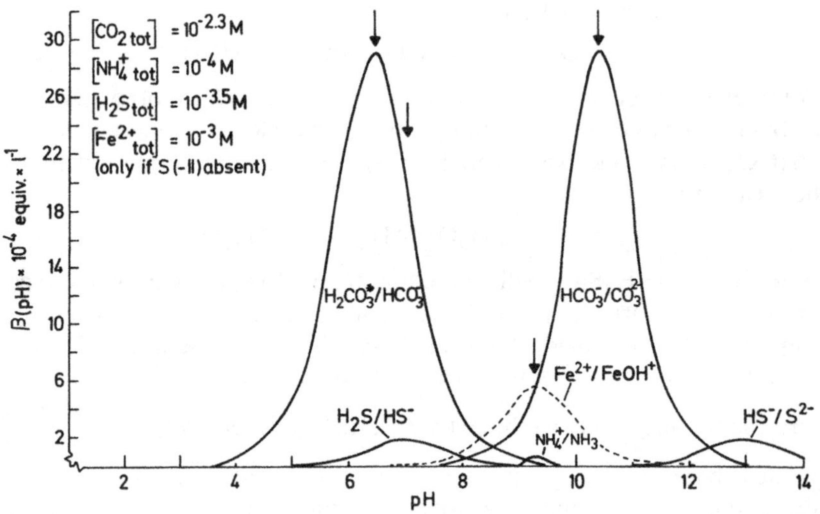

Fig. 4b. As 4a but due to common anoxic lacustrine conditions (autolysis of H_2O neglected)

acid-base systems (like NH_4^+/NH_3 or the even lower concentrated acid-base systems of Fe, P, Mn ions etc.) can usually be neglected.

The situation may change under anoxic conditions, e.g. in the hypolimnetic waters of a productive lake or in the pore water of its sediments. As shown in Fig. 4b the composition of pH influencing compounds may change due to the microbial reduction of NO_3^- (NO_2^-) to NH_4^+/NH_3, the possible microbial reduction of SO_4^{2-} to H_2S/HS^- (in the absence of Fe) or of Fe(III) to $Fe^{2+}/FeOH^+$ (or $FeHCO_3^+$).

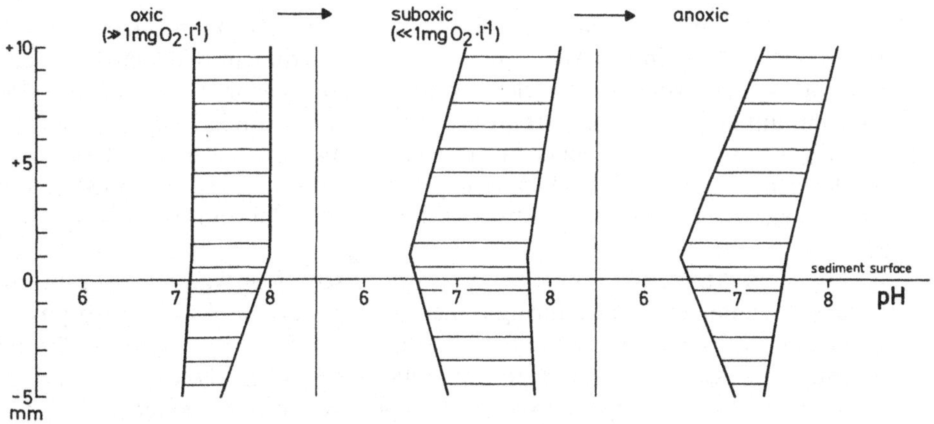

Fig. 5 a. pH values at the sediment-water interface of L. Constance (upper lake, 253 m depth) in a closed system developing from oxic to anoxic conditions

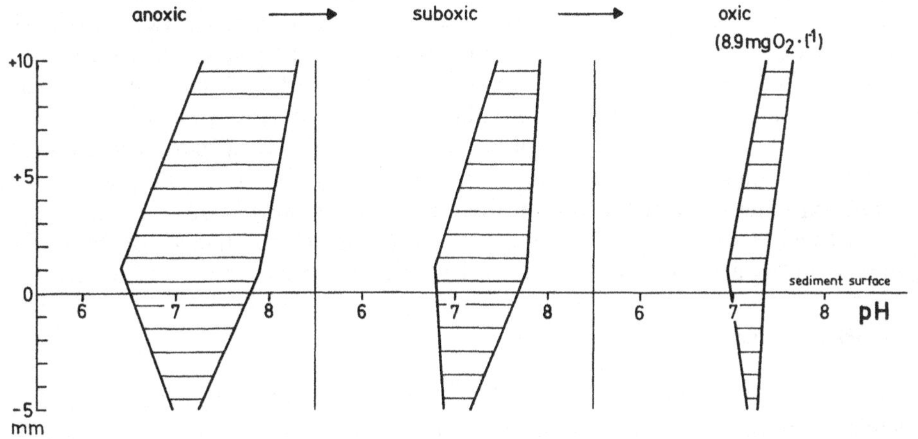

Fig. 5b. As 5a but in an open system developing from anoxic to oxic conditions by reaeration

Under conditions of acid deficit the pH in anoxic waters will, thus, hardly rise over 9.3–9.5 (arrow in Fig. 4b); under acidic conditions it will first drop to values at 7.0, and after further acid release at pH's between 7.0 and 6.4 (mostly ~6.7). Lower pH's are only possible if very high amounts of acid are released (mainly CO_2 and H_2S) which are not counterbalanced by dissolved HCO_3^-/CO_3^{2-}. The latter is common in the non-mixing hypolimnetic water layers of meromictic lakes ("monimolimnion") where accumulated organic carbon [46] is anaerobically oxidised by SO_4^{2-} reduction without mixing of the CO_2 and H_2S enriched water body,

$$2\,\mathrm{org\,C} + SO_4^{2-} + 2\,H_3O^+ \underset{\text{enzyme}}{\rightarrow} 2\,CO_2 + H_2S + 2\,H_2O\,. \qquad (6)$$

As $CO_2 + H_2O \rightleftarrows H_2CO_3^*$ and H_2S is a weaker acid than $H_2CO_3^*$, there will be a proton surplus corresponding to $H_3O^+(H_2CO_3^*)/H_3O^+(H_2S) = 2/1$.

Generally, the unbuffered anoxic pH range has to be more narrow than the corresponding pH range in oxic waters, i.e. 7.0–9.5 and 6.4–10.4, respectively.

This conceptual conclusion can again be confirmed by pH measurements in the overlying water, the sediment-water interface, and in sediment of a lacustrine sediment-water system using a glass electrode as proton sensor. It should be mentioned that, unlike $p\varepsilon$, detected pH values correspond to the conceptually calculated pH's, which is a consequence of the extreme high kinetics of acid-base reactions. Nevertheless, it should be kept in mind that pH measurements in sediments using glass electrodes may be less precise due to colloidal and ion diffusion effects.

In Fig. 5a and b the pH ranges in a calcareous sediment-water system during increasing D.O. depletion and during re-aeration are shown. As respiratory processes in a well-mixed water are predominant the pH values under oxic conditions are slightly depressed from the reference value ~ 8.4 (where $H_2CO_3^* \simeq CO_3^{2-}$). The restricted CO_2 exchange between sediment and overlying water leads additionally to the lower values in the sediment. Suboxic conditions in the overlying water lead to aerobic and (facultative) anaerobic metabolisms and consequently to a drop of pH to values around 7. There is no significant change under anoxic conditions. pH values below 6.4 do not occur and the suboxic and anoxic pH values reach their minima at the sediment surface. The latter is a clear consequence of the lacking mixing action and a local lack of acid neutralising capacity. Furthermore, sediment surfaces (like other density discontinuities in lakes) accumulate strata of detrital (e.g. organic) matter which locally increases the decomposition intensity and thus the release of acids (mainly CO_2) [47].

During reaeration of the system the opposite tendency is observed although the pristine pH values in the overlying water are not reestablished (probably because of a CO_2 accumulation in the overlying water during the experiment).

pK_{sp} and pH Values

Dissolved chemical compounds can be precipitated under the different hydrochemical conditions in rivers and lakes. This is mainly the case if – as a consequence of mixing and aeration of the runoff waters – the redox state of the dissolved species change and cations and anions recombine to less dissolvable compounds. For instance, Fe(II) hydroxide has a much higher solubility than Fe(III) hydroxide.

Reprecipitation may further occur in the deep hypolimnetic waters of productive lakes where the continuous depletion of D.O. leads to SO_4^{2-} reduction and subsequent metal sulfide precipitations (e.g. FeS, FeS_2, ZnS, Cu_2S etc.). Conceptually the solubility of less dissolvable compounds is defined by their solubility product $K_{sp} = [Me^{2+}]_{\text{equil.}} [S^{2-}]_{\text{equil.}}$ and $pK_{sp} \equiv -\lg K_{sp}$ [48]. However, it is obvious that pK_{sp} – although a "master variable" in the previously defined sense – is not sufficient in order to describe the real dissolution chemistry of the corresponding compounds [49]. Cations and anions may enter into acid-base or redox reactions and thus change their identity. Solubility products refer solely to the pronounced ion species.

As already mentioned, the pH value in epilimnetic and hypolimnetic waters will significantly be influenced by biotic action, mainly by the photosynthetic and the respiratory activity, respectively. The pertinent consequence drawn from that

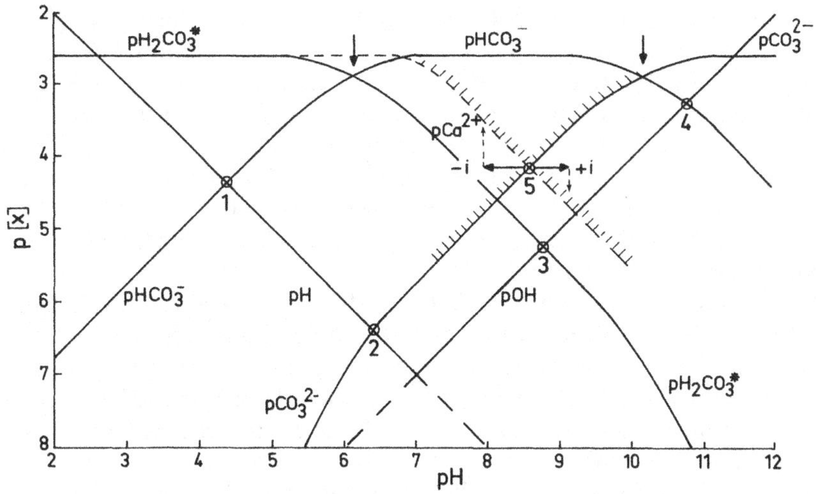

Fig. 6. pH stability diagram for the system $H_2CO_3^*/HCO_3^-/CO_3^{2-}$ under common lacustrine (calcareous) conditions. Arrows mark the $pK_{a,1}$ and $pK_{a,2}$ value; numbers of intercepts see text

is an indirect impact of organisms on the solubility of inorganic compounds leading to numerous "biohydrochemical" feedback mechanisms which are by far not fully understood [50]. [It should be mentioned that other (direct) impacts of organisms exist e.g. by selective incorporation which, however, cannot be discussed here.]

One of the most pertinent biohydrochemical interactions occurring in lacustrine environments in the *biogenic decalcification* [51]. This means the supersaturation and subsequent precipitation of $CaCO_3$ (mainly as calcite) as a result of photosynthetic rise of the pH. The precipitation occurs due to

$$Ca^{2+} + CO_3^{2-} \rightleftarrows CaCO_3(\text{calcite})\downarrow .$$

Figure 6 shows the pH stability diagram of the carbonic acid/carbonate system for a solution containing equivalent amounts of total carbonate and calcium ions corresponding to typical lacustrine concentrations ($10^{-2.6}$ *M*). The contour lines of the diagram indicate the pH-dependent change of the concentrations (activities) of the different dissolved ionic species.

Pertinent for dissolution and precipitation processes are the intersections of the lines. At intersect 1, $10^{-pH} = [HCO_3^-]$ which means that it represents the lowest possible pH value if $H_2CO_3^*$ is the only present acid in solution. Whereas intersect 2 (i.e. $10^{-pH} = [CO_3^{2-}]$) and intersect 3 (i.e. $10^{-pOH} = [H_2CO_3^*]$) mark only gradual boundaries of changes of the carbonate dissolution and precipitation intensity, intersect 4 and 5 are bound directly to the decalcification process; pH values rising over intersect 4 indicate the condition $10^{-pOH} \simeq [OH^-] > [HCO_3^-]$, i.e. where the reaction $HCO_3^- + OH^- \rightleftarrows CO_3^{2-} + H_2O$ will be shifted to the right side enhancing the $CaCO_3$ recombination [52]. Intersect 5 indicates the zero point of the calcite saturation index (i). At this point, where $[Ca^{2+}] = [CO_3^{2-}] = \sqrt{K_{sp}}$,the corresponding pH is defined by the calcite equilibrium pH value (pH_{calc}). If the measured pH of the solution is lower than pH_{calc}, i becomes negative, i.e. the solution has a calcite dissolving tendency. In the opposite case, i becomes positive, i.e. the solution has a calcite precipitating tendency (calcite supersaturation). Note that all marked intersections depend on the total carbonate concentration; changing carbonate concentrations will lead to different pH values corresponding to zero points of the calcite saturation index. (The intersections of $pH_2CO_3^*/pHCO_3^-$ and $pHCO_3^-/pCO_3^{2-}$ (arrows) and the pK_{sp} of calcite, however, are independent of the total carbonate concentration.)

We saw previously that the predominantly oxidizing conditions in catchment runoff and tributary waters of lakes have an acidifying effect. Corresponding waters from non-calcareous but also calcareous catchment areas should, therefore, be specified by i $\leqq$ O, i.e. by potential calcite undersaturation. However, in lakes the photosynthetic rise of pH in euphotic epilimnetic waters will shift the afternoon pH's much above pH_{calc} and thus to positive saturation (supersaturation) index values [53]. If not inhibited the solution can equilibrate by calcite precipitation depressing the dissolved Ca^{2+} and CO_3^{2-} concentration until they correspond again to the solubility product, i.e. to concentrations (activities) according to the pCa^{2+} and pCO_3^{2-} contour lines of the corresponding pH stability diagram (e.g. to the hatched lines in Fig. 6). In the metalimnetic as well as in the aphotic hypolimnetic waters i may become again negative due to respiratory CO_2 release reversing the precipitating into the dissolution process of calcite (lysocline). In highly calcareous lakes this re-dissolution process will not be quantitative. At least part of the participate will reach the sediment proving the decalcifying effect [54] and leading to permanent pH's$\geqq$7 in the lake.

The calcite precipitation from supersaturated solutions can kinetically be inhibited by interactions between Ca^{2+} ions and dissolved organic matter [55, 56]. Furthermore, co-precipitation of dissolved anions (like PO_4^{3-}) specified by similar low solubilities with Ca^{2+} ions can have a lowering effect on the photosynthetic activity and thus on the further increase of calcite supersaturation [57–60]. Co-precipitation refers also to different (often unknown) ion product values for the real precipitated compound [61]. As a whole these effects diminish the accuracy or (in the case of unknown K_{sp} values) even the precision of predictions about biogenic calcite precipitation or re-dissolution processes leaving still questions for further investigations. Nevertheless, if K_{sp} values are known complicated dissolution-precipitation patterns of multicomponent systems at equilibrium can be predicted by appropriate computerization [62].

Trophic Levels and Sediments

Changing occurrence of dissolved compounds as depending on the pH and/or pε(pe) of the solution may be crucial for the development of certain photosynthetic and respiratory organisms [63–67]. Although we cannot discuss the interaction between dissolved chemical plant nutrients and the planktonic (passively drifting) or benthic (sessile) organisms it is important to see that the biotic uptake of nutrient compounds competes to its abiotic precipitation. It is fascinating to understand "bio-hydro-chemical" competitions and to enlighten the positive or negative feedback control mechanisms between developing organisms and the hydrochemical constraints [68–70].

Chemical compounds with low solubilities (given pH, pe, complex ion concentrations etc.) but high nutritional pertinence (like phosphorus) will have the strongest influence on the *trophic level* of a lake, i.e. the annual average of photosynthetic (assimilative) production and respiratory (dissimilative) decomposition of organic matter. Lakes may be *oligotrophic* if their annual load in organic carbon (incl. particulate organic carbon) remains well below 60 g $C \cdot m^{-2}$; loads higher than 100 g . m^{-2} refer to *eutrophic* or even *hypertrophic* lakes; lakes with loads in between are usually considered as *mesotrophic* [2]. Note, however, that this nomenclature is neither precisely defined nor sufficiently accurate in interpreting the individual patterns of the algal assemblage of a lake.

Although the phosphorus concentration was found to be highly crucial for defining trophic levels in many (mainly phosphorus-poor) lakes [71, 72], it was also found that its availability to biota depends widely on the uptake and release rates by the biota itself [73], its chemical speciation (e.g. organic or inorganic bound phosphorus), and the relative abundance and residence time of the dissolved phosphorus fraction as referred to the average org. C/org. P ratio of the photosynthetic organisms [74, 75].

Nevertheless, since the pristine P concentrations in most lakes are extreme low (due to the low solubilities of Fe(III) hydroxophosphates and Ca-hydroxophosphates at the average lacustrine pH's) an artificial input of phosphorus into a lake ("phosphorus eutrophication") will lead to a higher annual photosynthetic production and, consequently, to an increased organic carbon load (increase of reducing equivalents) in the deeper hypolimnetic waters with predominantly respiratory metabolism of its biota. At this point we discover the pertinence of the morphometry of a lake and the geochemistry of its catchment area for its trophic level: Deep lakes (the volume of hypolimnion exceeds the volume of the euphotic epilimnion) with relative low shoreline developments but high mean depths (Z_m) and iron/carbonate rich catchment areas tend to be oligotrophic. Shallow lakes with small hypolimnetic D.O. or anoxic electron acceptor reservoirs and usually high shoreline developments tend to be eutrophic. Lakes with high mixing rates (dimictic or polymictic lakes) will tend to higher trophic levels since their nutrient load released by respiratory decomposition in the deeper aphotic waters is rapidly recycled into the euphotic photosynthetic active water layers. Temperature plays an important role for the trophic conditions. Metabolic activities are enhanced in warm lakes [76] leading usually to an accelerated recycling of essential nutrients from the decaying to the growing organisms. In fact, the latter effect may apparently indicate elevated eutrophic conditions although the annual organic carbon yield (org $C \cdot m^{-2} \cdot m^{-1}$) is relatively low. The inaccuracy of a simple trophic nomenclature is also confirmed if, naturally or artificially, toxic substances (e.g. heavy metals like Hg, Cu, Cd) are present in bioavailable species [77]. Trophic levels might therefore be defined by the presence as well as by the absence of certain components; it is thus obvious that a trophic scaling of lakes due to a single variable (unlike to the hydrochemical master variable) will be neither precise nor confirming the real conditions. However, it seems to be possible to obtain more

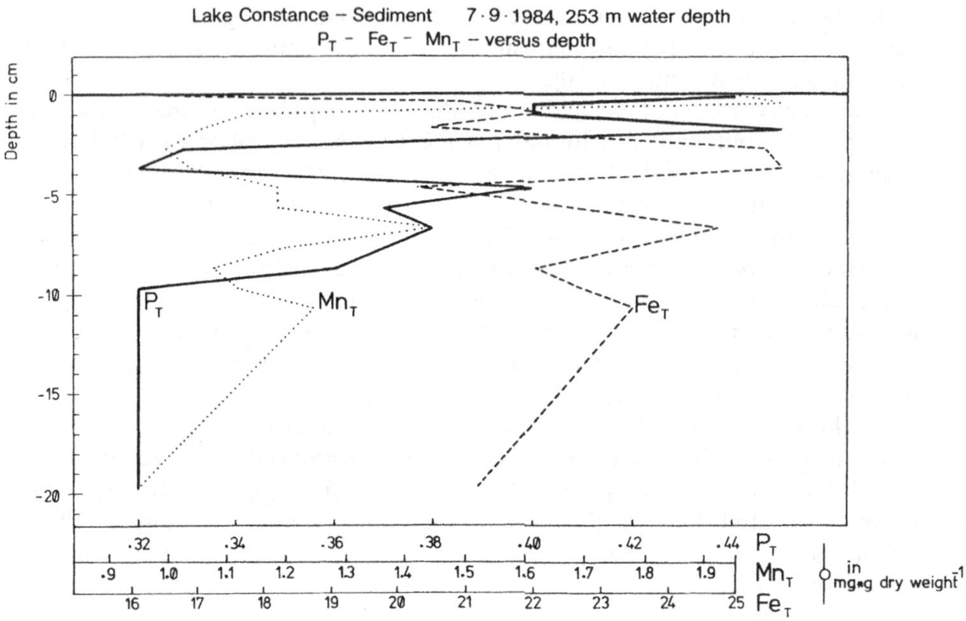

Fig. 7. Concentrations of total-phosphorus (P_T), total-iron (Fe_T) and total manganese (Mn_T) in the upper layer of sediment of L. Constance (upper lake, 253 m depth)

precise trophic notions by involving more physical and chemical variables into computerized lake models.

Many lakes suffer serious pollution, particularly by nutrient (mainly phosphorus) enriched tributaries. As a result former oligotrophic lakes develop to higher trophic levels [78, 79]. This involves higher organic carbon load [80], higher depletion of oxic or anoxic terminal electron acceptors with a subsequent climbing of the redoxocline from the upper sediment layers to the thermocline or even lower water depths [20]. The concomitant increase of organic carbon in the uppermost sediment layers leads to an enhanced diagenetic response particularly of redox-sensitive sediment components [34, 35, 38]. Increased concentrations in the sediment pore waters [38, 39, 42, 83] and subsequent diffusive dislocation and reprecipitation processes will result [81, 82, 84–87]. Sediments, if not repeatedly resuspended, are therefore excellent indicators for the trophic history of a lake [88]. This shall be exemplified by the vertical concentration profiles of total phosphorus (P_{tot}), total manganese (Mn_{tot}), and total iron (Fe_{tot}) in the dry weight of the sediment of Lake Constance (Upper Lake), a lake which was increasingly phosphorus polluted during the last 40 years (Fig. 7) [79]. This is clearly indicated by a concentration increase of P_{tot} from the deeper to the upper sediment layer. The upper 10 cm sediment layer reflecting the eutrophication process shows the typical alternative concentration pattern of P_{tot} and Fe_{tot}. Primarily are both sedimented as Fe(III) hydroxophosphate; the increased availability of reducing equivalents (organic carbon) in the anoxic sediment (~ -2 to -10 cm), however, enhances the Fe(III) reduction and subsequent SO_4^{2-} reduction, and thus phosphorus is released into the pore water. Concentration maxima of Fe_{tot} (mainly

FeS·aq) are concomitant to corresponding minima of P_{tot} (-2.5 to -4.5 cm). In the still oxic sediment upper layer (0 to -2 cm) Fe(III) hydroxides, precipitating dissolved phosphate, are still present; thus, Fe_{tot} and P_{tot} are concomitantly fixed to the sediment [89]. The uppermost drop in Fe_{tot} but increase in P_{tot} is due to the relatively higher concentration of organic matter (incl. organically bound phosphorus) in the sediment surface layer [80, 83].

The Mn_{tot} concentration profile indicates clearly the eutrophication as a process increasing the phosphorus *and* the organic carbon poising of lake and sediment [90]. As Mn(II) sulfides are improbable to be precipitated, Mn(II) is mainly released into the pore water by the reduction of MnO_2·aq with the organic carbon as primary electron donor. Mn(II)ions are mobile [91, 82, 92] thus explaining the extreme concentration minimum in the organic rich but anoxic layer between -2.5 to -4.5 cm, where Fe(II) was precipitated as FeS·aq. The straight concentration increase of Mn_{tot} to the sediment surface is again the result of a more positive redox intensity (pe value) in the uppermost oxic sediment layer where Mn(II) is oxidative reprecipitated in Mn(IV) compounds [82]. It is noteworthy that the (thermodynamic) redox sequence due to dropping $p\varepsilon$ values, resulting in a vertical sequence: Mn(IV), Fe(III) oxides – Fe(III) oxides – Fe(II) sulfides from oxic to deeper anoxic sediment layers is not reflected in the real concentration profiles. Irrespective of the different mobilities of Mn(II) and Fe(II) ions in the pore waters the latter is a result of the much lower pristine concentrations of Mn, and thus to its lower pe (redox) buffering capacity (per g sediment) compared to Fe and usually also SO_4^{2-}. As we already mentioned the most stable pe values (measured redox intensities) in the majority of lakes and sediments will be at the electron transfer levels of the redox couples D.O./H_2O, Fe(III)/Fe(II) and SO_4^{2-}/S(-II); in the absence of SO_4^{2-} also at the electron transfer level of the redox couple CO_2/CH_4 [93, 94].

Two Case Studies

In order to exemplify the annual development of the thermal and subsequent hydrochemical stratification in lacustrine water bodies (pelagial), some data of two lakes shall be reported. These lakes are similar concerning their morphometry and the geochemistry of their catchment areas (both calcareous) and sediments (clayey with carbonate concentrations up to 80% d.wt.) [95–97]. They differ significantly in their total-phosphorus concentrations, the sequence of annual algal developments and in their mean temperature and thermal stratification behaviour [98, 99].

I. Lake Constance (Gnadensee), a temperate dimictic and eutrophic-hypertrophic lake.

This lake is situated at 8°50′ E, 47°47′ N, 400 m MSL and is the highly eutrophic northeastern part of L. Constance-Lower Lake, however widely separated by a shallow water bank between a peninsula and an island, respectively. L. Gnadensee is postglacial with $Z_m = 21$ m, $\bar{Z} = 9.4$ m, and $\bar{Z}/Z_m = 0.45$, i.e. the lake basin corresponds close to an elliptic sinusoid [100]. It is fed by water of river Rhine and L. Constance (upper lake) bringing highly carbonate containing very fine

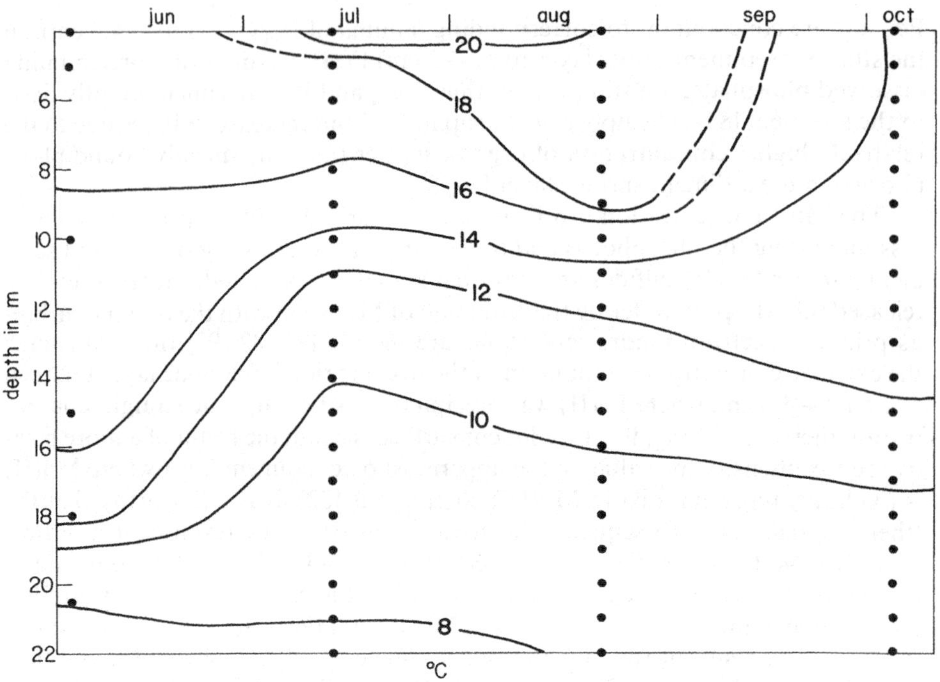

Fig. 8. L. Constance (Gnadensee) isotherms, June until October 1977

grained (clayey) sediments into the lake [95, 96, 101]. With its relative high shoreline development ($D_L = 3.2$ compared to $D_{circle} = 1.0$ of a circle shaped lake) and its relative high residence time (despite of a low relative depth Z_r of 0.5%), L. Gnadensee naturally tends to become eutrophic [102].

Heavy phosphorus pollution changed L. Gnadensee into a highly eutrophied lake with longterm and regular blooms of diatoms, bluegreen algae, and filamentous chlorophyta. The lake is dimictic and was repeatedly ice covered during January/February [103]. In Fig. 8 the isotherms of summer stratification 1977 are shown. After springtime overturn the lake is thermally stratified from late July until August; in September/October the autumn overturn of the water starts. The thermocline is at 8–10 m water depth and between 18 and 14 °C. As the development of the thermocline restricts the available D.O. in the deeper waters, the corresponding D.O. isoconcentration diagram (Fig. 9) reflects the thermal stratification with a delay according to the hypolimnetic D.O. reservoir and the biotic D.O. uptake (D.O. reduction rate), respectively. Similar to D.O. develops pe_7, the potentiometrically detected redox intensity refered to pH = 7. pe_7 drops from 6.5 to about 2.5 corresponding to ~ 10 ppm to ≦ 0.2 ppm D.O. (Fig. 10).

We know that NO_3^-–N and NO_2^-–N are important anoxic electron acceptors [32, 69, 104, 105]. Indeed, Figs. 11 and 12 (with NO_2^-–N concentrations in brackets) indicate that the pe_7 range between 5.0 and 2.5 seems to refer to the $NO_3^- \rightarrow NO_2^-$ reduction and the pe_7 range between 2.5 and 0 to the $NO_2^- \rightarrow NH_4^+$. The negative pe_7 value are accompannied by the occurrence of sulfide-sulfur which is

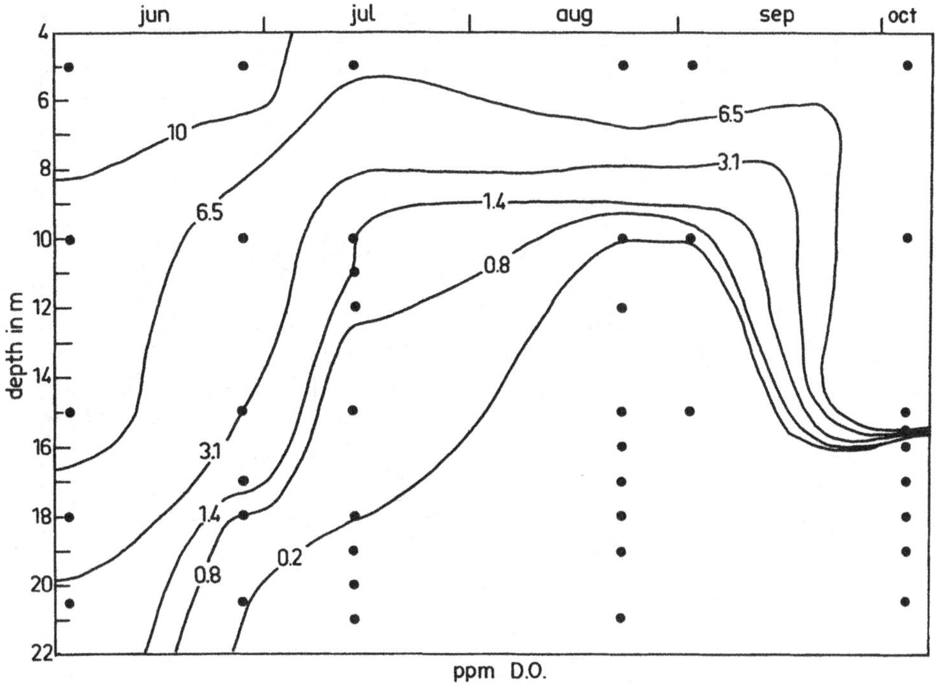

Fig. 9. L. Constance (Gnadensee) D.O. isoconcentration diagram, June until October 1977

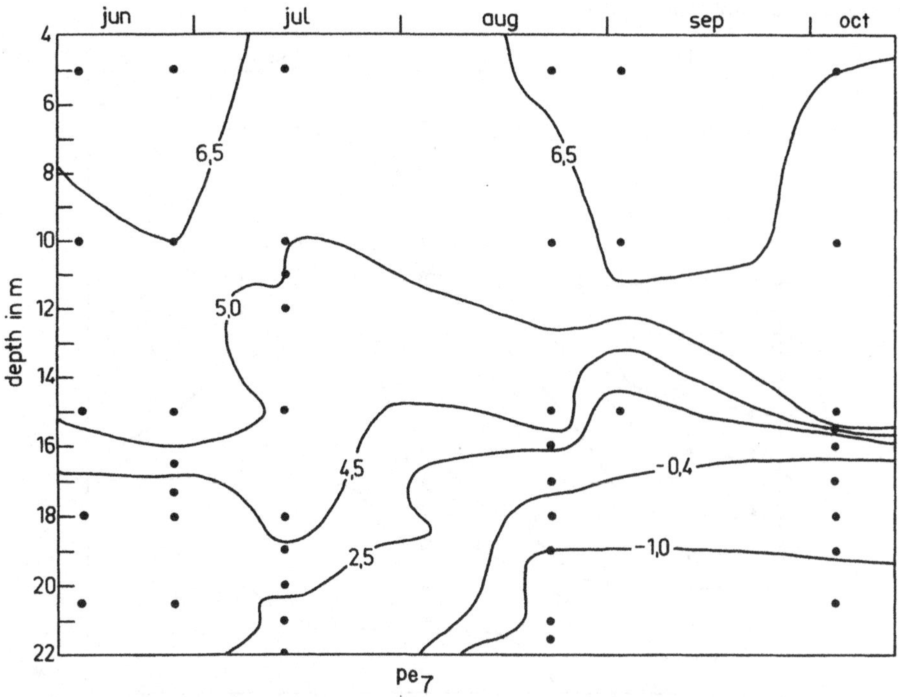

Fig. 10. L. Constance (Gnadensee) isoelectrogram (pe_7), June until October 1977

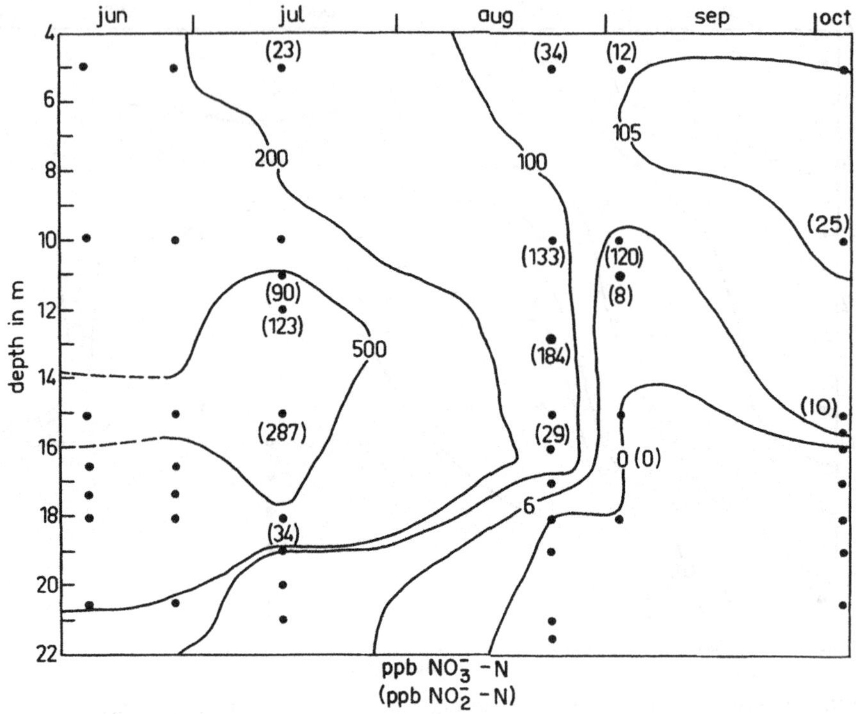

Fig. 11. L. Constance (Gnadensee) NO_3^-–N, NO_2^-–N isoconcentration diagram (NO_2^-–N in brackets), June until October 1977

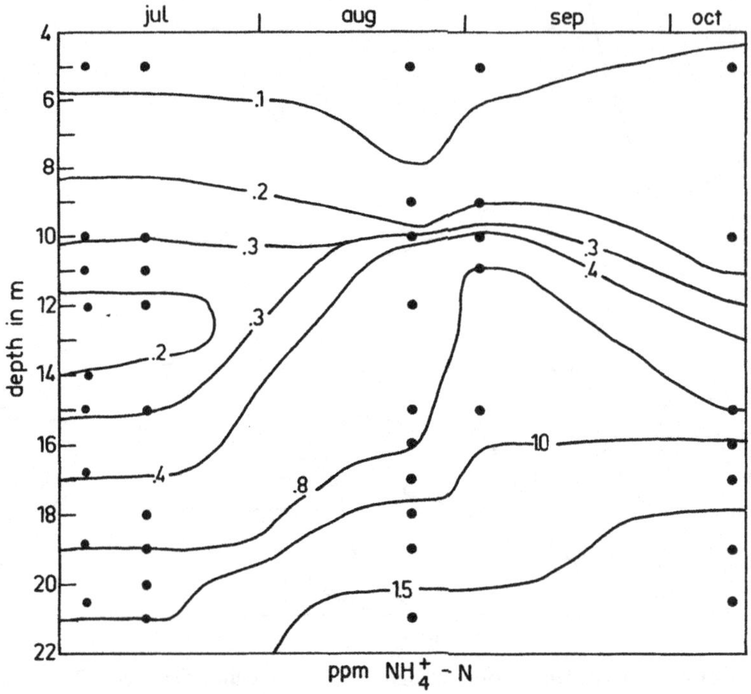

Fig. 12. L. Constance (Gnadensee) NH_4^-–N isoconcentration diagram, June until October 1977

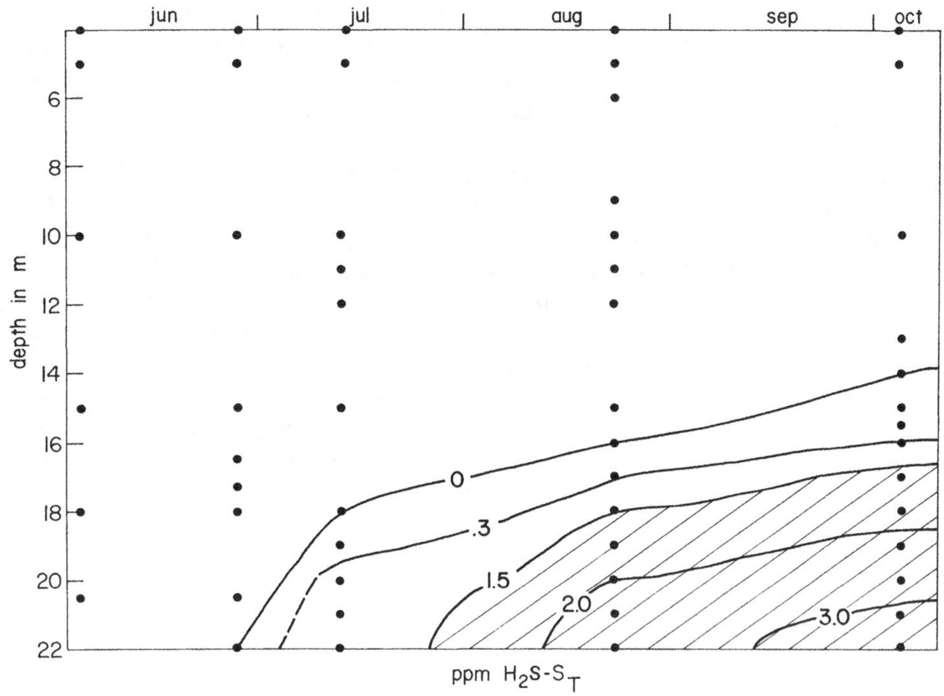

Fig. 13. L. Constance (Gnadensee) total sulfide-sulfur ($H_2S–S_T$) isoconcentration diagram, June until October 1977

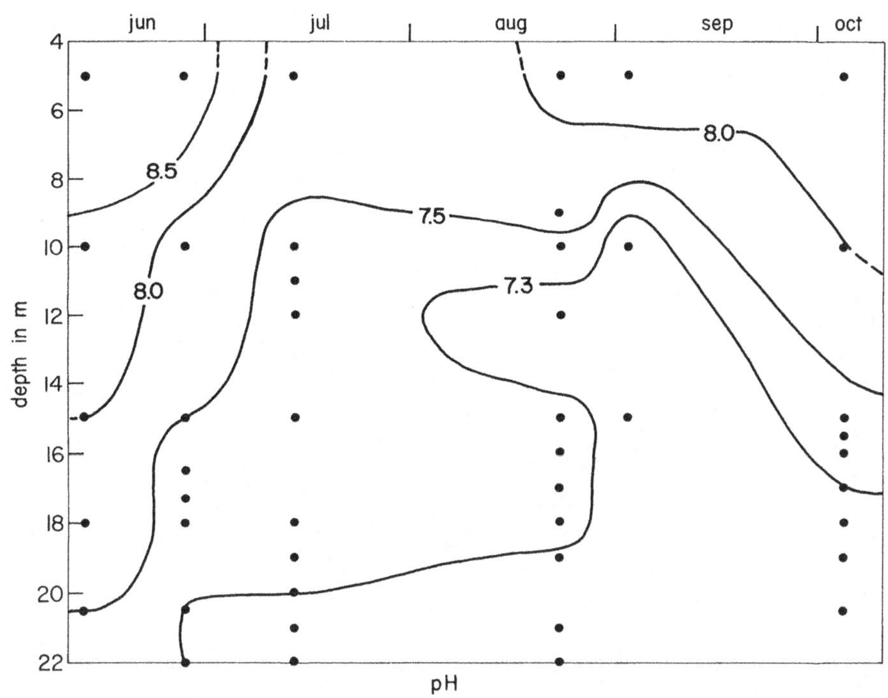

Fig. 14. L. Constance (Gnadensee) pH isogram, June until October 1977

shown in Fig. 13. Due to the calcareous buffering of the water (alkalinity: ~ 3.5 mequiv. · l^{-1}), the pH values (Fig. 14) range about 8.0 dropping to close to 7 in the deeper and H_2S-bearing waters and rising to 8.5 in the photosynthetic active epilimnion according to the buffering of the $H_2CO_3^*(H_2S)/HCO_3^-$ and the HCO_3^-/CO_3^{2-} systems.

The continuous high phosphorus input into L. Gnadensee is reflected by its extreme high total phosphorus concentrations throughout the year (Fig. 15). It is characteristic that the concentrations in the deep waters of the stratified lake are even higher. One reason for that may be the sedimentation of phosphorus bound to inorganic or organic particles (e.g. after coprecipitation with $CaCO_3$, or adsorption to clayey sediment [106]), and/or release of phosphorus from the sediment into the overlying water by the reduction of Fe(III)hydroxophosphates. Both processes accumulate phosphorus in the deeper waters as a potential nutrient reservoir for further photosynthetic production [107]. Reasonable proportions of the total phosphorus can only irreversibly be fixed into the sediment if the hypolimnetic waters never change from oxic to anoxic conditions with the corresponding drop of pe.

II. Lake Kinneret, a subtropical monomictic and oligotrophic-mesotrophic lake.

This lake is situated at 35°35′ E, 32°48′ N in the northern part of the Afro-Syrian Rift Valley at −210 m MSL. The lake is the deepest situated freshwater

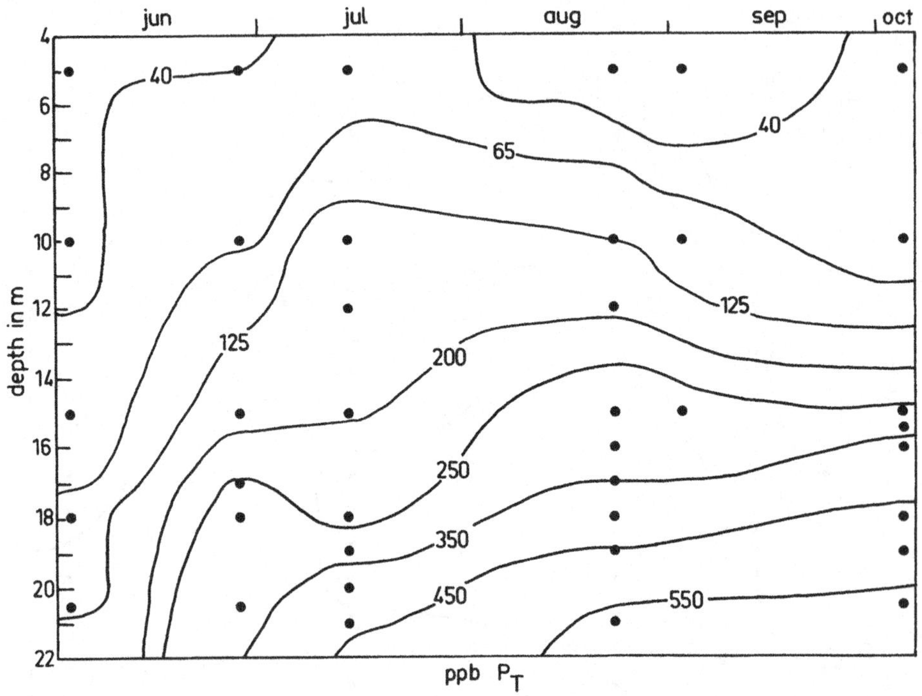

Fig. 15. L. Constance (Gnadensee) total-phosphorus (P_T) isoconcentration diagram, June until October 1977

body on earth with $Z_m = 43$ m, $\bar{Z} = 26$ m, and $Z_m/\bar{Z} = 0.6$ corresponding to a trench-like lake basin [99]. River Jordan the main tributary of L. Kinneret enters the lake in the north and leaves it in the south draining in its major part a calcareous and basaltic catchment area. The sediments of the lake are thus carbonate rich and clayey [97]. Usually once a year during the rainy winter season the lake water is overturned and with the relative low shoreline development and relative depth of $D_L = 1.15$ and $Z_r = 0.27\%$ the lake should be oligotrophic. Unlike L. Constance (Gnadensee) L. Kinneret is much less polluted with phosphorus which, however, does not prevent the regular blooming of the dinoflagellate *Peridinium cinctum fa. westii* from April until June [108].

In Fig. 16 the isotherms of summer stratification 1981 are shown. Unlike L. Gnadensee the stratified period is longer and at a temperature about 10 °C higher than in Gnadensee. Subsequently, the D.O. depletion in the hypolimnion starts already in March (Fig. 17) and no drop of the thermocline with subsequent mixing of the overlying water can be observed until October. Although there was no pe_7 monitoring as with L. Gnadensee, the pe data from March and October in Tab. 6 indicate the same development as observed in L. Gnadensee except for the sharper gradient in L. Kinneret with more positive values in the oxic epilimnion and more negative ones in the H_2S-bearing hypolimnion.

Typical for L. Kinneret are the high NO_3^-–N concentrations after winter rain (Fig. 18) but the relative low concentration and shortterm occurrence of NO_2^-–N (Fig. 19). We see that the redoxchemical cut from full oxic to full anoxic (H_2S-bearing) conditions is not only two months earlier but also more straight than in L. Gnadensee. NH_4^+–N occurring together with sulfide-sulfur (Fig. 20) is lower concentrated than in L. Gnadensee which is likely to be the result of intense denitrification ($NO_3^-/NO_2^- \rightarrow N_2\uparrow$) [109]. The accumulation of NH_4^+–N can have very serious consequences on the biota of a lake. If such waters are overturned and come into contact we waters of increased pH values (>8.3) NH_4^+ will partly or quantitatively (pH > 9.3) converted into ammonia which can initiate fish dying.

Table 6. pH, pe and pe_7 values in waters of Lake Kinneret (maximum depth 43 m); pe_7 means pe refered to pH 7

Date	Depth	pH	pe	pe_7
22.03.81	0	8.56	6.4	7.9
	5	8.54	5.6	7.1
	10	8.36	5.5	6.9
	25	8.26	5.8	7.0
	41	7,85	5.7	6.6
10.10.81	0	8.48	5.7	7.0
	5	–	–	–
	10	8.58	5.5	7.0
	16	8.57	5.3	6.9
	17	7,88	5.2	6.1
	18	7.70	4.2	4.9
	30	7.33	−2.6	−2.3
	41	7.33	−2.7	−2.4

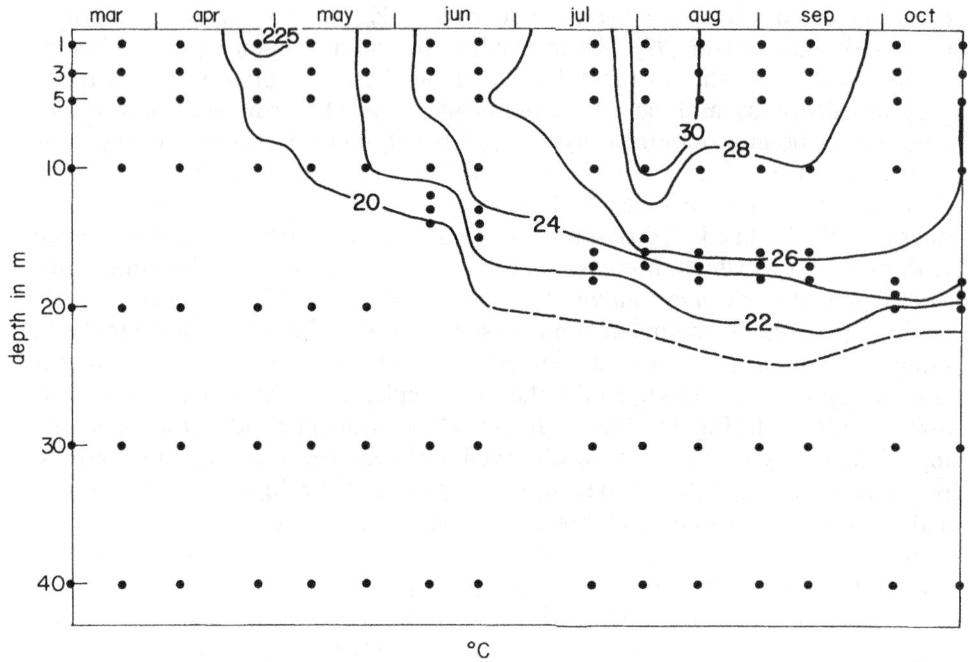

Fig. 16. L. Kinneret isotherms, March until October 1981

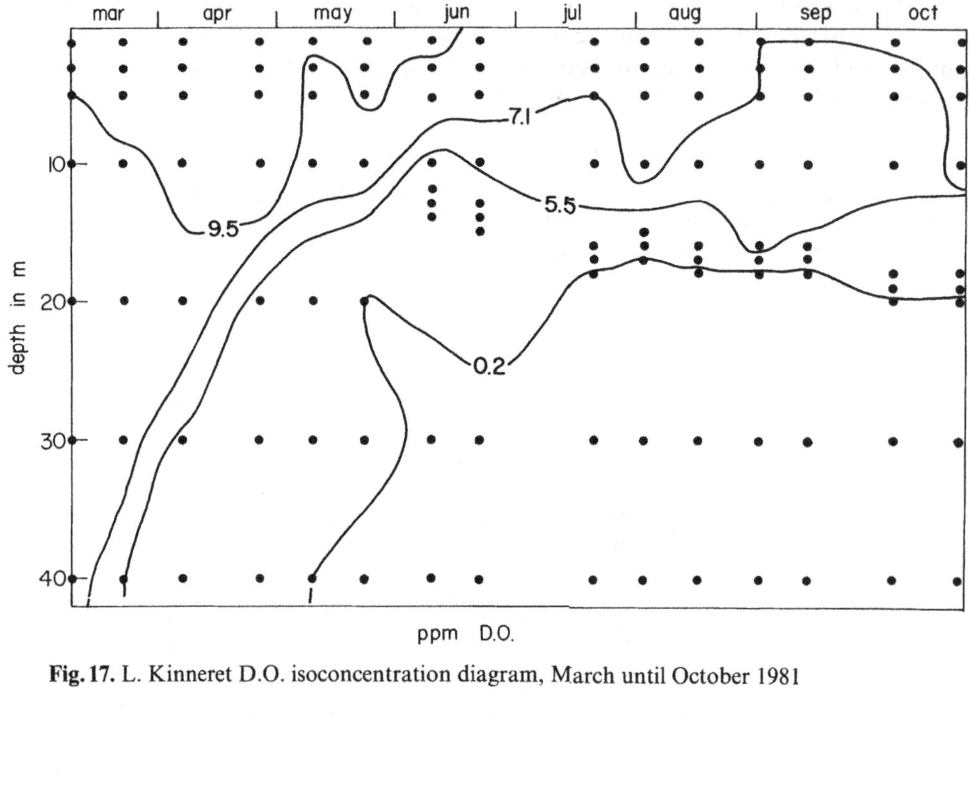

Fig. 17. L. Kinneret D.O. isoconcentration diagram, March until October 1981

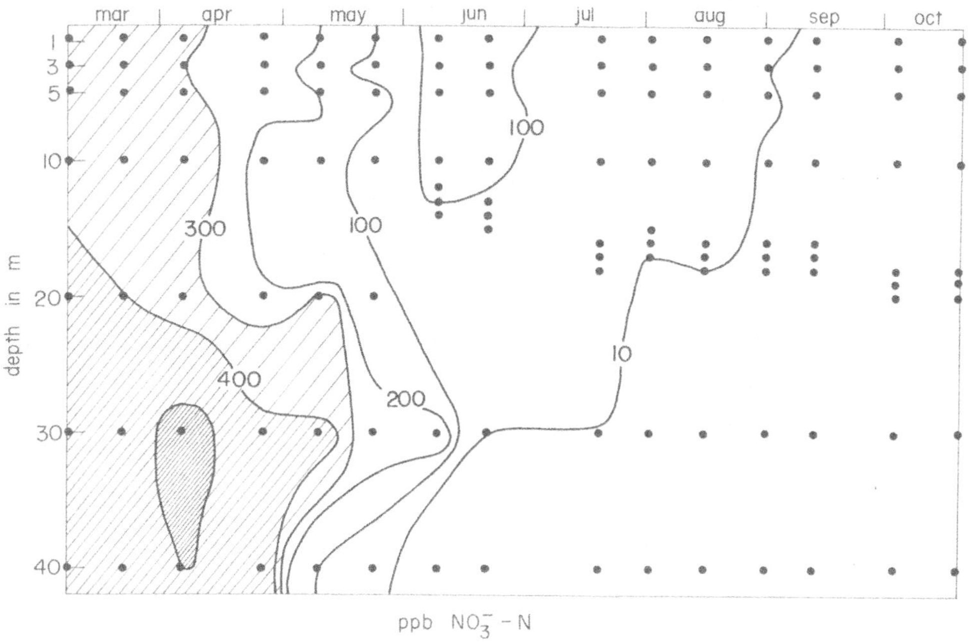

Fig.18. L. Kinneret NO_3^-–N isoconcentration diagram, March until October 1981 (dark hatched area = over 500 ppb)

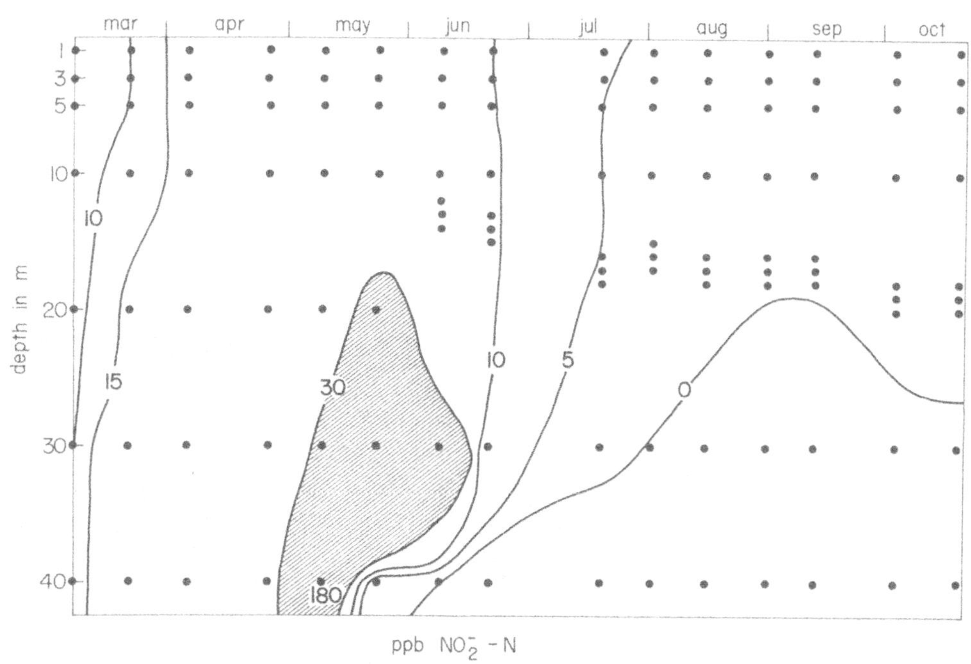

Fig.19. L. Kinneret NO_2^-–N isoconcentration diagram, March until October 1981 (hatched area = 30–180 ppb)

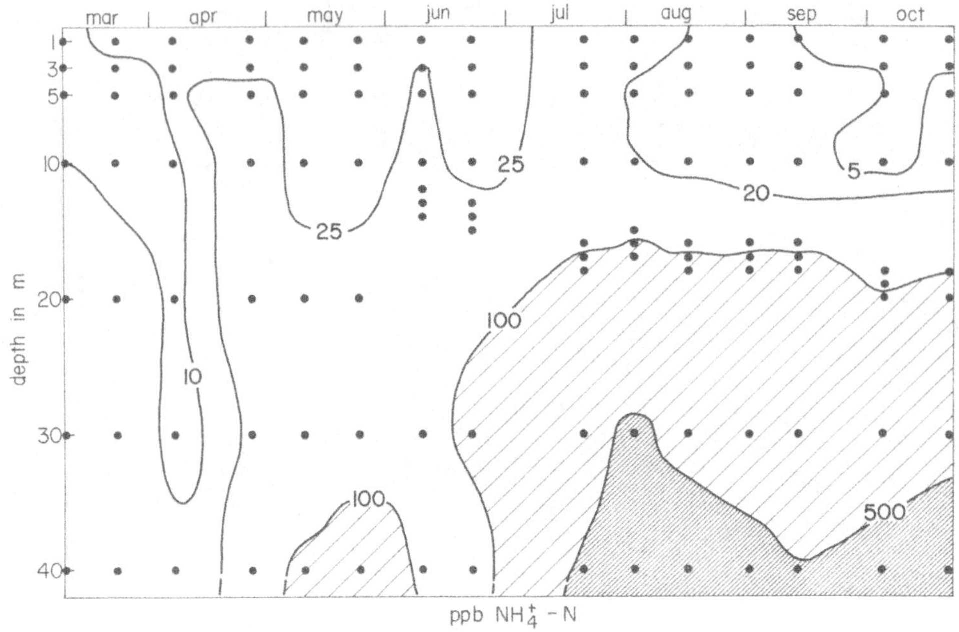

Fig.20. L. Kinneret NH_4^-–N isoconcentration diagram, March until October 1981 (dark hatched area = over 500 ppb)

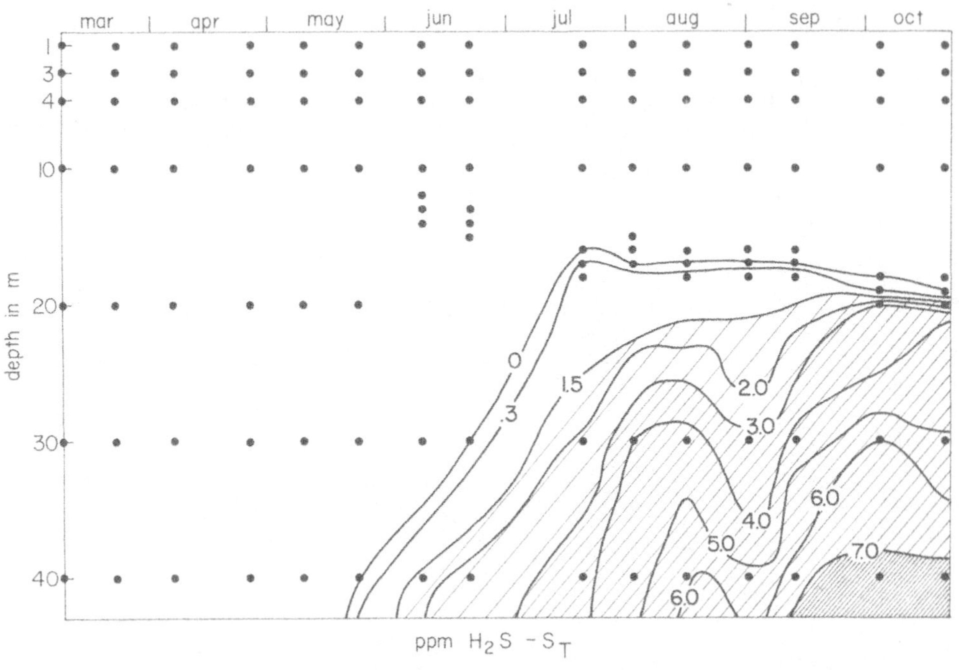

Fig. 21. L. Kinneret total sulfide-sulfur (H_2S–S_T) isoconcentration diagram, March until October 1981 (dark hatched area = 7–9 ppm)

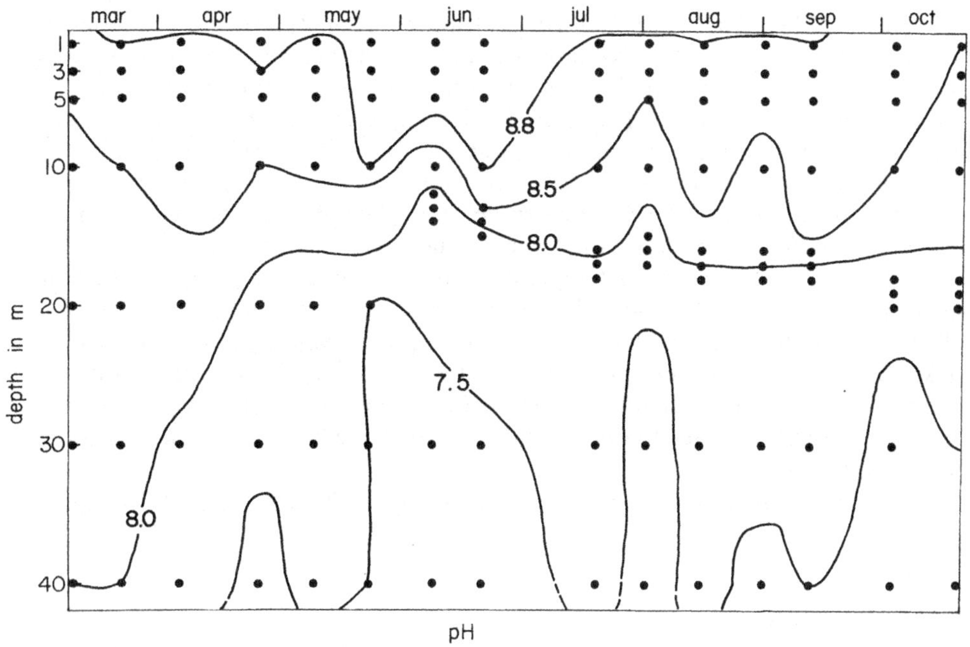

Fig. 22. L. Kinneret pH isogram, March until October 1981

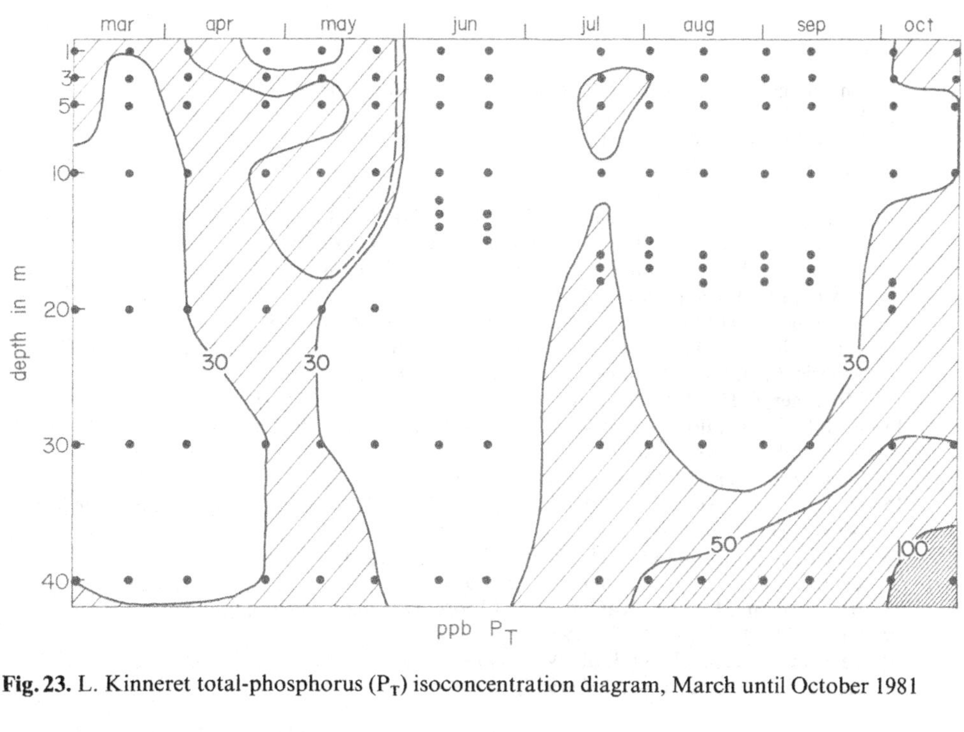

Fig. 23. L. Kinneret total-phosphorus (P_T) isoconcentration diagram, March until October 1981

The sulfide-sulfur (H_2S–S_T) concentrations (Fig. 21) fit to the concentrations of the other redox components insofar as the SO_4^{2-} reduction starts soon after D.O. disappeared. It is interesting to see that in both lakes the same redox processes occur but in L. Kinneret in an obvious accelerated way. Thus intermediate redox compounds (e.g. NO_2^-) play a minor role and concentration gradients are more steep in the subtropical (L. Kinneret) compared to the temperate lake (L. Gnadensee) [110]. The pH value is predominantly carbonate buffered and shows the same annual development (Fig. 22).

Beside of the higher water temperature and its higher NaCl content L. Kinneret is mainly much less phosphorus eutrophied than L. Gnadensee (Fig. 23). This is clearly indicated by the total phosphorus (P_T) concentrations in the both lakes. Generally the concentrations in L. Kinneret are $^1/_{10}$ the ones of L. Gnadensee and concomitant to oligotrophic-mesotrophic conditions; thus, under temperate conditions L. Kinneret could be a low productive lake with oxic conditions in its deep waters all over the year.

A comprehensive review on warm lakes was given by Serruya and Pollingher [111].

Acknowledgments

Thanks are due to the Deutsche Forschungsgemeinschaft and the Israel Oceanographic and Limnological Research Ltd. (Kinneret Limnological Laboratory) for their financial support of my field research on L. Constance and L. Kinneret. I am further grateful to H. Bernard and G. Schulze for preparing the drawings.

References

1. Thienemann, A.: Limnologie. Jedermanns Bücherei, Breslau 1926
2. Wetzel, R.G.: Limnology, 2nd ed. W.B. Saunders Co., Philadelphia 1981
3. FAO: The State of Food and Agriculture (C77/INF/19.11.77)
4. Berger, F.: Carinthia II, Sonderheft 31, 29 (1971)
5. Birge, E.A.: Trans. Americ. Microsc. Soc. *25,* 5 (1904)
6. Bloß, G., Grigull, U.: Z. Wasser/Abwasserforsch. *4,* 121 (1974)
7. Mortimer, C.H.: Verh. Internat. Verein. Limnol. *14,* 79 (1961)
8. Patalas, K.: Verh. Internat. Verein. Limnol. *22,* 97 (1984)
9. Bowen, I.S.: Phys. Rev. *27,* 779 (1926)
10. Mortimer, C.H.: Proc. Roy. Soc. London (Ser. B) *236,* 355 (1952)
11. Wunderlich, W.O.: In: Hall, G.E. (ed.), Reservoir Fisheries and Limnology. Spec. Publ. 8, Americ. Fish. Soc. Washington D.C., 219 (1971)
12. Mortimer, C.H.: Verh. Internat. Verein. Limnol. *20,* 124 (1974)
13. Endoh, S., Okamoto, I.: In: Mori, S., An introduction to the limnology of Lake Biwa. SIL Spec. Publ. August 1980, 10
14. Stewart, R.: Limnol. Oceanogr. *9(4),* 606 (1964)
15. Csanady, G.T.: In: Lerman, A. (ed.), Lakes, Chemistry, Geology, Physics. Springer-Verlag, New York, Heidelberg 1978
16. Likens, G.E., Bormann, F.H.: Science *184,* 1176 (1974)
17. Boström, K.: Adv. Chem. Ser. *67,* 286 (1967)
18. Baas-Becking, L.G.M., Kaplan, J.R., Moore, D.: The J. of Geology *68(3),* 243 (1960)
19. Olson, J.S.: Ecology *14,* 322 (1963)
20. Lerman, A.: Hydrol. Sci. Bull. *19,* 25 (1974)
21. Horio,T.: J. Biochem. *45,* 267 (1958)

22. Gorden, R.W., Beyers, R.J., Odum, E.P., Eagon, E.C.: Ecology *50,* 86 (1969)
23. Golterman, H.L.: Verh. Internat. Verein. Limnol. *22,* 23 (1984)
24. Abd-el-Malek, Y., Rizk, S.G.: J. Appl. Bacteriol. *26,* 7 (1963)
25. Sillén, L.G.: In: Stumm, W. (ed.), Equilibrium Concept in Natural Water Systems, Adv. Chem. Ser. *67,* 45 (1967)
26. Stumm, W.: Proc. Int. Water Pollut. Res. Conf., 3rd, Munich 1, *283* (1966)
27. Morris, J.C., Stumm, W.: Adv. Chem. Ser. *67,* 270 (1967)
28. Morgan, J.J.: In: Stumm, W. (ed.), Equilibrium Concept in Natural Water Systems, Adv. Chem. Ser. *67,* (1967)
29. Whitfield, M.: Limnol. Oceanogr. *19,* 857 (1974)
30. Frevert, T.: Schweiz. Z. Hydrol. (in press)
31. Javornichy,P., Prokesova, V.: Intern. Rev. Ges. Hydrobiol. *48,* 335 (1963)
32. McCarty, P.L.: In: Faust, S.J., Hunter, J.V. (eds.), Organic Compounds in Aquatic Environments. M. Dekker, New York 1971
33. Vallentyne, J.R.: Ann. N. Y. Acad. Sci. *108,* 342 (1963)
34. Doyle, R.W.: Americ. J. Sci. *266,* 840 (1968)
35. Einsele, W.: Arch. Hydrobiol. *33,* 361 (1938)
36. Kuznetsov, S.J., Romanenko, U.J.: Akad. Nank SSSR Dokeady *151,* 181 (1963)
37. Gorham, E.: Limnol. Oceanogr. *3(3),* 291 (1958)
38. Gorham, E., Swaine, D.I.: Limnol. Oceanogr. *10,* 268 (1965)
39. Mortimer, C.H.: J. Ecol. *29,* 280 and *30,* 147 (1941/1942)
40. Frevert, T.: Arch. Hydrobiol./Suppl. *55(3/4),* 278 (1979)
41. Berner, R.A.: Geochim. Cosmochim. Acta *27,* 563 (1963)
42. Emerson, S.: Geochim. Cosmochim. Acta *40,* 925 (1976)
43. Bagander, L.E.: Ambio Spec. Report *5,* 147 (1977)
44. Hargrave, B.T.: Verh. Internat. Verein. Limnol. *19,* 1073 (1975)
45. Berner, R.A.: Principles of Chemical Sedimentology. McGraw-Hill, New York 1971
46. Tessenow, U.: Verh. Internat. Verein. Limnol. *19,* 1251 (1975)
47. Hallberg, R.O., Bågander, L.E., Engvall, A.G., Lindström, M., Odén, S., Schippel, F.A.: Contrib. Askö Lab., Univ. Stockholm *2(1),* 1 (1973)
48. Sillén, L.G., Martell, A.E.: Stability constants of Metal-Ion Complexes. The Chemical Society, London 1964, 1971
49. Davison, W.: Geochim. Cosmochim. Acta *44(6),* 803 (1980)
50. Briand, F., Trucco, R., Ramamoorthy, S.: J. Fish. Res. Bd. Canada *35(11),* 1482 (1978)
51. Kelts, K., Hsü, K.J.: In: Lerman, A. (ed.), Lakes, Chemistry, Geology, Physics. Springer-Verlag, New York, Heidelberg 1978
52. Wiechers, H.N.S., Sturrock, P., von Marais, G.: Wat. Res. *9,* 835 (1975)
53. Schäfer, A., Stapf, K.R.: Natur. Mus. *102(8)* (1972)
54. Schöttle, M., Müller, G.: In: Müller, G., Friedman, G. (eds.), Recent Developments in Carbonate Sedimentology in Central Europe. Springer-Verlag, Heidelberg, Berlin 1968
55. Suess, E.: Geochim. Cosmochim. Acta *34,* 157 (1970)
56. Allen, H.E., Unger, M.T.: Z. Wasser/Abwasserforsch. *13(4),* 124 (1980)
57. Arnold, P.W.: Trans. Faraday Soc. *46,* 1061 (1950)
58. Clark, I.S., Turner, R.C.: Can. J. Chem. *33,* 665 (1955)
59. Otsuki, A., Wetzel, R.G.: Limnol. Oceanogr. *17,* 763 (1972)
60. Hsu, P.H.: Environ. Lett. *5,* 115 (1973)
61. D'Anglejan, B.F.: Can. J. of Earth Science *5,* 81 (1968)
62. Truesdell, A.H., Jones, B.F.: J. Res. US Geol. Surv. *2(2),* 233 (1974)
63. Goldman, C.R.: Verh. Internat. Verein. Limnol. *15,* 365 (1964)
64. Gächter, R.: Schweiz. Z. Hydrol. *30(1),* 1 (1968)
65. Shannon, E.E., Brezonik, P.L.: Environ. Sci. Technol. *6,* 719 (1972)
66. Imboden, D.M.: Limnol. Oceanogr. *19,* 297 (1974)
67. Akashi, K., Ikeda, S., Shibai, H., Kobayashi, K., Hirose, Y.: Biotechnol. and Bioeng. *XX,* 27 (1978)
68. Stumm-Zollinger, E.: Arch. Mikrobiol. *83,* 110 (1972)
69. Aminuddin, M., Nicholas, D.J.D.: Biochim. Biophys. Acta *325,* 81 (1973)
70. Almgren, T., Hagström, T.: Wat. Res. *8,* 395 (1974)

71. Vollenweider, R.A.: Techn. Rep. O.E.C.D. Paris DAS/CSI/68, 27 (1968)
72. Thomas, E.A.: In: Griffith, E.J., Beeton, A., Spencer, J.M., Mitchell, D.T. (eds)., Environmental Phosphorus Handbook. J. Wiley, New York 1973
73. Lane, P., Levins, R.: Limnol. Oceanogr. *22*, 454 (1977)
74. Lee, G.F., Rast, W., Jones, R.A.: Environ. Sci. Technol. *12*, 900 (1978)
75. Janus, L.L., Vollenweider, R.A.: Verh. Internat. Verein. Limnol. *22*, 179 (1984)
76. Reimann, K.: gwf (Wasser/Abwasser) *118(1)*, 10 (1977)
77. Singer, P.L. (ed.): Trace Metals and Metal Organic Interactions in Natural Water. Ann Arbor Sci. Publ., Inc., Ann Arbor, Michigan 1973
78. Wagner, G.: gwf (Wasser/Abwasser) *111*, 485 (1970)
79. Elster, H.-J.: gwf (Wasser/Abwasser) *123(6)*, 277 (1982)
80. Wagner, G., Bührer, H., Ambühl, H.: Intern. Gewässerschutzkomm. Ber. *17*, 1 (1976)
81. Ohle, W.: Vom Wasser *13*, 87 (1938)
82. Ostendorp, W., Frevert, T.: Arch. Hydrobiol./Suppl. *55(3/4)*, 255 (1979)
83. Frevert, T.: Arch. Hydrobiol./Suppl. *55(3/4)*, 298 (1979)
84. Tessenow, U.: Arch. Hydrobiol./Suppl. *38*, 353 (1972)
85. Tessenow, U.: Arch. Hydrobiol./Suppl. *47*, 1 (1974)
86. Kamp-Nielsen, L.: Arch. Hydrobiol. *73(2)*, 218 (1974)
87. Holdren, G.C. jr., Armstrong, D.E., Harris, R.F.: Wat. Res. *11*, 1041 (1977)
88. Züllig, H.: Schweiz. Z. Hydrol. *18*, 487 (1956)
89. Fitzgerald, G.P.: Limnol. Oceanogr. *15(4)*, 550 (1970)
90. Kamp-Nielsen, L.: Verh. Internat. Verein. Limnol. *19*, 1057 (1975)
91. Childs, C.W.: Nature Phys. Sci. *241*, 119 (1973)
92. Frevert, T.: Arch. Hydrol./Suppl. *59(4)*, 319 (1982)
93. Chen, R.L. et al.: J. Envir. Quality *1*, 155 (1972)
94. Wolfe, R.S.: Adv. Microbiol. Phys. *6*, 107 (1971)
95. Müller, G.: Naturwiss. *10*, 237 (1966)
96. Schäfer, A.: N. Jahrb. Min. Abh. *117*, 118 (1972)
97. Avnimelech, Y.: Wat. Air Soil Poll. *14*, 451 (1980)
98. Schröder, R.: Symp. Biol. Hung. *15*, 21 (1975)
99. Serruya, C. (ed.): Lake Kinneret. Dr. W. Junk bv Publ., The Hague, London 1978
100. Penck, A.: Morphometrie des Bodensees. Festschrift der geogr. Gesellschaft München 1894
101. Müller, G.: gwf (Wasser/Abwasser) *105(30)*, 810 (1964)
102. Schröder, R., Schröder, H.: Arch. Hydrobiol. *82(1/4)*, 240 (1978)
103. Kiefer, F.: Naturkunde des Bodensees, 2nd ed. J. Thorbecke, Sigmaringen 1972
104. Bailey, L.D., Beauchamp, E.G.: Can. J. of Soil Science *51*, 51 (1971)
105. Graetz, D.A., Dennis, K.R., Aspiras, R.B.: Limnol. Oceanogr. *18(6)*, 908 (1973)
106. Edzwald, I.K., Toensing, D.C., Lenug, M.Chi-Y.: Environ. Sci. Technol. *10*, 485 (1976)
107. Barica, J.: Arch. Hydrobiol. *73,3*, 334 (1974)
108. Pollingher, U.: In: Serruya, C. (ed.), Lake Kinneret. Dr. W. Junk bv Publ. The Hague, London 1978
109. Cavari, B.Z.: In: Serruya, C. (ed.), Lake Kinneret. Dr. W. Junk bv Publ. The Hague, London 1978
110. Bailey, L.D., Beauchamp, E.G.: Can. J. of Soil Science *53*, 213 (1973)
111. Serruya, C., Pollingher, U.: Lakes of the Warm Belt. Cambridge University Press. Cambridge 1983

Environmental Microbiology

W. D. Grant

Department of Microbiology, University of Leicester
University Road, Leicester LE1 7RH, Great Britain

P. E. Long

Department of Botany, University of Leicester
University Road, Leicester LE1 7RH, Great Britain

Summary

Microorganisms were the first living things on this planet. Microbial activity has contributed in a major way to the formation of an essentially hospitable biosphere from extremely inhospitable beginnings, and continues to ensure that the present day biosphere is an equable place for higher forms of life. Present day industrialized society owes a further major debt to microorganisms of the past, being largely dependent on microbially-derived fossil fuels and ores. In biogeochemical terms, microbial activity at the present time is clearly evidenced by the continuing contribution of microorganisms to soil formation and fossil fuels and by microbially-mediated interconversions of different valence states of certain elements. Microbial activities of these kinds are a natural consequence of particular physiologies possessed both by individual microorganisms and by microbial communities. In order to appreciate the impact of microorganisms both past and present it is necessary to be aware of the complexities of microbial community structure and the great diversity of microbial form and function.

The Microorganism – a Part of the Natural Environment

Defining the Boundaries of Environmental Microbiology

The term "biosphere" is often used to describe the global composite of all living organisms and their abiotic surroundings. *Microbial Ecology* or *Environmental Microbiology* is the study of microorganisms and their interrelationships in the biosphere – interrelationships among microorganisms, interrelationships between microorganisms and higher forms of life, and more significant with regard to this particular series, interrelationships between microorganisms and their abiotic environment. This last facet of environmental microbiology is sometimes called *Microbial Biogeochemistry* or *Geomicrobiology*.

Microorganisms are to be found in almost any natural environment on earth. It is also certain that microorganisms were the first living things on this planet. The fossil record has indications of microbial activity which date from 3·5 Ma [1] years ago [1–3], close to the age of the oldest rocks on earth. Any microbiologist will confidently assert that microorganisms have played major roles in the formation of the present day biosphere, and that the continued equability of the planet as an environment for higher forms of live is due to continuing microbial activity. Assertions of that kind are inevitably based on intuition rather than incontrovertible fact – environmental microbiology is a complex area of study because of the number of variables involved, and indeed there is often argument as to the boundaries of the area of study itself [4]. Nevertheless, it is clear that microbially-related transformations of various kinds enmesh in complicated and sometimes less than obvious ways with significant consequences for the environment. The present day environmental microbiologist attempts the difficult task of trying to unravel these interrelationships by combining the strategy of highly defined laboratory-based experimentation with observation and measurement from the less than highly defined natural environment. The non-microbiologists may be less convinced of the importance of the past and present micro-flora – we seek here to persuade them otherwise.

Prokaryotes, Eukaryotes, and Archaebacteria

By definition, all microorganisms are small, individuals generally being invisible to the unaided eye. The discipline of microbiology is largely defined by the microscopic size of the subject matter and by the methodology used in manipulation and culture. The microbiologist is thus concerned with a diverse group of different organisms including bacteria, protozoa, fungi, and algae. Some microbiologists would include viruses, but we consider that these entities, albeit microscopic at a different order of magnitude, are acellular and therefore differ in so many fundamental ways that they fall outside the definition of the subject matter alluded to above. Furthermore, although viruses clearly have an impact on organisms, and therefore on interrelationships between organisms and the environment, the effect is indirect.

1 Ma = Milliard (billion, 10^9) years

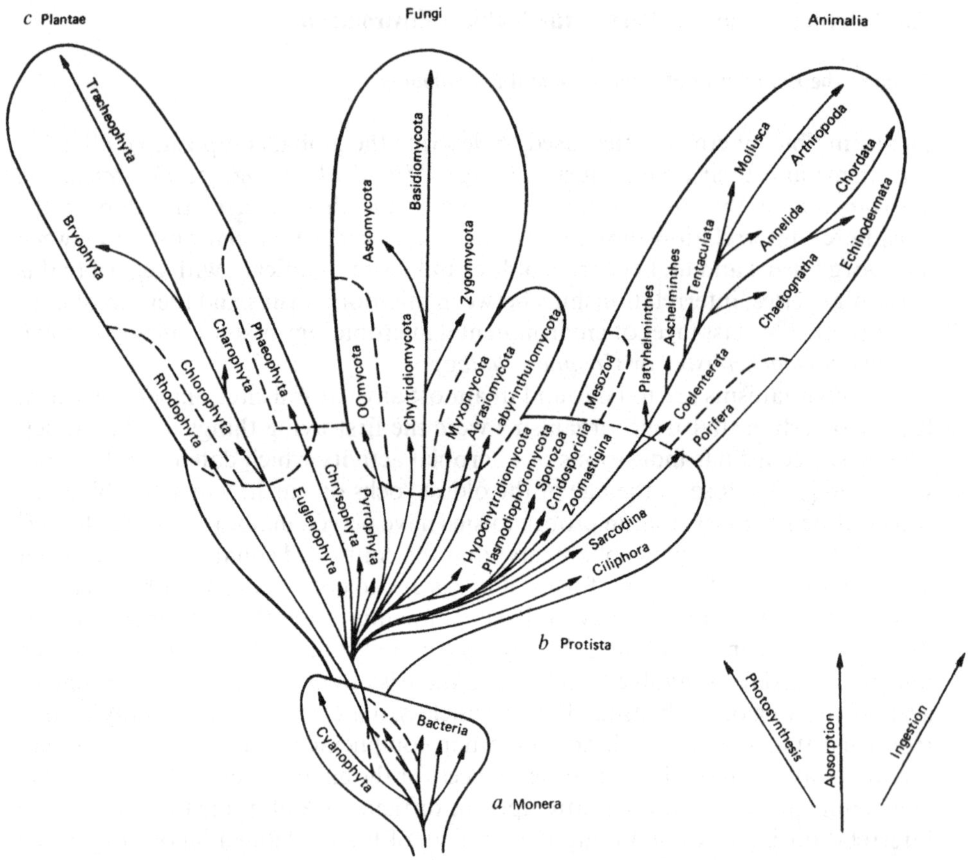

Fig. 1. A five kingdom hierarchy based on three levels of organization: *a* prokaryotic bacteria and cyanobacteria (Monera); *b* undifferentiated eukaryotes (Protista); *c* differentiated eukaryotes (Plantae, Fungi, Animalia). From [5] with permission

The lack of tissue differentiation even in multicellular microorganisms has been used as the basis for classifying microorganisms separately from plants or animals. In Whittaker's five kingdom classification (Fig. 1), algae and protozoa (The Protista) constitute one kingdom, the fungi a second kingdom, and the bacteria including the distinct blue-green algae or cyanobacteria a third kingdom (The Monera), each of these kingdoms being distinct from the plant and animal kingdoms. A further distinct group, the didemnid endosymbiont *Prochloron* spp. [6] should now be added to the Monera. Over the years a major evolutionary and structural discontinuity has become apparent between the cell structure of microorganisms in the Monera and that of all organisms in all other kingdoms. Moneral cells are in general an order of magnitude smaller in size compared to the other cells, but more importantly have a much simpler cell structure, lacking any membrane-bounded nucleus or intracellular organelles comparable to the chloroplasts and/or mitochondria seen in cells from the other groups. These fundamental structural differences were first noted last century, but with the advent

Table 1. Comparison of the prokaryotic and eukaryotic cell[a]

	Prokaryotes	Eukaryotes
Nuclear structure and function		
Nuclear membrane	Absent	Present
DNA	Single molecule, not complexed with histones	in several or many chromosomes, usually complexed with histones
Division	No mitosis	Mitosis: mitotic apparatus with microtubular spindle
Sexual reproduction	Fragmentary process: no meiosis; usually only portions of genetic complement reassorted	Regular process: meiosis: reassortment of whole chromosome complement
Cytoplasmic structure and organization		
Internal membranes	Relatively simple; mesosomes	Complex: endoplasmic reticulum: Golgi apparatus
DNA-containing membrane-bounded organelles	Absent	Present: chloroplasts, mitochondria
Ribosomes	70S in size	80S except for ribosomes of mitochondria and chloroplasts which are 70S
Flagellar movement	Flagella of submicroscopic size, each flagellum composed of one fibre of molecular dimensions	Flagella or cilia: microscopic size; composed of microtubular elements arranged in a characteristic pattern of nine outer doublets and two central singlets
Size	Generally small: usually <2 μm diameter	Usually larger: 2 μm to >100 μm diameter

[a] Modified from [8]

of the electron microscope are now beyond dispute. The evolutionary implications of such a major structural discontinuity have been extensively discussed over the last twenty five years particularly since a seminal review paper by Stanier and van Niel [7] highlighted these structural and biochemical differences between cells with a true nucleus (*eukaryotes*) and those without (*prokaryotes*). The differences between eukaryotic and prokaryotic cells have been catalogued many times – Table 1 provides a list of the main differences. The structural discontinuity between prokaryotes and eukaryotes is particularly highlighted by differences between eukaryotic and prokaryotic ribosomes which differ both in size and antibiotic sensitivity patterns. The eukaryotic cell is, however, a complex compartmentalised structure and, remarkably, eukaryotic organelles have their own complement of ribosomes which turn out to be prokaryotic rather than eukaryotic. These organelles also possess some independent genomic complement. There are thus clear structural similarities between eukaryotic organelles and prokaryotic cells, leading to continued speculation as to the origins of such organelles.

The fossil record indicates that the first organisms on earth were microscopic prokaryotes, the oldest authenticated prokaryotic fossils being dated at 3·1 Ma ago [9], although there is older evidence of microbial (presumably prokaryotic) activity [1–3]. Eukaryotic fossils do not appear until 1·3–1·4 Ma ago [10, 11]. The order of appearance of the two cell types, and the prokaryotic features of eukaryotic organelles has led to suggestions that the original eukaryote was a chimaera of prokaryotic cells, i.e. a now extinct engulfing species with large "eukaryotic" cytoplasmic ribosomes in association with an ancester of chloroplasts and/or mitochondria containing small "prokaryotic" ribosomes. This view has been developed and expanded particularly by Margulis [12] to include the suggestion that the eukaryotic flagellum is derived from an ancestral flexuous prokaryote. Oxygenic phototrophic prokaryotes such as cyanobacteria are imagined to be similar to the chloroplast ancestor(s), whereas an ancestor of *Paracoccus denitrificans* which has an electron transport chain (see Section on Microbial Energetics and Geochemistry) very similar to certain mitochondria could be a candidate as a protomitochondrion [13]. However, eukaryotes are likely to have evolved under conditions quite different to the present day and inferring evidence as to the origin of eukaryotes from extant prokaryotes must remain highly speculative.

The fossil record is of little help in deducing any relationship(s) between eukaryotes and prokaryotes, but in recent years it has been argued that another kind of fossil record exists in genetic sequence – in essence an organism's evolutionary history is enshrined in its genome, which is in itself a reflection of a mutational clock. Accordingly, it should be possible to deduce evolutionary relationships amongst organisms (both prokaryote and eukaryote) by comparing appropriate genetic sequences (directly as in DNA, or indirectly as in protein or ribosomal RNA), assuming of course that the mutational clock has run at the same rate in the molecules examined. A number of so-called "chronometric" macromolecules have now been examined, notably the highly conserved 16 S ribosomal RNAs which are particularly appropriate molecules for study in that they are highly conserved in view of their vital role in the biosynthetic machinery of the cell. Ribosomal RNAs are also of universal occurrence, including being present in the prokaryotic ribosomes of eukaryotic organelles.

Studies of 16 S rRNA sequence homologies by comparing hexanucleotides in common between pairs of organisms (the S_{AB} coefficient) notably by Woese and his colleagues [14–16], initially with a view to constructing a general prokaryote phylogeny, have shown that present day choroplasts appear to share a common ancestor with mitochondria and certain present day bacteria [17]. The larger but phylogenetically equivalent 18 S rRNA of the eukaryotic ribosome, however, does not appear to be related to any 16 S rRNA examined to date. This work would seem to support chimaeric origins for the original eukaryote(s).

Woese's work is also notable for the demonstration that two major lines of prokaryotic descent are present, rather than the monolithic prokaryote lineage previously supposed. In the course of analysing many different 16 S rRNAs with a view to constructing a phylogeny it became clear that certain specialised methanogenic bacteria, although clearly prokaryotic, had rRNAs quite unrelated to all the other prokaryote RNAs examined, and indeed these bacteria were as distinct from other bacteria as they were distinct from eukaryotes [15]. Since that time,

Table 2. Comparison of eukaryotes, eubacteria and archaebacteria [a]

	Archaebacteria	Eubacteria	Eukaryotes
Cell size	Usually <2 μm	Usually <2 μm	Usually 2–100 μm
DNA-containing Cellular Organelles	Absent	Absent	Present
Nuclear Membrane	Absent	Absent	Present
Cell Wall	Variety of types: none incorporates Muramic Acid	Variety within one type: all incorporate Muramic Acid	No cell wall in animal cells: variety of types in other phyla
Membrane Lipids	Ether-linked branched aliphatic chains	Ester-linked straight aliphatic chains	Ester-linked straight aliphatic chains
Transfer RNA's: Thymine in "common" arm	Absent	Present in most transfer RNA's of most species	Present in most transfer RNA's of all species
Dihydrouracil	Absent in all but one genus	Present in most transfer RNA's of all species	Present in most transfer RNA's of all species
Amino acid carried by initiator transfer RNA	Methionine	Formylmethionine	Methionine
Ribosomes	70S	70S	80S except for mitochondria and chloroplasts which are 70S
Translation-Elongation factor	Reacts with Diphtheria toxin	Does not react with Diphtheria toxin	Reacts with Diphtheria toxin
Sensitivity to Chloramphenicol	Insensitive	Sensitive	Insensitive
Sensitivity to Anisomycin	Sensitive	Insensitive	Sensitive
Sensitivity to Kanamycin	Insensitive	Sensitive	Insensitive
RNA Polymerase	9–11 subunits	4 subunits	12–15 subunits

[a] Modified from [19] and [21]

particular thermophilic and halophilic bacteria have also been found to be unrelated to other bacteria, but related to each other and the methanogenic bacteria [18, 19]. The implication of this work is that there are three major prokaryotic lines of descent from the first living organisms on earth, the majority of present day bacteria (the Eubacterial line), the now extinct original engulfing species in the first chimaeric eukaryotes (the Urkaryote line), and these distinct methanogenic, halophilic, and thermophilic prokaryotes (the Archaebacterial line). Although the three lines were originally defined primarily on the basis of the S_{AB} coefficient, there is now a wealth of biochemical detail supporting this tripartite lineage (Table 2). Archaebacteria have some eukaryotic characteristics as well as prokaryotic characteristics, but characteristically have membrane lipids based on

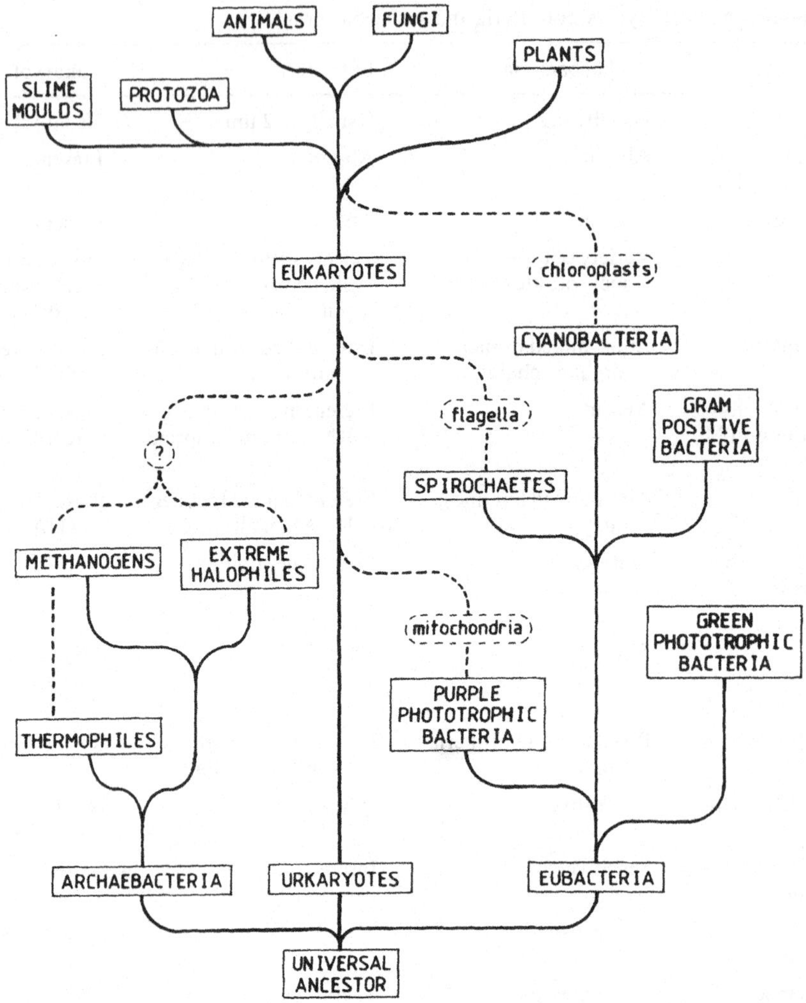

Fig. 2. The three primary lines of descent. It is supposed that the archaebacteria, the eubacteria, and an urkaryote stemmed from a common ancestor. Eukaryotes evolved when the urkaryote engulfed organisms from the other line(s), producing a stable symbiosis. Modified from [19]

C_{20} isoprenoid glycerol ethers rather than the fatty acid glycerol esters seen in eubacteria and eukaryotes [19, 20].

The name *archaebacteria,* first coined by Woese, denotes antiquity, and archaebacteria may have first colonised the surface of the earth, the extant examples being relics from the past lurking in what, by present day standards, are extreme environments but which would have been unexceptional 3·5 Ma ago. However, it is not possible to determine from the S_{AB} values the order of appearance of the three lines, or whether they arose simultaneously from a common precursor cell line. The importance of this work lies in the delineation of the second prokaryote line, whenever it arose, and indeed the presence of eukaryotic traits in archaebacteria has given rise to the interesting speculation that the original eukaryote may have been a chimaera of eubacterial, urkaryote, and archaebacterial lines (Fig. 2).

Major Microbial Groups

Chemotrophs and Phototrophs

It is outside the scope of this Chapter to attempt any detailed taxonomic survey of the major microbial groups. What is more important is how particular physiological groups react with the environment in relation to their modes of nutrition and energy generation.

Organisms may be usefully grouped according to their source of carbon and energy. Traditionally, animals have been considered to be *heterotrophs,* obtaining both carbon and energy from reduced organic compounds, whereas higher plants are *autotrophs* (or photoautotrophs), obtaining carbon from the reductive fixation of CO_2 and energy from light. In both of these cases the mode of energy generation is invariably associated with a particular mode of carbon nutrition. Eukaryotic microorganisms usually fit either of these two categories, but prokaryotic microorganisms are less easy to place in either of these two groups – here there is no requirement that a particular type of energy-generating mechanism is exclusively associated with a particular type of carbon source. Accordingly it is more useful to categorise microorganisms with respect to mode of energy generation alone.

There are two major divisions with respect to energy generation. Organisms harnessing light energy are *phototrophs,* whereas *chemotrophs* harness chemical bond energy by oxidizing a high energy electron donor, electrons being transferred from the donor to a lower energy electron acceptor. Chemotrophs may be subdivided depending on whether the chemical bond energy derives from organic electron donors (*chemoorganotrophs*) or inorganic electron donors (*chemolithotrophs*). *Chemomixotrophs* are capable of using both organic and inorganic electron donors. Animals are clearly chemoorganotrophs in this system of nomenclature. Chemolithotrophy seems to be an exclusive property of prokaryotes.

Chemoorganotrophs may use an internally derived organic molecule as an electron acceptor, when the mode of energy generation is said to be *fermentative.* However, it is energetically far more favourable for chemoorganotrophs and chemolithotrophs to use molecular oxygen as the terminal electron acceptor (see Sect. on Microbial Energetics and Geochemistry). Organisms which utilize O_2 in this way are clearly dependent on oxic conditions under normal circumstances. Such organisms are *aerobic* or *microaerophilic* depending on the oxygen concentration required for growth. Prokaryotes differ from eukaryotes in that electron acceptors such as NO_3^-, SO_4^{2-}, Fe^{3+}, and CO_2 may be used by certain groups in place of O_2. Evidence is now beginning to accumulate that certain eukaryotes are capable of utilizing NO_3^- as a terminal electron acceptor [22], so the use of electron acceptors other than organic compounds and O_2 is no longer held to be an exclusively prokaryotic property. Organisms capable of using other than O_2 as an electron acceptor can clearly grow under anoxic conditions and may be *obligatively anaerobic* or *facultatively anaerobic,* depending on whether O_2 can also be used and whether the alternative electron acceptor is stable in oxic conditions.

Phototrophic organisms may be categorised in a similar way, based on the nature of the electron donor ultimately required to reduce CO_2 to cell material, should CO_2 fixation occur [23]. Thus *photoorganotrophs* have organic electron donors, *photolithotrophs* inorganic electron donors. *Photomixotrophy* is also possible. Water serves as the electron donor for all eukaryotic phototrophs, and, additionally, the prokaryotic cyanobacteria and *Prochloron* group. The photolysis of water results in the production of oxygen so these phototrophs are *oxygenic*. The other major group of phototrophic prokaryotes, the "classical" green and purple phototrophic bacteria (see Sect. on S Oxidation) are *anoxygenic* because they utilize electron donors other than water, such as reduced organic compounds and/or reduced sulphur compounds.

The carbon source used by microorganisms in each of these energy classes can be organic compounds, CO_2 or a mixture of both, depending on organism and there is no *a priori* reason to link a particular carbon source with a particular energy type, although an obligate lithotroph is clearly likely to be dependent on CO_2 as its major source of carbon.

Microbial metabolism therefore enmeshes in a complicated way with the carbon cycle and with the cycling of different valence states of various inorganic compounds.

The Prokaryotes

The current taxonomy of prokaryotes is phenetic, i.e. based on present day morphological and metabolic characteristics. Future classifications may well seek to incorporate at least some of the purportedly phylogenetic information generated by sequence analysis of chronometric macromolecules [14], but at the present time this work is at a preliminary stage. However, this work has already indicated that some of the important parameters in the current classification are poor indicators of phylogenetic relatedness. Cell shape in particular, although a major taxonomic determinant in the current classification, seems to have little relevance in phylogenetic terms [17].

Bergey's Manual of Determinative Bacteriology periodically summarizes the state of bacterial taxonomy. The 8th edition, now some 10 years old, lists 19 "form groups" of bacteria [24]. Some groups contain what are believed to be relatively closely related bacteria, but more often than not these groups contain essentially unrelated bacteria categorised for convenience rather than to demonstrate any real taxonomic veracity. The 9th edition of Bergey is currently in preparation, and will be substantially different in approach. A recent compendium, "The Prokaryotes" treats genera described at the time of its publication (1981) in both taxonomic and ecological terms [25]. Cyanobacteria are not included in Bergey's Manual, although recent treatments of their taxonomy and ecology are now available [25–27]. Table 3 lists prokaryote "form groups." Some of the distinguishing features of these groups have obvious ecological significance. Certain groups are composed of individuals that have an immediate impact on the biosphere because of their mode of nutrition and energy generation. The effect is

Table 3. Prokaryote form groups [a]

Form group	Morphology	Examples of genera	Comments
Phototrophic bacteria	Spiral, spheres, ovoid	*Chromatium Chlorobium, Rhodopseudomonas*	Phototrophic; anoxygenic; bacteriochlorophylls
Gliding bacteria	Rods which may aggregate and differentiate to form a fruiting body	*Myxobacter, Cytophaga*	Chemoorganotrophic; hydrolyse many polymers
Sheathed bacteria	Rods in chains within sheaths	*Sphaerotilus, Leptothrix*	Chemoorganotrophic; some deposit iron and/or manganese oxides
Budding and/or appendaged bacteria	Cells may bear buds or appendages	*Caulobacter, Pedomicrobium*	Chemoorganotrophic; some deposit iron and/or manganese oxides
Spirochaetes	Flexible helices	*Leptospira, Treponema*	Chemoorganotrophic; many animal parasites
Spiral and curved bacteria	Rigid, helically curved cells	*Spirillum, Campylobacter, Azospirillum*	Chemoorganotrophic
Gram negative [b] aerobic rods and cocci	Straight or curved rods: a few may form branches, others are cocci	*Pseudomonas, Halobacterium, Azotobacter*	Chemoorganotrophic; many soil isolates; includes archaebacterial halophiles
Gram negative [b] facultatively anaerobic rods	Rods	*Escherichia, Vibrio, Salmonella*	Chemoorganotrophic; many animal parasites
Gram negative [b] anaerobic bacteria	Rods	*Bacteroides, Desulfovibrio*	Chemoorganotrophic; some sulphate reducers
Gram negative [b] cocci and coccobacilli	Cocci or coccobacilli	*Neisseria, Paracoccus*	Chemoorganotrophic; many animal parasites
Gram negative [b] anaerobic cocci	Cocci	*Veillonella*	Chemoorganotrophic
Gram negative [b] chemolithotrophic bacteria	Rods and cocci	*Nitrosomonas, Thiobacillus, Sulfolobus*	Chemolithotrophic; oxidise N or S, or deposit iron and/or oxides; archaebacterial examples
Methane-producing bacteria	Rods or cocci	*Methanobacterium, Methanococcus*	Strictly anaerobic producing methane as a result of CO_2 reduction; all archaebacteria
Gram positive [b] cocci	Cocci	*Micrococcus, Streptococcus*	Chemoorganotrophic; many animal parasites
Endospore-forming rods and cocci	Rods or cocci	*Bacillus, Clostridium*	Chemoorganotrophic; produce endospores
Gram positive [b] asporogenous rod-shaped bacteria	Rods	*Lactobacillus*	Chemoorganotrophic; often in soils, vegetation
Actinomycetes and related organisms	Irregular rods, filaments, branching	*Arthrobacter, Streptomyces*	Chemoorganotrophic; major inhabitants of soils

Table 3 (continued)

Form group	Morphology	Examples of genera	Comments
Rickettsias	Rods, cocci	*Rickettsia, Chlamydia*	Obligate parasites
Mycoplasmas	Highly pleomorphic: outer layer is triple layered membrane	*Mycoplasma, Thermoplasma*	Chemoorganotrophic; archaebacterial examples known
Cyanobacteria	Cocci, rods, filaments	*Synechococcus, Gloeocapsa, Oscillatoria, Spirulina, Fischerella*	Phototrophic; oxygenic; chlorophyll *a*. Many forms from unicells to complex filamentous thalli.
Prochlorophytes	Cocci	*Prochloron*	Phototrophic; oxygenic; chlorophyll *a* and *b*; endosymbionts of didemnids

[a] Compiled from [24, 25]

[b] Gram positive cells retain a particular stain in the presence of decolourizers, whereas Gram negative cell are decolourized

indirect with other bacteria that are largely parasitic on higher forms of life (Table 3). The important attribute of being able to reductively "fix" and assimilate nitrogen (see Sect. on Nitrogen fixation) is to be found in individuals from many of the form groups.

Archaebacteria are not considered separately in this treatment, examples of archaebacteria being found in several "form groups" containing both lithotrophs and organotrophs. It is likely that future versions of prokaryotic taxonomy will separately categorise these phylogenetically distinct bacteria even in the absence of a complete phylogeny of prokaryotes.

The Fungi

The fungi are chemoorganotrophic spore-bearing organisms, in the main obligately aerobic, although facultatively anaerobic fermentative types are known. Fungal classification is largely based on the morphology of sexual and asexual spores and of the spore-bearing structures. The microbiologist is concerned largely with the ubiquitous microfungi that can be easily grown and induced to spore under laboratory conditions. Unlike most microfungi, the fungi that form large "bracket" or "mushroom" sporing structures often have pronounced geographical distributions. The very size of the sporing structures of these fungi has tended to place them within the province of the botanist rather than the microbiologist. Any one of a number of general texts describing the biology and taxonomy of the fungi can be consulted for detailed taxonomic information [28–30].

Fungi can be divided into slime moulds and true fungi. The slime moulds lack cell walls except in spores and in many ways are similar to protozoa, having unicellular amoeboid phases in their characteristic life cycles. There are those who believe that the slime moulds are sufficiently different from all other forms of life to merit the taxonomic status of a separate kingdom (see Fig. 1). Many of these organisms have phases which presumably feed on bacteria, and it is possible that many soil amoebae are developmental stages of slime moulds. However, it is difficult to estimate the effects of this group in view of the difficulties encountered in precisely identifying the amoeboid stages. There are four groups of slime moulds: members of the Acrasiomycota grow vegetatively as amoebae (myxamoebae) but under certain conditions the myxamoebae aggregate together to form a multicellular mass known as a pseudoplasmodium which goes on to produce a fruiting body called a sorocarp; members of the Labyrinthulomycota form a net-like pseudoplasmodium in aquatic habitats; members of the Myxomycota produce a multinucleate true plasmodium which behaves like a great multinucleate amoeba with synchronous nuclear divisions, eventually giving rise to fruiting bodies containing myxamoebal gametes; the Plasmodiophoromycota are obligate parasites, producing plasmodia inside host cells of plants, algae, and other fungi, eventually producing spores which differ from other slime mould spores in that they germinate to produce flagellate cells.

The true fungi have chitinaceous or cellulosic cell walls and have a vegetative mode of growth which consists of branching tubes or filaments (hyphae). The true fungi also have four main groups, the members of one group having non septate hyphae and therefore resembling some slime moulds in having a multinucleate form of vegetative growth. The Phycomycetes, whose members include many aquatic fungi, have aseptate hyphae and the group is divided into five main subgroups: the Chytridiomycota, the Hyphochytridiomycota, the Oomycota, the Zygomycota, and the Trichomycota: none of these groups produces sexual spores in or on a complex fruiting structure, and the subgroups are defined by the type of spore. Ascomycetes, whose members include most of the yeasts and truffles and many of the fungal partners in the lichen symbiosis have septate hyphae and produce a specialised sporing structure known as an ascus which contains ascospores; further subdivision depends on the type of ascus and the subtending structure. Basidiomycetes have septate hyphae and members of this group are the most complex and differentiated fungi including mushrooms and bracket fungi, all producing an often large sexual structure known as a basidium which subtends basidiospores: further subdivision depends on the structure of the basidium and the basidiospores. The fourth group known as the Deuteromycetes or Fungi Imperfecti have in general not yet been shown to produce a sexual stage, and these largely microfungi are identified by the morphology of their asexual sporing structure (conidiophore).

Fungi have a major impact on the global carbon cycle in view of their mode of nutrition, their abundance in soils in particular, and the capacity of large numbers of types to degrade relatively recalcitrant carbonaceous material like cellulose and lignin (see Sect. on Decomposition of Cellulose and Lignin). The environmental microbiologist is likely to most commonly encounter members of the fungi imperfecti since these are readily cultured in the laboratory. Examples of the genera *Aspergillus, Penicillium, Geotrichum, Trichoderma,* and *Alternaria* are easily isolated since they produce copious numbers of conidiospores. However, many different ascomycetes are also present in soils, and soil also contains a large group of so far non-culturable mycorrhizal fungi.

The Algae

Eukaryotic algae are oxygenic aerobic phototrophs containing chloroplasts which house chlorophyll(s) and accessory light-harvesting pigments such as carotenoids or phycobilins. The organisms in this group range in structure from relatively simple unicells to very large macroscopic types representing borderline cases between algae and higher plants. The algae are classified into nine or ten major groups based on morphology, position, and type of flagellum if present, types of pigments and the nature of any storage material. Some of these groups are depicted as phyla in Fig. 1, although there is continuing dispute over the precise taxonomic ranking of several of these groups. A number of treatments of algal taxonomy and biology are available and should be consulted for detailed information [31–33].

The Phaeophyta or brown algae comprise the common brown algae of the seashore, the majority of examples being wholly marine. These are the most complex organisms classified as algae and may form very large structures. The brown colour is due to brown carotenoids of the fucoxanthin series which mask chlorophylls *a* and *c*. Members of the red coloured Rhodophyta are also conspicuous on the seashore, most of these organisms being macroscopic and confined to marine and brackish ecosystems. These organisms are considered to be among the most primitive algae, lacking any flagellate stage in the life cycle, and exhibiting a photosynthetic apparatus similar to that seen in cyanobacteria with phycobilins as accessory pigments.

The microbiologist is more likely to concern himself with other groups of largely unicellular algae, particularly the green algae or Chlorophyta which comprise an ecologically dominant extremely large class of chlorophyll *a* and *b*-containing algae, ranging from unicellular types to multicellular siphonaceous types common in soils and waters, particularly fresh waters. The Euglenophyta share similar habitats and have similar pigments but are characteristically unicellular flagellates lacking a cell wall. The more restricted range of algae in the Charophyta were once considered to be part of the Chlorophyta, but differ in their mode of sexual reproduction. The Xanthophyta form a much smaller, largely fresh water group once also classified with the Chlorophyta but commonly with silicification of the cell wall and lacking chlorophyll *b*.

The mainly unicellular Chrysophyta or golden algae are common components of plankton from a number of different aquatic habitats. These algae have chlorophyll *a* and golden brown accessory carotenoid pigments. Members of the Chrysophyta together with brown and red coloured chlorophyll *a* and *c*-containing members of the Bacillariophyta (diatoms), Pyrrophyta (desmids and dinoflagellates), and the unusual phycobilin-containing Cryptophyta constitute major components of marine phytoplankton in particular.

Algae are the major primary producers in the aquatic environment and thus make a major contribution to the carbon cycle. Algae with calcareous or silica-impregnated cells clearly interact in other ways with the biosphere. The diatoms in particular feature extensively in the fossil record and past fluctuations in the silica concentration of the world's oceans can be precisely correlated with the appearance of this group [34].

The Protozoa

Protozoa are unicellular chemoorganotrophic microorganisms ranging in size from 10 μm to 2–3 mm. Most are obligately aerobic, but anaerobic fermentative types are known to inhabit the anoxic intestines of animals. Anaerobic types are also to be found in anoxic sediments. Most protozoa ingest particulate food material including bacteria, and many possess specialised mouth regions. Protozoa are divided into four groups based on their mode of locomotion (Fig. 1) being motile either by fluid cytoplasmic extensions (pseudopodia), flagella, or the shorter cilia usually arranged in rows or tufts. Details of the biology and classification can be obtained from any one of a number of general texts [35–37].

The Mastigophora are flagellate protozoans, and some classifications include members of the phototrophic Euglenophyta. The organisms in this group include attached (sessile) types, but the majority are free swimming inhabitants of soils and waters. The group contains a number of parasitic members, including members of the genus *Giardia* which produce gastroenteritis in humans. Representatives of the genus *Bodo* are commonly to be found in polluted, highly anoxic environments where they obtain energy by fermentation.

The Sarcodina are typically amoeboid, the form of pseudopodia being of significance in further classification. Some classifications include the slime moulds in the Sarcodina. The Amoebida which have normal pseudopodia are largely inhabitants of soils and fresh waters. However, many members of the Sarcodina are marine organisms with silica or calcareous endo- or exoskeletons through which pass specialised pseudopodia. Skeletal members of the Foraminiferida and Radiolaria may form a significant fraction of marine plankton.

The Ciliophora constitute the largest group of protozoa. These microorganisms are ciliated and often have specialised mouth regions. Most members are free living but a few are parasites of earthworms, crustaceans, and amphibians. The entodiniomorphs are fermentative anaerobes occupying the rumen of herbivores: some can digest cellulose. Members of the Peritrichida are typically sessile, commonly attached to a substrate by a contractile stalk. Sessile peritrichid protozoa are known to have an important function in grazing bacteria in biological sewage treatment processes [38, 39].

The fourth major group, the Sporozoa, have a less direct impact on the environment in that members of this group are parasites of vertebrates and invertebrates. The sporozoa have no obvious organelles for locomotion, usually produce spores, and often have complex life cycles. Examples include the malarial parasites.

Protozoans clearly constitute a significant trophic level in the foodchain, since many consume bacteria, and there are important predator-prey relationships between protozoans and bacteria in sewage treatment systems [39], the rumen [40–42] and detrital breakdown in aquatic environments [43].

Microbial Energetics and Geochemistry

For the environmental microbiologist, in particular the geomicrobiologist, the study of microbial interactions with the environment is to a large extent the study of microbial energetics. The conservation of energy by all living things generally involves the harnessing of oxidation-reduction reactions. The basis of an oxidation-reduction reaction is the transfer of reducing equivalents (usually electrons or electrons accompanied by H^+) from a donor molecule to an acceptor molecule that is therefore reduced. Thus both the electron donor and the electron acceptor are essential as part of the oxidation-reduction reaction. During the oxidation of AH_2 to A, B is reduced to BH_2

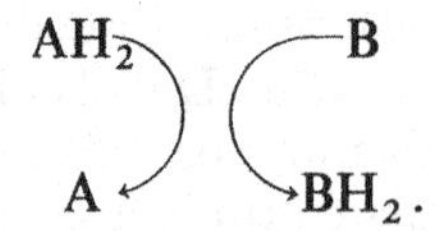

This four component system consists of two oxidation-reduction *couples*, A/AH_2 and B/BH_2. By convention, the oxidized form is placed first. Substances vary in their tendency to give up reducing equivalents and thus become oxidized, or alternatively to accept reducing equivalents and become reduced. When two oxidation-reduction couples are combined as an oxidation-reduction pair, the net flow of reducing equivalents is determined by the tendency of each couple to donate or accept reducing equivalents. This tendency can be quantified as the oxidation-reduction (*redox*) potential measured electrically in mV by reference to a standard redox couple. The standard redox couple is that present at the hydrogen electrode where hydrogen gas is in contact with H^+ in solution at 25° and at pH 0. A couple which is more reducing than the standard couple is credited with a negative redox, whereas a couple that is more oxidizing is credited with a positive redox.

Many redox processes involve only the transfer of electrons, but commonly in biological systems the simultaneous transfer of H^+ occurs. Under these conditions since H^+ is a participant in the redox couple, the redox potential is affected by the concentration of H^+ i.e. the pH of the system. Thus at pH 7 the redox potential of the standard hydrogen electrode is -420 mV. The symbol E'_θ is used for redox potentials measured under these conditions, which are referred to as standard since, additionally, other solute components are present in standard amounts (1 mol dm^{-3}). Redox measurements in the natural environment are seldom made under standard conditions and the symbol E_h is used for the redox potentials of redox couples measured under non-standard conditions.

A redox couple of lower (i.e. more reducing) potential will always donate reducing equivalents to a redox couple of higher redox potential. This transfer of reducing equivalents is accompanied by the release of free energy ($\Delta G'_\theta$). The amount of free energy released under standard conditions is related to the difference in redox potential between the two couples as given by

$$\Delta G'_\theta = -_n F \cdot \Delta E'_{\theta (kJmol^{-1})},$$

where n is the number of reducing equivalents transferred, F is the Faraday constant and $\Delta E'_\theta = E'_\theta$ oxid-E'_θ red. Under non-standard conditions $\Delta E'_\theta$ and $\Delta G'_\theta$ are replaced by ΔE_h and ΔG respectively.

In both eukaryotes and prokaryotes the transfer of reducing equivalents from an electron donor to a final electron acceptor occurs via the participation of intermediate carrier compounds which can exist in the oxidized or reduced form.

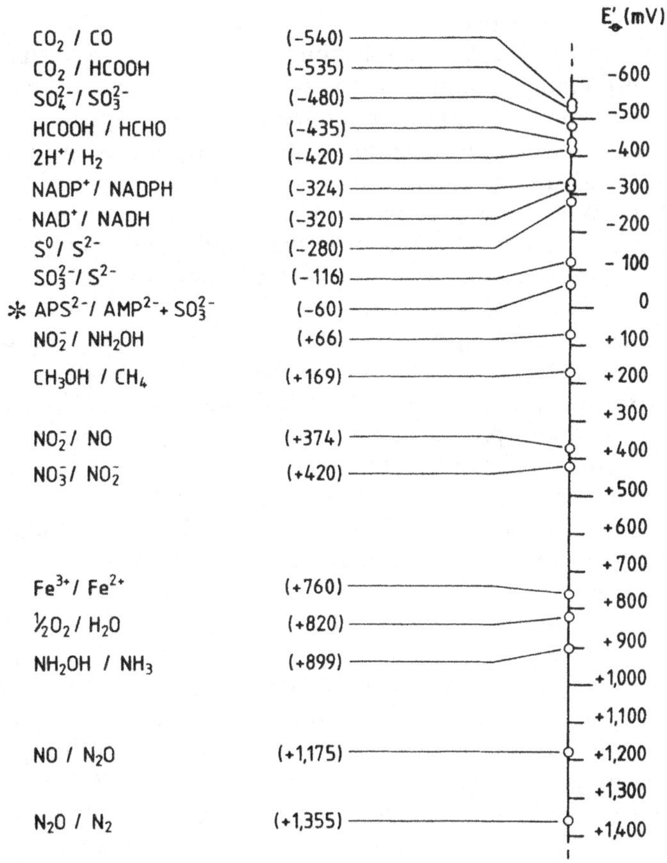

* See Sect. on S Reduction

Fig. 3. The redox column. Modified from [8]

Organisms therefore possess an *electron transport chain* composed of a series of components with closely spaced redox potentials, and as a result energy is released in relatively small packets and hence can be conserved more efficiently. In phototrophs, this multistep electron transport process is potentiated at one or more points by light energy. Pyridine nucleotides (NAD and NADP), flavoproteins, and quinones are hydrogen carriers whereas compounds like ferredoxins and cytochromes are electron carriers. The chemistry of these components, their organization in chains, and the nature of the energy conservation is dealt with in a number of texts [44–47].

The redox column illustrated in Fig. 3 depicts in a convenient way the possibilities for energy generation by oxidation-reduction reactions involving a variety of organic and inorganic compounds. The column ranks redox couples from the most energetic at the top of the tower to the least energetic at the bottom. Reducing equivalents flow naturally in the downwards direction and reducing equivalents from a redox couple in any position can be captured by a redox couple at any lower level, the amount of energy obtained from the transfer being proportionate to the distance between the two levels for any one reducing equivalent transferred. If a reducing redox couple donates more than one reducing equivalent, then correspondingly more energy is available per mol oxidized.

Chemoorganotrophs and photoorganotrophs oxidize organic compounds by preliminary dehydrogenation reactions with high E_0' values, generally around –300 – 400 mV, and are thus capable of donating reducing equivalents directly to NAD or NADP. When aerobic respiration occurs, these reducing equivalents are then ultimately transferred via the electron chain in the direction of more positive redox potential finally reacting with O_2 as the final electron acceptor. The couple $\frac{1}{2}O_2/H_2O$ has an E_0' of +820 mV and O_2 is the most energetically favourable acceptor known, although $\frac{1}{2}O_2/H_2O$ is not the most electropositive redox couple known to be utilized by some microorganisms (Fig. 3). Viewed globally, the end products of oxygenic photosynthesis are aerobically consumed by prokaryotic and eukaryotic organotrophs, resulting in the continued cycling of carbon.

The basis for the capacity of prokaryotes in particular to occupy an extensive array of unusual environments, and to exert effects on the biosphere beyond the carbon cycle, lies in the capacity of lithotrophs to use a variety of inorganic compounds as electron donors and the capacity of certain chemoorganotrophs to use inorganic compounds as electron acceptors.

In the middle of the redox column (Fig. 3) several redox couples, e.g. NO_3^-/NO_2^-, S^0/S^{2-}, can act as electron acceptors or donors, although clearly the amount of energy available is considerably reduced compared to chemoorganotrophic growth with O_2 as terminal electron acceptor, since the redox span between potential donors and acceptors is less. Accordingly, lithotrophs generally utilize O_2 as terminal electron acceptor although the NO_3^-/NO_2^- and Fe^{3+}/Fe^{2+} couples are energetically satisfactory for growth on some sulphur compounds. However, many prokaryotic chemotrophs replace O_2 with alternative acceptors (both organic and inorganic). The use of an acceptor from the middle of the redox column dictates that organisms that perform this kind of *anaerobic respiration* are generally chemoorganotrophic. Oxyanions of nitrogen and sulphur are most commonly used as electron acceptors and there is now evidence that certain eukaryotes are capable of anaerobic respiration of this type [22]. Fe^{2+} and CO_2 can also be used as terminal electron acceptors by certain groups of bacteria. Microorganisms may be *obligate* users of particular electron acceptors, or they may be *facultative,* preferring O_2 should this be available. Thus NO_3^- is used only under anoxic conditions by many groups of bacteria, O_2 being the preferred acceptor, whereas SO_4^{2-} is used obligately by specific sulphate-reducing bacteria. Obligate and facultative lithotrophs are also known.

Obligate chemolithotrophs utilizing donors of redox potential more positive than –320 mV have a problem in addition to the general problem of a relatively small redox span between acceptor and donor, in that the donor is incapable of donating reducing equivalents directly to pyridine nucleotides. Reduced pyridine nucleotides are required for the reduction of CO_2 to cell material. Accordingly, energy has to be expended to achieve reverse electron flow. As a consequence these organisms must transform very large amounts of substrate to sustain other than a very slow growth rate [48]. In view of the nature of the electron donors this clearly has a considerable significance for the biosphere. Methanogenic bacteria also have problems in using CO_2 as a terminal electron acceptor in view of the extremely negative redox potentials of couples involving CO_2 – how these odd bacteria overcome this difficulty is discussed later (see Sect. on Methanogenesis and Methylotrophy).

In the natural environment the overall redox potential actually measured (E_h) is influenced by both the pH and the relative concentrations of the oxidizing and reducing components making up the various redox couples present. O_2 participates in one of the most oxidizing redox couples, so clearly the concentration of dissolved oxygen has an effect on the E_h of the environment. It follows, therefore, that a change in overall E_h for whatever reason will have an effect on the concentration of reactants in individual redox couples, which in turn exerts selective pressure on the type of microorganisms in a particular habitat and on their mode of energy generation. The concentration of dissolved O_2 has an additional effect in that the facultative use of some alternative electron acceptors such as NO_3^- is sometimes controlled by O_2 concentration rather than E_h *per se* [49]. Very strict anaerobes may also be confined to totally anoxic conditions because they lack protective mechanisms against certain toxic O_2 derivatives such as superoxide anion ($O_2^{\cdot -}$) which is produced by procedures effecting a one electron reduction of O_2 [50, 51].

Aerobic respiration and oxygenic photosynthesis thus drive the carbon, hydrogen, and oxygen cycles. The ingenuity of microorganisms in harnessing a wide range of redox couples, means that microorganisms have a profound effect on other element cycles including sulphur, and nitrogen with concomitant geochemical effects in the biosphere, either directly, or by providing inorganic compounds at a valence state appropriate for higher forms of life. Microbially-mediated element cycles thus enmesh in complicated and sometimes less than obvious ways in the environment. As an example, the interrelationships of the carbon, oxygen, nitrogen, and sulphur cycles in a sediment in relation to E_h are depicted in Fig. 4. It would, of course, be possible to draw up similar interrelationships between these cycles in other kinds of environment, underlining the importance of a microbial activity in the biosphere.

The Limits of Microbial Colonization

Microorganisms are to be found colonizing almost every naturally occurring environment on earth. Many of these environments may be unsuitable for most higher forms of life because of one or more environmental stresses, but nevertheless microorganisms are to be found surviving and even growing under these so-called "extreme" conditions. Generally, as conditions become more and more demanding, the diversity of microbial types diminishes and prokaryotes in particular seem to exclusively colonize the most extreme environments. In essence, the difference between a "normal" and an "extreme" environment is simply one of diversity of population [4, 52]. Given such a definition then it is clear that a wide range of selective pressures such as high radiation levels, low nutrient levels, high barometric pressure and even extreme redox potentials are capable of producing extreme environments, although it is more common to discuss the environments

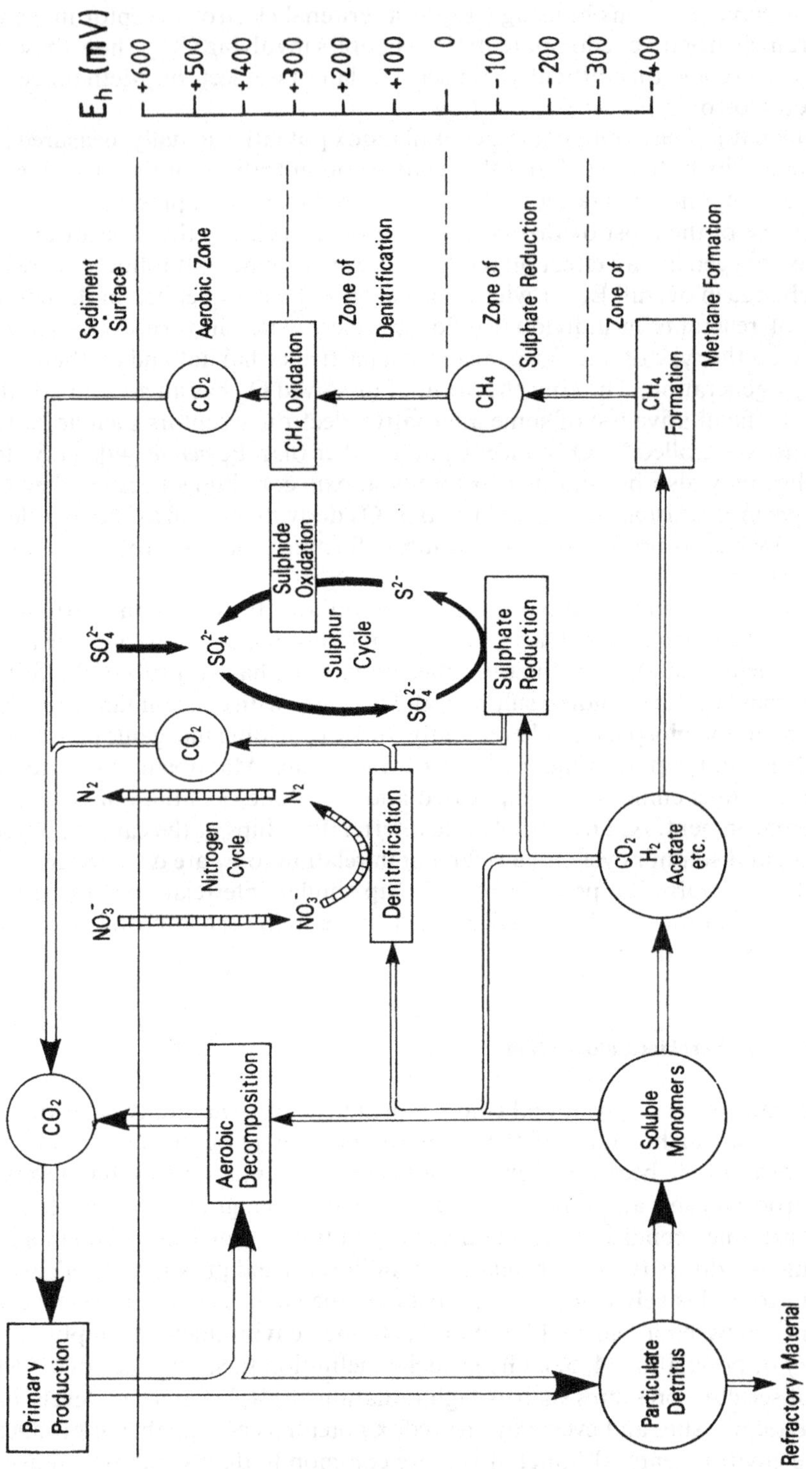

Fig. 4. Pathways of carbon during decomposition in an aquatic sediment: relationship with sulphur, nitrogen, and redox. From [4] with permission

imposed by extremes of pH, water activity (a_w) and temperature. Detailed information on a variety of extreme environments is available from a number of recent texts [4, 53, 54]. Here we concentrate on the stressed environments produced by extremes of pH, temperature, and water activity (a_w). Elsewhere in this chapter brief consideration is given to the strategies employed by microorganisms to combat low nutrient (oligotrophic) situations.

Extremes of Temperature

The stress imposed by extremes of temperature is unique in that it is impossible for a small cell to maintain its internal environment in a more equable condition since there is no way of preventing the internal temperature from equilibrating with the exterior. *Thermophiles* are considered to have a temperature optimum >45 °C, being incapable of growth in the *mesophilic* range (20–45°). Organisms with temperature optima in the mesophilic range, but capable of growth at higher temperatures are referred to as *thermotrophs*. *Psychrophiles* are considered to be incapable of growth and are killed by temperatures >20 °C, usually have temperature optima around 0 °C, whereas *psychrotrophs*, although capable of growth at low temperature, have temperature optima in the mesophilic range.

It is now clear that the maintenance of membrane stability is a vital aspect of life at extremes of temperature. Organisms are able to modulate the relative proportion of high melting point and low melting point lipids in the cell membrane in order to maintain a functional sol-gel system [53, 54]. Psychrophiles and thermophiles represent extreme examples of this capacity. It is also clear that proteins and nucleic acids can be stabilised with respect to temperature by the addition or removal of extra hydrogen bonds or salt bridges, so there is no need to postulate a mechanism for thermophiles that involves the rapid resynthesis of heat-denatured macromolecules [53].

A few years ago a microbial ecologist interested in thermophiles would have confined his attention to those high temperature environments not exceeding the temperature of boiling water (91°–101 °C depending on altitude). The discovery of microbial activity at temperatures of 300 °C in pressurized superheated "black smoker" vents in the floor of the East Pacific Rise has led to the realisation that the upper temperature for microbial growth has not yet been defined. Complex mixed prokaryote communities have now been cultured from these "black smokers" in pressurised containers at 250 °C [55]*. These communities produce CH_4 as well as other gases and therefore presumably include archaebacteria. These discoveries go a long way towards verifying Brock's originally controversial prediction [52], that life is likely to be present in any boiling habitat that contains liquid water. The diversity of microorganisms that have been cultured at temperatures close to 100 °C is considerable, although eukaryotic activity is slight at temperatures >70 °C. Table 4 lists some of the recorded genera found at high temperatures, but does not include the as yet unidentified "black smoker" bacteria. However, other slightly pressurised thermal sea floor environments have recently been described which have yielded new archaebacteria capable of growing at 105 °C [56]. These archaebacteria carry out sulphur transformations in remarkable consortia comprised of several different archaebacterial types [56–61].

* This work has now been called into question [Nature *307*, 737 (1984)]

Table 4. Thermophilic microorganisms [a]

	Approximate optimum temperature (°C)	Approximate maximum temperature (°C)
Eukaryotes		
15–20 genera of fungi including *Rhizopus, Penicillium, Coprinus, Humicola, Dactylaria,* etc.	40–50 Depending on species	50–60 Depending on species
A few genera of protozoa including *Cercosulcifer, Naegleria*	40–45	50–55
Cyanidium [b] (red algae)	45	55
Prokaryotes		
8–10 genera of cyanobacteria including *Synechococcus, Mastigocladus, Oscillatoria*	40–65 Depending on species	50–75 Depending on species
15–20 genera of bacteria including archaebacteria: *Thermoanaerobium, Chloroflexus, Bacillus, Thiobacillus* [b], *Thermoplasma* [b], *Thermothrix* [b], *Thermus, Sulfolobus* [b], *Thermomicrobium, Thermoactinomyces, Thermoproteus* [b], *Thermodiscus, Pyrodictium, Thermofilum, Desulfurococcus, Thermococcus, Methanobacterium, Methanococcus*	40–105 Depending on species	50–>105 Depending on species

[a] Compiled form [4, 56–62]
[b] Genus contains acidophilic thermophiles

Psychrophiles grow over a more restricted span of temperature compared to most thermophiles, and their biochemical problems are not so acute. Presumably, like thermophiles, psychrophiles are restricted by the availability of liquid water, but at the other end of the temperature spectrum. Accordingly, microbial activity is much greater in the oceans of the world at temperatures below 0 °C than in frozen soils at comparable temperatures [53]. Despite the abundance of low temperature habitats compared to high temperature habitats, relatively few real psychrophiles are known compared with many psychrotrophic types. Most are classified in common soil and marine genera, although unique psychrophilic fungi and bacteria have been described from Antarctic regions [63, 64].

Deep sea psychrophiles and thermophiles clearly have to be capable of growing under hyperbaric conditions. Most appear to be *barotrophic* rather than *barophilic,* although to a large extent technical problems in maintaining deep sea samples under constant pressure, and pressure may have prevented the survival of many barophiles. However, a number of barophiles have now been cultured, mostly from the guts of deep sea animals [65, 66]. Such bacteria have more rapid growth rates under pressure [67] and do not survive under normal barometric pressure [68]. The physiology of barophiles has been reviewed [53, 69]. It is clear that, excepting the thermal vent areas in the sea floor (See Sect. on S Oxidation) microbial activity in the abyssal deeps is at a very low level due to a combination of hyperbaric low nutrient and low temperature conditions [70].

Extremes of Water Activity (a_w)

Water activity (a_w) is generally used to describe the amount of *available* water in any environment, and this largely reflects the amount bound by solutes and/or interfaces. The water potential (ψ) is another related term [4] commonly applied to the soil environment (See Sect. on Specific Habitats: Soils).

Many different types of microorganisms capable of tolerating low a_w values have now been described (Table 5). It is much easier to subject an organism to a controlled a_w stress in a homogeneous medium by the addition of particular solutes, so most information about a_w stress has been obtained in this way. However, there is no reason to suppose that an equivalent a_w stress achieved by capillarity effects at interfaces would cause a microbial population to react any differently. Occasionally microorganisms react differently to ionic and non-ionic solutes [53].

The environment produced by high concentrations of salt (NaCl) is most studied. *Halotrophic* microorganisms will grow both in the presence and absence of salt; *halophilic* microorganisms have a specific requirement for salt. Marine microorganisms are considered to be slight halophiles (0.2–0.5 *M* NaCl), moderate halophiles require 0.5–2.5 *M* NaCl, and extreme halophiles (requiring at least 2.5 *M* NaCl) are to be found as salt concentrations approach saturation. Naturally occurring high salt environments are quite common, and the red colouration in soda evaporation ponds, and salt lakes in equatorial regions, is due to the

Table 5. Microorganisms capable of growth under high solute concentrations[a]

	Approximate lower limit of a_w for growth
Eukaryotes	
10–12 genera of fungi including *Mucor, Penicillium, Aspergillus, Saccharomyces, Xeromyces*	0.75–0.80 for most, 0.60–0.70 for *Xeromyces* and *Saccharomyces* spp. (in sugar)
A few genera of green algae including *Dunaliella*[b]	0.75–0.80 (in salt)
Prokaryotes	
A few genera of cyanobacteria including *Aphanothece*[b], *Dactylococcopsis*[b]	0.75–0.80 (in salt)
12–15 genera of bacteria including archaebacteria: *Bacillus, Staphylococcus, Vibrio, Ectothiorhodospira*[b], *Actinopolyspora*[b], *Halococcus*[b], *Halobacterium*[b], *Natronobacterium*[b], *Natronococcus*[b]	0.75–0.80 depending on species (in salt)

[a] Compiled from [4, 71]

[b] Moderate or extreme halophiles

domination of particular archaebacteria that possess red-coloured carotenoids. The genera *Halobacterium* and *Halococcus* are characteristically found in neutral salt environments often in association with halophilic cyanobacteria such as *Dactylococcopsis salina,* and the odd eukaryotic alga *Dunaliella salina* [53]. The genera *Natronococcus* and *Natronobacterium* are found in alkaline (soda) salt lakes usually associated with bacterial phototrophic bacteria of the genus *Ectothiorhodospira* [71, 72]. The main difference between soda lakes and neutral salt lakes apart from pH is in the concentration of Ca^{2+} and Mg^{2+} which are essentially absent from soda lakes due to precipitation as carbonates.

Most of the non-ionic low a_w environments are man-made such as foods, although nectaries and honeycombs provide similar environments. These environments are dominated by *osmotrophic* fungi and many of these fungi are to be found in dry soils where capillarity effects produce low a_w values [53].

A consideration of the thermodynamics of high solute environments indicates that it is impossible for a microorganism to maintain an interior significantly more dilute than the exterior. It is now clear that microorganisms (and other organisms) achieve an internal a_w which balances that of the exterior. Archaebacterial halophiles may be unique in that they exclude Na^+ and accumulate K^+ [53]; certain cyanobacteria and eubacteria accumulate a mixture of Na^+ and K^+ [53, 54]. Other microorganisms synthesize high concentrations of osmotically active compounds known as *compatible solutes*. These are often polyols [54], sometimes amino acids such as proline, and occasionally odd substituted amino acids like betaine [73].

Archaebacterial halophiles are clearly different from other halophiles in a variety of ways, not least in possessing the usual archaebacterial traits not specifically related to halophily. The rod-shaped examples also have unusual cell walls with structural glycoproteins [74] that require a high concentration of salt for stability. Under microaerophilic conditions most synthesize areas of *purple membrane* containing a photoactive pigment *bacteriorhodopsin* which is part of a light-mediated proton-pumping energy-generating mechanism. Accordingly this group is unique in being the only known organism capable of non-chlorophyll-mediated phototrophy [75].

Extremes of pH

Microorganisms are to be found growing at pH values approaching 0 and at pH values above 11. Organisms with an obligate requirement for low pH are *acidophiles,* whereas those with an obligate requirement for high pH are *alkaliphiles.* Alkalitrophic and acidotrophic organisms are also known, where, although growth occurs at high or low pH, the pH optimum for growth is around neutrality. Table 6 illustrates some of the microorganisms known to grow at extremes of pH.

Fungi dominate moderately acid environments, whereas prokaryotes tend to dominate very acid environments. Sulphureous geothermal springs and sulphide-containing mining effluent are generally very acid (pH 0–2) and often relatively hot. The low pH is generated by chemolithotrophic sulphur-oxidizing prokaryotes that generate H_2SO_4 (See Sect. on S Oxidation). Accordingly, these habitats are commonly dominated by chemolithotrophic, acidophilic, often thermophilic

Table 6. Microorganisms capable of growth at extreme pH [a].

Growth under acid conditions (pH <3)	Growth under alkaline conditions (pH >10)
Eukaryotes 6–8 genera of fungi including *Saccharomyces, Cephalosporium, Penicillium*	Eukaryotes A few genera of fungi including *Aspergillus, Fusarium, Penicillium*
Cyanidium (red algae), *Chlorella, Chlamydomonas, Euglena* (green algae)	diatoms including *Nitzchia*
Prokaryotes	*Prokaryotes* Cyanobacteria including *Anabaenopsis, Spirulina, Microcystis, Plectonema*
12–14 genera of bacteria including archaebacteria: *Thiobacillus* [b], *Metallogenium, Sulfolobus* [b], *Bacillus* [b], *Sulfobacillus* [b], *Thermoplasma* [b], *Thiomicrospira, Leptospirillum, Thermoproteus* [b], *Desulfurococcus* [b]	A few genera of bacteria including archaebacteria, *Flavobacterium* [c], *Agrobacterium, Ectothiorhodospira* [c], *Bacillus* [c], *Natronobacterium* [c], *Natronococcus* [c]

[a] Compiled from [4, 56, 61, 71]
[b] Genus contains acidophiles
[c] Genus contains alkaliphiles

prokaryotes of this type, both eubacterial and archaebacterial [52, 76]. However, non-lithotrophic eubacteria and even a few eukaryotes are also to be found in certain of these environments [52, 53].

Very alkaline environments are relatively rare, presumably because of the extensive buffering capacity of the atmospheric CO_2. It is likely that naturally occurring alkaline environments do not exceed pH 11.5 for this reason, although certain man-made industrial wastes exceed this value. Alkaline soda lakes are found in many equatorial parts of the world, usually in relatively inaccessible areas, and these lakes (pH 10–11) have characteristic, mainly prokaryotic, populations of microorganisms [67]. The more dilute lakes (pH 10.0) harbour cyanobacteria such as *Spirulina* spp. and phototrophic bacteria of the genus *Ectothiorhodospora*. The very saline soda lakes (pH 11.0) have populations of the haloalkaliphilic archaebacterial genera *Natronococcus* and *Natronobacterium*, along with different haloalkaliphilic *Ectothiorhodospora* spp. [66].

Organisms that grow at extremes of pH maintain an internal pH that does not deviate much from neutrality and indeed a deviation of more that 2 pH units is usually lethal [72].

The study of some of the inhabitants of extreme environments may well continue to provide us with profound insights into the limits of life not only on this planet but perhaps also in a wider context.

Microbial Community Structure

General Aspects

Terminology

The study of microbial communities is in its infancy. Microbiologists are endeavouring to free themselves from the confines of Koch's pure culture approach [78–80], but our knowledge of microbial behaviour in natural environments remains rudimentary. Many of our current concepts derive, necessarily, from laboratory models which may represent oversimplifications of natural situations. Pure cultures are extremely unlikely in most parts of the biosphere, most microorganisms being engaged in a wide spectrum of interactions with other organisms and their products. These interactions are difficult to observe because of the minuteness of the participants and the heterogeneity of their habitats. Only the more obvious, tightly coupled systems have received extensive attention, for example the nitrogen-fixing root nodule associations between angiosperms and prokaryotes [81, 82].

Ecology centers around interactions both among the members of the biosphere and between them and their abiotic environment. Van der Ploeg [83] has presented a succint account of basic ecological concepts in an earlier volume in this series, to which the reader is referred.

The majority of bacteria are unicellular or at most assembled into simple multicellular filaments. Optical and scanning electron microscopy suggests that most unicellular bacteria are present in natural environments as dispersed cells or micro-colonies on suitable interfaces; colonies of the size seen in laboratory culture are exceptional although visible films may appear. The prokaryotic actinomycetes and the eukaryotic fungi have both evolved filamentous, branched hyphal systems which can ramify through the interspaces between sediment particles or living structures, forming complex networks in which it is difficult to delimit individuals. Myxomycetes (see Sect. on Fungi) exacerbate this problem, compounding it by being migratory like the gliding bacteria (myxobacteria). Consequently, is often becomes difficult to define or quantitatively describe a microbial habitat or niche.

Wimpenny [80] has introduced the terms *"compartment"* and *"domain"* with the aim of overcoming such problems, although these expressions have yet to gain wide acceptance. His definitions are:

(1) "The *compartment* is a functional physiological unit having a discrete boundary separating distinct physio-chemical spaces."
(2) "The term *domain* represents a region of physio-chemical influence. This influence may originate in the environment (*habitat domain*) or may be due to compartments in that habitat (*activity domain*)."

Relationships between compartments and domains are illustrated in Fig. 5. A compartment can be synonymous with a cell but it may be applied to smaller or larger units ranging from the active centres of enzymes to microbial colonies or even metazoan organs or organisms. The limits of domains represent points at which particular activities fall below predetermined values. Domains can be either source or sink domains for particular solutes, depending on flux direction. Domains can have positive, negative or neutral effects on any compartments that

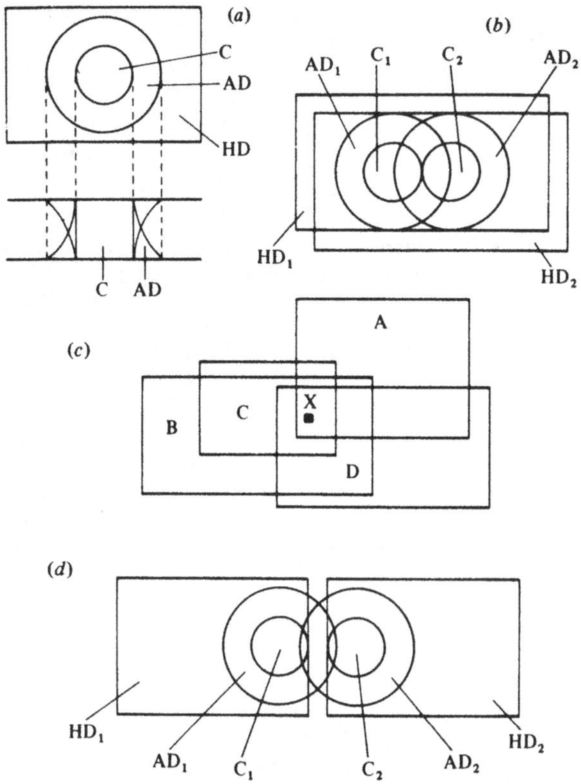

Fig. 5 a–d. Compartments (C), activity domains (AD), and habitat domains (HD) in microbial ecology. **a** A single compartment with (lower part) source or sink domains for particular solutes. **b** Interacting compartments with overlapping habitat and activity domains. **c** A mixed microbial community growing under homogeneous conditions at one point (X) in their common habitat domains; each organism shares a common activity domain. **d** Interacting compartments with exclusive habitat domains. From [80] with permission

lie within them. Interactions occur, in Wimpenny's view, when there is overlap of the activity domains of compartments within the same or different habitat domains.

There have been various reviews of microbial interactions which include attempts at classification [84–86]. For clarity, most schemes confine themselves to interactions between two distinct populations. These classifications reflect Odum's [87] matrix of interactions, and the terms commonly applied to the outcome of these interactions are illustrated and defined in Table 7. *Symbiosis* requires qualification; it is frequently employed when discussing *mutualistic* interactions as in this chapter, but some authors [88] apply it in De Bary's [89] original broad meaning of "a common life" to cover interactions ranging from antagonism to mutualism. The complexities possible when more than two populations are present have deterred much analysis. Relatively stable communities can be established where either complex carbon and energy sources are being degraded or where marked physico-chemical gradients develop. If such communities are

Table 7. Descriptions of interactions between two populations, based on effects

Class of Interaction	Subclasses	Description
1. Neutralism		No interactions between populations
2. Commensalism		One population benefits, the other is not affected
3. Mutualism		Both populations reciprocally benefit
	a) Synergism Protocooperation	Facultative and transient mutualism; combined populations achieve metabolic transformations beyond their separate abilities
	b) Symbiosis	Frequently obligate long-term mutualism between specific partners, often in direct contact
4. Parasitism/Predation		One population attacks and feeds on the other. Parasites are usually smaller and predators larger than the food organism.
5. Amensalism		Growth of one population is restricted by the second unaffected population.
6. Competition		Both populations are restricted by a shared need for a growth-limiting factor.

tightly coupled and have only a few members they may be referred to as *consortia*.

The Microbial Habitat

Many chemolithotrophic and chemoorganotrophic bacteria of biogeochemical importance seem almost world-wide in their distribution, crossing the boundaries between major ecosystems. This impression of ubiquity arises because:

(i) Bacterial distributions relate to the presence of suitable microhabitats [4] which can occur in many combinations in varied ecosystems. The range of microhabitats in soil is a prime example.

(ii) Bacterial taxonomy is often not adequate to permit easy discrimination between populations in different ecosystems, so homologies may be more apparent than real.

Eukaryotic microorganisms appear to have narrower environment tolerances and are more obviously restricted to particular major ecosystems. The microhabitat concept has had a strong hold on the imaginations of microbial ecologists although the detection and study of minute environments in technically difficult without gross disturbance. Microhabitats can be as small as the habitat/activity domain [80] that can exist around individual bacterial cells or microcolonies. Domains capable of supporting anaerobes, aerobes, chemoorganotrophs, and chemolithotrophs can exist in a small volume of soil as a result of the irregular distribution of the various organic and inorganic components, their association as aggregates and related variations in pore size, water, and gas content. The selective effect of these microhabitats relate to microbial responses to both *interfaces* and *gradients*.

Interfaces between aqueous and non-aqueous phases are thought to play critical roles in determining the micro-scale distribution of bacteria, as well as having practical implications in the development of fouling communities on submerged surfaces [90, 91]. The importance of such interfaces for fungi and other eukaryotic protists is not so well documented except with regard to colonization of plants and animals. Non-aqueous phases forming interfaces of microbiological significance range from gases, including the atmosphere, through hydrophobic liquids such as hydrocarbons to solids varying from mineral grains to the complex organic surfaces of other organisms. At all these interfaces microorganisms must remain in contact with the aqueous phase in order to be metabolically active.

Most experimental studies on interfacial effects have employed solid surfaces of metals, glass or plastics to follow bacterial attachment and subsequent film formation. The initial, reversible attachment to solid surfaces is thought to arise from combinations of Van der Waal's forces and electrostatic attraction. The latter results from interactions between negatively charged microbial surfaces and positive countercharges that have accumulated on the solid surface. Consequently, adsorption of organic macromolecules such as proteins upon the solid surface can greatly modify the initial attachment process. Permanent attachment follows an irreversible, time dependent process requiring the presence of microbial extracellular polymer adhesives, either glycoproteins or polysaccharides [92]. Once attachment has become irreversible cell division can give rise to a monolayer of bacteria from which more complex bacterial films can arise. In a fully developed film bacterial cells are enveloped in an extensive, often fibrillar, extracellular adhesive matrix, termed the *glycocalyx* [93]. Some bacteria produce more defined adhesive structures including fimbriae, holdfasts, and stalks.

Microbial growth at solid/liquid interfaces is enhanced particularly in nutrient limited habitats. With solid/liquid interfaces this effect is often stated to be the result of preferential adsorption of organic molecules on the solid surface [94]. An extension of this concept is the idea that the charged polysaccharides of the glycocalyx act as ion exchangers with enhanced surface areas and thus accumulate nutrients [93, 95]. Ellwood et al. [96] have speculated that bacterial film formation has survival value because it is energetically advantageous. This could arise from the trapping of leaked protons between bacterial cells, or between the cells and the solid surface (Fig. 6) as a result of impaired diffusion in the film. The consequent concentrations of protons could either contribute to cell positioning by causing local polarity gradients, or lead to a gain in energy if there was re-uptake of protons via a membrane-bound ATPase. Whatever the nutritional or energetic gain, microbial adhesion to surfaces has clear advantages in turbulent aquatic environments. Air/water interfaces are also environmentally important but much less studied. The surface *neuston* (pleustron) films of lakes and oceans are sites of organic accumulation and contain relatively large numbers of bacterial cells [97, 98].

Gradients of physical and chemical factors are another widespread feature of natural environments strongly influencing microbial distributions and activities [99]. Whereas interfaces are precisely locatable phase boundaries, gradients vary enormously in scale and complexity. At the smallest scale, gradients of electrons, protons, inorganic ions, and metabolites regulate and integrate the activities of

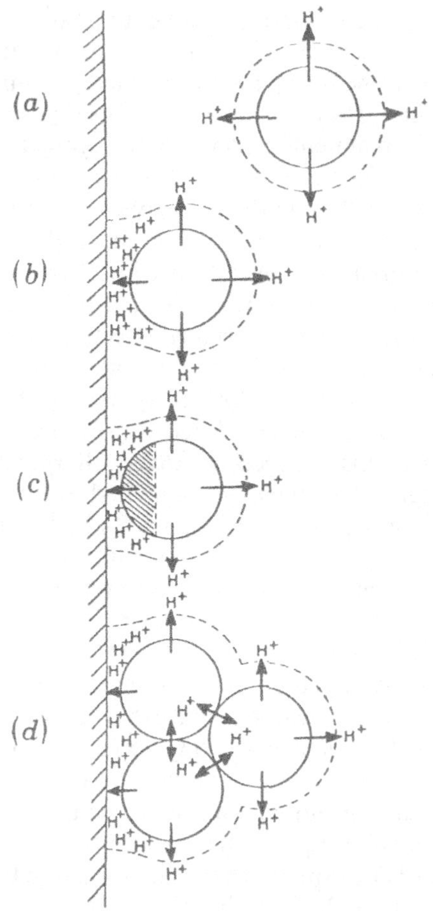

Fig. 6 a. Postulated chemiosmotic energetic advantage for a microbial cell at a surface. **a** Proton gradients generated by free suspended cell. **b** Interaction of a cell's domain with a surface establishing a localized higher concentration of extruded protons. **c** Localized proton flux and concomitant ATP synthesis bringing about increased metabolic activity and polarity of the cell to drive adhering process in the same region. **d** Development of a microcolony with shared proton gradients between members. From [96] with permission

compartments within the cell. The domains of individual cells [80] are defined by the gradients produced by the cells acting as sources or sinks for solute molecules. Solute gradients elicit chemotactic responses in flagellate or gliding bacteria [100] and chemotropic responses from filamentous fungi.

Much larger scale physical and chemical gradients profoundly affect the distribution of physiologically specialized microorganisms. The most striking examples occur in aquatic environments, including sediments, where varying degrees of stratification are common. Stratification partly arises from the restriction of photosynthesis to the upper illuminated *euphotic* zone. Solar heating of water surfaces is another important cause, producing a warmed upper region separated by a short steep temperature gradient, the *thermocline,* [101, 102] from the lower

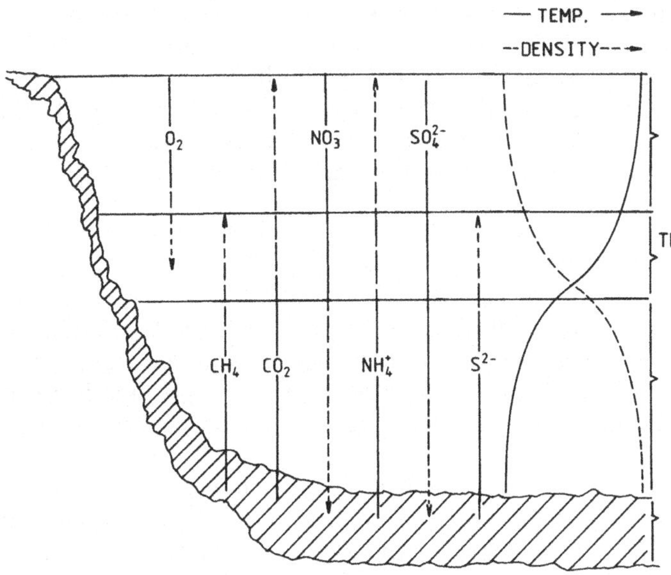

Fig. 7. Physical and biogenic chemical (*arrows*) gradients that can develop in a seasonally stratified lake

cooler part of the water column. These thermally separate water masses persist when turbulence and current velocities are low and can become chemically differentiated as a result of biological processes because solute transport is limited to slow diffusion across the thermocline. The thermocline thus also forms a chemical boundary, the *chemocline*. The seasonal development of the *epilimnion* and *hypolimnion* of eutrophic lakes is a good example of this process (Fig. 7). Once fully established the epilimnion is a warm, oxygenated, upper zone depleted of inorganic nitrogen sources and other plant nutrients by photosynthetic plankton blooms. Much of the organic detritus derived from these blooms accumulates in the hypolimnion and its underlying sediments, and oxygen becomes depleted as decomposition commences. Anaerobic decomposition becomes established with the concomitant reduction of NO_3^- and SO_4^{2-} and the production of CH_4. When fully stratified such a lake possesses a variety of opposing physical and chemical gradients (Fig. 7). Specific microbial groups are distributed in the water column according to their physiological optima, phototrophic sulphur-oxidizing bacteria in the illuminated sulphide-rich part of the hypolimnion, non-heterocystous microaerophilic cyanobacteria in or near the chemocline, methane oxidizers just above the chemocline, and so on. This banding is even more accentuated in the relatively rare permanently-stratified systems such as salt-stabilized lakes, intertidal sediments and enclosed brine-rich ocean basins. Banded microbial growth can be simulated in the laboratory by bacterial seeding of thick gel layers containing opposing oxygen and nutrient gradients [103].

Thus, when considered at the level of the individual cell, microbial habitats are extremely heterogeneous. The interplay of factors controlling the distribution of specific physiological groups produces highly localised sites with favourable environments for particular organisms. As a result an apparently homogeneous habitat can contain many dissimilar compartments. Many of these microhabitats recur in different major ecological systems resulting in biogeochemically important microorganisms being very widely distributed in soils, sediments, and water masses. Unfortunately, *in situ* investigation of microhabitats and their inhabitants is beyond the resolution of most presently available techniques.

Specific Habitats

Soils

Soils are part of the biosphere, derived by weathering from the terrestrially exposed surface of the lithosphere. They are characteristically capable of supporting microbial, plant, and animal life, and their structure and composition is greatly modified as a result. Details of soil chemistry are given in an earlier volume of this series [104].

The variety of microhabitats to be found in a small volume of soil is probably unmatched. This arises from the diverse and irregularly distributed organic and inorganic constituents of soils and their assembly, partly by action of microbes and other soil inhabitants, into a complex of aggregates and pores giving a characteristic structure to a mature, undisturbed soil (see Sect. on General Aspects of Soil and Sediment Formation). Two interfaces which are especially critical to microbial distribution and activities are the surfaces of clay minerals and organic complexes. Attachment of ions, solutes or cells to these charged surfaces profoundly affects the formation of microhabitats by evoking local differences in pH, accumulating or sequestering potential nutrients, and inducing micro-colony formation. The presence of clay minerals also provides enormous internal surfaces within soils: 1 cm^3 of a clay loam (44.3% clay by weight) presents 6.7 m^2 of internal surfaces, 98% of which is due to the clay [105]. Thus, in a small soil volume, microbial cells and colonies can be widely separated. Gradients developed within soil aggregates add another dimension to microhabitat complexity; even in well aerated soils, aggregate cores may be anoxic and permit the growth of anaerobes. The isolation of thermophilic bacteria from both arctic and temperate soils hints at the possibility of localised, transient heating of added organic matter to temperatures of 50°–60 °C, well above the normal soil range and introducing yet another set of microhabitats [53].

Within soils, the timing and extent of microbial activity is regulated seasonally by soil temperatures but, more importantly, in the short term, by fluctuations in soil moisture and energy supplies. For the dominant chemoorganotrophs energy derives from the irregularly available additions of utilizable organic carbon. The roles of soil moisture content are complex and require amplification as there are direct and indirect effects on microbial activity.

Soil water is in three forms, *gravitational, capillary,* and *hygroscopic* [104]; the last is tightly bound to soil particles and is not available to plants or soil microorganisms. Gravitational water fills the soil pores and is lost by rapid drainage and its effects on microbial activity are indirect although it may be important in the dispersal of motile stages. When soil pores become occluded by gravitational water, gaseous exchange between the soil and the atmosphere is impaired. Respiration by soil organisms and plant roots depletes the O_2 but increases the CO_2 content of soils up to c 0.1 atm, a level inhibitory to some microorganisms. Prolonged waterlogging can lead to anoxia and a switch to nitrate or sulphate respiration if these electron acceptors are available (see Sect. on Microbial Engergetics and Geochemistry). Capillary water, in fine pores and held by particles at suction pressures between 0.1 and 10 MPa, is the form most readily available for microbial metabolism.

The availability of water to living organisms can be expressed in terms of *water potential* (ψ), which is a measure of the amount of work done by an organism to obtain water. Water potential is conventionally 0 for pure water and is usually expressed as a negative value. Three components contribute to water potential (ψ), viz:-

$\psi = \psi_p - \psi_s - \psi_m$, where
ψ_p = the potential as reduced by hydrostatic pressure
the pressure component (negligible in soils),
ψ_s = the solute component (potential as reduced by solutes),
ψ_m = the potential as reduced by interfacial effects, the *matric potential.*

The matric potential is the most critical component for soil inhabitants. It is largely generated by the surface tension forces at menisci of water columns within the soil capillaries. In aquatic habitats the solute and pressure components may become significant. An alternative term which is useful to microbiologists is *water activity* (a_w), which relates to water potential thus:–

$$\psi = RT \ln a_w ,$$

where R is the gas constant and T the absolute temperature. The concept of water activity (a_w) was introduced to microbiologists by the food industry and has been used particularly where water availability relates to solute concentration. For pure water, $a_w = 1$, in all other cases $a_w < 1$.

Soil bacteria generally need ψ above -10 MPa (a_w 0.92) for activity. Soil fungi are less sensitive, some *Aspergillus* and *Penicillium* species being osmotrophic, growing optimally at ψ of -5 MPa ($a_w = 0.96$), but continuing down to -40 MPa or -50 MPa ($a_w \simeq 0.70$). Bacteria in soils are affected by water stress in an indirect way initially since they require a surface film of moisture in which to move. Accordingly, the first effects noted in drying soils relate to the cessation of bacterial motility as pores drain, rather than a direct effect on metabolism *per se*. Filamentous microorganisms avoid this problem since filaments can bridge air gaps. Sequential bacterial activities like nitrification are affected first since different populations of bacteria require access to the same site. The effect of water stress on microbial activity is extensively discussed by Griffin [106].

Waters

The hydrosphere is the largest microbial habitat, dominated by the oceans which occupy 71% of the earth's surface and contain 97% of its water. Fresh waters account for a mere 0.0091% of the global total. The residue is located in glaciers, ice-sheets and ground waters which are not major sites of microbial activity. Marine waters are distinguished by their moderately high ionic strength ($\sim$0.7 *M*) due to 35 g l^{-1} salts, mainly Na^+ and Cl^-, and by their narrow pH range (7.8–8.2). The ionic content of fresh water varies from 0.3 g l^{-1} for hard water to 0.065 g l^{-1} for soft water, while pH can be anywhere between 1.5 and 11.0. Both types of surface waters contain carbon as dissolved organic carbon (DOC) and particulate organic (POC), much of which may be recalcitrant organic material. Levels of these organic constituents are generally low, DOC being 1–5 mg l^{-1} in the open ocean, but up to 40 mg l^{-1} in productive near-shore areas. Lakes have 1–50 mg l^{-1} DOC depending on whether they are low nutrient and poorly productive (*oligotrophic*) or high nutrient and highly productive (*eutrophic*). Nutrients such as nitrogen, phosphate, sulphur, and silica are commonly growth-limiting in aquatic environments, particularly when stratification has developed. Stratification often also leads to O_2 limitation in parts of the water column. The converse, O_2 saturation and the generation of toxic oxygen radicals such as superoxide ($O_2^{\cdot -}$) can occur in the uppermost well-lit parts of the euphotic zone during phytoplankton blooms. Further details of the chemistry of the hydrosphere is provided elsewhere in this series [101].

Physical features of aquatic environments that profoundly affect the microbial population include light penetration, temperature stability, and hydrostatic pressure. Light penetration determines the location of phototrophic primary production and varies according to the particulate and DOC content of the water. In pure water, light intensity decreases by 50% for each metre depth. Once light levels have fallen to 1% of the incident light at the surface, the *compensation point* is reached where photosynthesis is balanced by respiration. This point marks the base of the *euphotic* zone within which primary production is possible. The compensation point lies at 200 m in clear waters such as Crater Lake (USA) or the Sargasso Sea; is only 20 m in the turbid productive North Sea and can be as shallow as 8 m in estuaries. Changes in spectral composition with depth result in a preponderance of violet-blue (430–470 nm) light at the compensation point in clear lakes and seas. The increase in particles and DOC shifts the spectral peak upwards towards green (c 575 nm) in coastal waters. The nature of the phototrophic population at depth is often related to spectral composition [107].

Thermal stability derives from the high thermal capacity of water, which permits considerable exchanges of heat energy with little change in temperature. In fresh waters, water approaching 4 °C sinks, so the bottoms of large ponds, lakes and their sediments are rarely frozen. In oceans 90% of all waters do not exceed 4 °C and thus a psychrophilic or psychrotrophic existence is obligatory for inhabitants of the greater part of the hydrosphere. Hydrostatic pressure significantly slows biological reactions at above 300 atm at temperatures prevailing in ocean depths; in nearly 25% of marine waters, pressures exceed 500 atm. To survive in deep sea water microorganisms thus need to be *barophilic* or *barotrophic* (see Sect. on Extremes of Temperature) as well as adapted to low temperatures. An exceptional situation occurs at deep sea hydrothermal vents where the pressure-induced elevation of the boiling point allows bacteria to grow at temperatures well in excess of 100 °C. Microbial habitats within the hydrosphere are listed in Table 8. The water-column and the sediment constitute the major habitats.

Table 8. Aquatic microbial habitats

Habitat	Definition	Community terms	Sub-divisions
1. Neustonic	The surface film (air/water interface)	Neuston	
2. Planktonic	The water column; its free floating inhabitants of limited motility	Plankton	Potamoplankton (= river plankton)
3. Benthic	In and on permanently submerged sediments	Benthos	a) Psammon or Hydropsammon (interstitial water between sand grains) b) Episammon (on sand grains) c) Epipeplon (on mud)
4. Epibiotic	On living and non-living surfaces large enough to support a mixed microbial community	Periphyton (all attached organisms)	a) Epiphyton (on plants) b) Epizoon (on animals) c) Epilithic (on rocks and stones)
5. Enteric-faecal sestonic	Within animals guts and continuing on faeces; also on any floating particulate organic matter	Enteric Faecal Seston	
6. Endobiotic	Within tissues or cells of other organisms	Parasites Symbionts	

Many of the properties of the water-column have been discussed in the Sect. on the Microbial Habitat. At the microhabitat level it provides an homogeneous environment for free-floating (*planktonic*) organisms, although conditions will obviously vary from the surface downwards according to the degree of stratification present. Biogenic particles (*seston*) derived from the plankton, often as faeces, introduce heterogeneity into the water column of large lakes and oceans. Similarly, suspended sediments and fragments of the surrounding or submerged vegetation provide surfaces for microbial growth in shallow lakes, rivers, and coastal waters and increase the variety of microhabitats. In fast flowing streams and tidal coastal waters the surfaces of submerged vegetation, animal exoskeletons or rocks and stones provide an important microbial habitat exposed to the water mass.

Sediments are a key habitat for biogeochemical transformations. They resemble soils in providing large internal interfacial areas but differ in their origins and detailed structure and are defined as "solid materials that settled down from a state of suspension in a liquid" [108]. Details of their properties and microbiology are to be found in Nedwell & Brown [109]. Unlike soils, their formation is by surface accretion and they accordingly increase in compaction with depth; biological activity is mainly in the upper 1 m. Three classes of material contribute to sediment formation: detritus from erosion of the exposed lithosphere (mostly aluminosilicates), *biogenic* substances and *authigenic* material formed *in situ* within the sediments. There are three main biogenic inputs: calcite ($CaCO_3$) mainly from foraminifera and coccolithophores, opal (SiO_2) from diatoms and radiolarians, and organic matter. Authigenic components include precipitated mineral phases such as pyrite (FeS_2), vivianite [Fe_3 $(PO_4)_2 \cdot 8H_2O$) and hauerite (MnS_2); their formation is closely linked to organic decomposition and the resultant redox gradients. Organic input is essentially from the surface plankton although benthic organisms obviously contribute. Even deep ocean sediments receive phytoplankton residues within a few weeks [110] although there is considerable C and N loss during sinking. The amount of organic material reaching the sediment determines the development of redox and electron acceptor gradients of microbial origin. In shallow eutrophic lakes the oxic/anoxic boundary can be well above the sediment surface but deep sea sediments may remain oxygenated [111]. The classical zonation from an aerobic surface layer to the lowest methanogenic zone can also be modified by water chemistry and by animal burrowing (bioturbation) which permits oxygenated water to penetrate below the sediment surface. In marine sediments the high SO_4^{2-} content may restrict methanogenesis through competing with CO_2 for electrons (see Sect. on S-Reduction).

Sediments differ from soils in other respects. Particle aggregation is less. There is no equivalent of soil crumb structure although faecal pellets may provide local anoxic domains within the oxygenated zone. Most sediments are permanently waterlogged, lacking an appreciable gas phase, so sediment microbes are not subject to stress through fluctuating matric potentials.

The Microbial Population

Methodology

Laboratory studies on isolated organisms or communities, analyses of samples and *in situ* observations have all contributed to the characterization of the microbial inhabitants of natural environments and the identification of their geochemical activities. Some of the techniques employed in field studies are listed in Table 9. Their applications and limitations are dicussed in [112] and [122]. In general, methods providing taxonomic or population data involve destructive sampling while *in situ* activity measurements do not permit taxonomic identification of the participants. Consequently it is rarely possible to follow the *in situ* interactions of individuals or micro-colonies with their abiotic or biotic environment. This immediately puts microbial ecologists at a disadvantage compared to those with higher plants or animals, and puts constraints on our understanding of events at the microhabitat level. Indeed, the present state of *in situ* studies of soil microbiology has been compared to analysing the ecology of a tropical rain forest from 5,000 m up in an aeroplane [113].

Population Dynamics

A single microorganism will grow exponentially where all the growth resources are in unlimited supply and physiochemical conditions are favourable and stable. The behaviour of the resultant population can be described by the four *basic growth equations* [114]. The first defines the size of the population after a period

Table 9. Methods in microbial ecology

Aim	Method	Applicability			
		Soil	Surfaces	Sediments	Water
Isolation	Direct: Micromanipulation	+	+	+	+
	Size fractionation and dilution	+	+	+	+
	Pour or spread plates of nutrient media	+	+	+	+
	Indirect:				
	Enrichment with selective media	+	+	+	+
Enumeration	Total counts:				
	Light microscopy	+	+	+	+
	Epifluorescence microscopy (EFM)	+	+	+	+
	Scanning electron microscopy (SEM)	+	+	+	+
	Viable counts:				
	Plating on nutrient media	+	+	+	+
	Most Probable Number techniques	+		+	+
	Autoradiography + EFM		+		+
	Cell division inhibitors + EFM				+
	Electron transport chain activity			+	+
Biomass (Standing crop)	Total: Calculations from microscopic counts	+	+	+	+
	ATP assays	+	+	+	+
	$CHCl_3$ fumigation	+			
	Specific groups:				
	Calculations from microscopic counts	+	+	+	+
	Assays for group specific substances	+	+	+	+
Distribution	Direct:				
	Microscopy of sections and surfaces	+	+	+	+
	Indirect:				
	Recovery of buried slides, capillaries, nylon mesh, etc.	+		+	
Growth and activity	Changes in microbial communities:				
	Alteration in number or biomass	+	+	+	+
	Frequency of dividing cells				+
	Colonization of buried slides, etc.	+		+	
	Metabolic changes:				
	O_2 uptake	+	+	+	+
	CO_2 evolution	+	+	+	+
	Substrate utilization	+	+	+	+
	Product appearance	+	+	+	+
	Lithotrophic CO_2 fixation		(+)	(+)	+
	Organotrophic CO_2 fixation		(+)	(+)	+
	Nucleotide incorporation				+
	SO_4^{2-} incorporation into proteins				+

of growth x_t as

$$x_t = x_o \cdot 2^{t/td}, \tag{1}$$

when x_o is the initial population size, *t* is the elapsed time and *td* is *doubling time* of the population (sometimes referred to as the *mean generation time*). This expression can be converted into a linear form by taking the natural logarithms of both sides of the equation 1

$$\ln x_t = \ln x_o + \ln 2 \cdot t/td. \tag{2}$$

Most laboratory cultures start with an asynchronously dividing population rather than from a single cell; the change in population size with time can then be expressed as

$$\frac{dx}{dt} = \mu x_0, \tag{3}$$

where μ represents a proportionality constant, the *specific growth rate* which is a measure of the amount of new biomass produced per unit of existing biomass per unit time. As with the doubling time, the specific growth rate depends on the specific organism and the conditions under which it is growing. A maximum value for μ is reached with all growth requirements in excess, μ then becomes the *maximum specific growth* rate, μ_{max}. Solving equation 3 gives

$$x_t = x_o \cdot e^{\mu\tau}, \tag{4}$$

the final basic growth equation which is equivalent to equation 1, and can also be converted to a linear form

$$\ln x_t = \ln x_o + \mu t. \tag{5}$$

By comparing equations 2 and 5 an expression, permitting calculation of the specific growth rate can be derived:

$$\mu = \frac{\ln 2}{td} = \frac{0.693}{td}, \tag{6}$$

In practice many environments are closed and at least one growth substrate becomes rate limiting; in these conditions Monod [115] applied an expression analogous to Michaelis-Menton kinetics to define the specific growth rate

$$\mu = \mu_{max} \frac{S}{(K_s + S)}, \tag{7}$$

where S is the concentration of the growth-limiting substrate. The term K_s represents the *saturation constant* and has been defined as the growth-limiting substrate concentration which allows the organism to grow at half its maximum specific growth rate (μ_{max}). It is the third basic growth parameter, the others being the doubling time td and the specific growth rate.

The *observed growth yield Y* is another important parameter, defined as that quantity of biomass produced per unit time as the result of the utilization of limiting substrate in the same time.

Yield can be represented as

$$-\frac{dx}{ds} = Y, \tag{8}$$

while the rate of substrate utilization is given by

$$-\frac{ds}{dt}=\frac{\mu}{Y}\cdot x, \tag{9}$$

By substituting the Monod relationship for μ in 7 this can be represented as

$$-\frac{ds}{dt}=\mu_{max}\frac{S}{(K_s+S)}\cdot\frac{x}{Y}. \tag{10}$$

However not all the substrate energy input is consumed in growth since a proportion is needed to maintain cell structure and function. Thus:

Overall substrate utilization = Rate of consumption for growth + Rate of consumption for maintenance

$$\frac{\mu x}{Y}=\frac{x}{Y_g}+\frac{ax}{Y_g}, \tag{11}$$

where Y_g is the theoretical growth yield assuming no energy consumption in maintenance and a is the specific maintenance rate (h^{-1}). Where the biomass, energy input and maintenance rate are known, the microbial growth rate can be calculated using the relationship between growth rates and substrate utilization:

$$\frac{dx}{dt}+ax=Y\frac{ds}{dt}, \tag{12}$$

These laboratory-derived relationships have been used by microbial ecologists to gain insights into microbial growth in soils and aquatic systems. Most studies relate to chemoorganotrophs, particularly bacteria, and there is a dearth of information on lithotrophs. Furthermore, the field information obtained relates to mixed populations, usually under nutrient limitations, whose maintenance requirements and growth efficiencies may deviate considerably from those obtained in pure cultures.

Estimates of the growth rates of temperate soil bacteria based on the energy input into the soils, the observed biomass, an experimental figure for maintenance energy and another for yields, suggest doubling times of 40–50 h or longer, even when discounting the presence of fungi [116, 117]. As fungi may account for the greatest proportion of the living microbial biomass in a soil [118], these bacterial generation times may be overestimates. Growth rates of unattached planktonic bacteria can be inferred from the frequency of dividing cells [119], uptake of sulphate S into proteins [120], precursor incorporation into RNA [121] or less directly by use of diffusion cultures and chemostats. In sea-water most estimates for doubling times fall in the range 10–100 h [122] but may be down to 4 h near subtropical phytoplankton blooms or well in excess of 1,000 h in cold seasons or at depth. Within the highly eutrophic temperate Lake Mendota there was a seasonal range of bacterial turnover times, from 1.3 h to 28 h or 4 h to 15 h, depending on the method of calculation [123]. There are no comparable figures for microbes on particles or within sediments although these may be more active than the free-living inhabitants of the water columns. Doubling times for chemoorganotrophic bacteria within major environments are thus at least an order of magnitude slower than those possible in pure cultures, indicating that natural populations are generally under severe carbon and energy limitations. High growth rates are presumably locally possible on plant or animal debris and other favourable microhabitats, but are not readily measured.

This view of natural habitats as scenes of famine briefly relieved by feasts is supported by the high proportion of microbial cells that appear to be dead, dormant or inactive in soil or water. A mere 0.4% of the bacterial biomass of an English oakwood soil was classed as "living" [118]. Less than 1% of the chemoorganotrophic bacteria in sea-water can form colonies in conventional media although the metabolically active proportion might average 30% of the total count [122]. Data for fungi and eukaryotes is sparse and scattered. About 20%–30% of fungal mycelium in a temperate woodland soil appeared viable [124]. The proportion of soil fungal biomass represented by dormant spores is unknown although it is likely that most colonies on soil plates originate from these propagules [125].

Oligotrophy

The nutrient-limited status of many natural environments means that their chemoorganotrophic inhabitants are rarely exposed to more than millimolar concentrations of carbon and energy sources and often have to survive on nanomolar amounts. Chemoorganotrophic microbes that are able to grow with very low nutrient fluxes are considered *oligotrophic*. One definition of an oligotroph is that it can be isolated on media containing 1–15 mg C l^{-1} [126], a C concentration 1,000 fold less than that provided by conventional culture media. This may be too restrictive a view as it is likely that there are both facultative and obligate oligotrophic bacteria [127], while the distinction between oligotrophy and survival under starvation as "growth-precursor" cells [128, 129] is blurred. Poindexter [130] has introduced the opposing term *copiotroph* to describe organisms whose survival typically depends on C levels at least on 100 times greater than those demanded by oligotrophs. Most chemoorganotrophic bacteria employed in genetics, biochemical and physiological studies are copiotrophic.

Recent reviewers [131, 127] have formulated the properties of an idealized oligotroph. Key attributes include a large surface to volume ratio and constitutive high affinity (low K_s) but low specificity transport systems to facilitate substrate uptake into the cells. Other features might be the rapid and efficient conversion of substrates into reserve materials such as polyhydroxy butyrate (PHB); low maximum growth rates (μ_{max}) and possession of an oxygen dependent respiratory system to maximise efficient energy production (see Sect. on Microbial Energetics and Geochemistry). The study of obligate oligotrophs is difficult because current isolation methods may not select typical members of the group [127], while cell yields in culture may be too low for normal biochemical and physiological measurements [132]. Aquatic bacteria that seem to possess oligotrophic properties include isolates of *Vibrio, Spirillum,* and *Achromobacter* [133, 134], some (*Caulobacter* and *Hyphomicrobium*) but not all appendaged bacteria [127], and possibly the marine *ultramicrobacteria* [135]. The latter group characteristically pass through a 0.3 μm filter [136] but this dwarfing could be a phenotypic response to starvation by production of a dormant state [128, 137].

The concept of oligotrophy has mainly developed from studies on aquatic bacteria, but oligotrophs have been isolated from soil [138] and the characteristic soil genus *Arthrobacter* has many oligotroph attributes [127]. Oligotrophy in fungi has not been established as such, but is highly likely because of their ability to scavenge carbon sources in inorganic media [134] and within rock surfaces [140, 141].

Soils

Soil microbes play key roles in all terrestrial biogeochemical cycles as decomposers of natural and synthetic manufactured organic matter (see Sect. on the Carbon Cycle) and as participants in the oxidation and reduction transformation of N, S, P, and other elements. In practical terms microorganisms oxidize to CO_2 most of the enormous variety of organic molecules that arrive or are added to soils, have potent effects on soil fertility and plant nutrition, and help maintain soil structure by promoting aggregate formation. This range of activities reflects the diversity of the soil microbiota and its genetic potential, which is amplified by the possibilities of natural gene transfer between dissimilar organisms [142]. An additional sometimes overlooked complication is that most soils are extensively penetrated by plant roots and other underground organs which produce a *rhizosphere* effect in their vicinity, modifying, and frequently increasing microbial activities as well as having a selective effect on the composition of the microflora. A detailed review of soil microbiology is provided by Alexander [143], shorter accounts are to be found in Grant and Long [4] and Atlas and Bartha [144].

The soil microbiota is dominated by chemoorganotrophic bacteria and fungi; the phototrophic and chemolithotrophic prokaryotes implicated in elemental cycles other than C are a minor part of the soil biomass. Soils also contain a variety of eukaryotic algae, slime molds and protozoa whose population sizes and quantitative contributions to soil processes remain largely unknown. The aerobic and anaerobic chemoorganotroph population is concentrated in the upper organic rich soil layers, including the litter layer where present. Although bacteria may dominate viable counts of soil, fungi constitute the largest "living" or "active" fraction in terms of biomass, particularly in woodland soils (Table 10). Use of appropriate isolation methods on soil samples can produce the majority of the non-parasitic bacteria described in Bergey's Manual [24] but it is likely that further soil bacteria remain undescribed or are not readily assignable to existing taxa. The commonest aerobic chemoorganotrophic soil bacteria are Gram-positive. Up to 70% of soil isolates may be loosely identified with the potentially oligotrophic coryneform genus *Arthrobacter*, other important Gram-positive genera include *Micrococcus*, *Corynebacterium*, and the spore-forming *Bacillus*. Actinomycetes, particularly *Streptomyces* are widespread soil inhabitants responsible for the earthy smell. *Clostridium*, also Gram-positive and spore-forming, is the most commonly isolated anaerobe. Strains of *Pseudomonas* and the ill-defined genus *Flavobacterium* are among the most frequent Gram-negative genera; other genera commonly identified include *Acinetobacter*, *Agrobacterium*, *Alcaligenes*, and *Cellulomonas*.

Table 10. Living microbial biomass on the floor of a temperate deciduous wood (Meathop Wood, Cumbria, U. K.)[a]

Resource or soil horizon	Microbial biomass ($Kg\ ha^{-1}$ dry wt.)		Bacteria and actinomycetes
	Basidiomycetes	Other fungi	
Woody debris	30.5	7.3	2.6
L (litter) horizon	3.1	0.5	37.3 (L, Oh + Ah, A and B horizons combined)
Oh + Ah (organic) horizon	8.9	3.4	
A horizon	1.0	26.4	
B horizon	1.0	31.4	
Dead roots	228.0	65.1	8.0
Total	271.5	134.1	47.9

[a] Compiled from [118]

Soil fungi include members of all major classes, but isolation plates are frequently dominated by heavily sporing imperfect fungi such as *Aspergillus, Penicillium, Cephalosporium,* and zygomycetes including *Mucor, Mortierella,* and *Zyorrhynchus.* Direct observation of fungal mycelium in soils, particularly where woody debris is common, indicates that basidiomycetes are an important and underestimated component of the fungus flora. The diversity of soil fungi was clearly illustrated by an intensive study of two German wheatfield soils, which produced 23,500 isolates of 209 species from washed organic and mineral particles [145]. Many soil fungi appear cosmopolitan but studies are beginning to reveal links between particular fungal species and soil or vegetation type [146].

Soil microbiology has been plagued by uncertainty about what can be regarded as indigenous members of the soil population. It is clear that a large proportion of the biomass is inactive at any one time and Winogradsky [147] early on introduced the concepts of *authochthonous* and *zymogenous* components to describe persistently or transiently active members of the community. In modern terms this division may relate to oligotrophic and copiotrophic survival strategies respectively. The problem is compounded by the widespread dormancy of many microbial propagules in soils, partly due to the fungistatic or bacteristatic effects of the soil microbiota [148–150]. Furthermore, the soil receives spores and other survival structures of a rich variety of plant and animal commensals, symbionts, and parasites.

Waters

Aquatic microorganisms resemble the soil community in their participation in decomposition and element cycling but differ in their greater importance in primary production. The scale of their environment, particularly in the oceans, has hindered adequate sampling of the chemoorganotrophic population. Accounts of the microbiology of aquatic environments are provided by Rheinheimer [151], Sieburth [135], and Caldwell [152].

Phototrophs are the dominant microbes within the illuminated euphotic zone. Their quantitative contribution to primary production is discussed in the Sect. on

General Aspects and Primary Productivity. Anoxygenic phototrophic bacteria [153] are confined to zones containing appropriate electron donors within the anoxic region of stratified water masses or sediments and are of only local importance. These phototrophs comprise green S-oxidizing genera such as *Chlorobium*, purple S oxidizers including *Chromatium, Thiocystis*, and *Ectothiorhodospira*, and the non-sulphur purple genera *Rhodospirillum, Rhodopseudomonas*, and *Rhodomicrobium* which require organic acids as electron donors.

The cyanobacteria are by far the most important phototrophic prokaryotes, characterized by possession of a complete oxygenic photosynthetic system utilizing H_2O as the electron donor. It is becoming apparent, however, that many cyanobacteria can switch to anoxygenic photosynthesis with H_2S as the electron donor [154]. Thus the group has a unique dual capability for colonizing both oxic and anoxic illuminated sites and is often found producing blooms in regions of fluctuating oxygen levels as on intertidal sediments or within the metalimnion of lakes. An important ancillary activity is biological N_2 fixation carried out by filamentous heterocystous aerobes typified by *Anabaena* and *Nostoc*, and anaerobic N_2 fixation by non-heterocystous filaments including *Plectonema*. Cyanobacteria are a major component of freshwater phytoplankton, capable of producing massive blooms in eutrophic lakes, commonly with contributions from at least 20 genera including *Microcystis, Aphanizomenon, Anabaena*, and *Nostoc*.

Another major region of cyanobacterial diversity is the marine littoral zone where complex surface mats or even stromatolites develop especially in the tropics. It was considered that cyanobacteria were a minor part of the marine phytoplankton except for tropical blooms of the filamentous *Trichodesmium* (*Oscillatoria*) but small (<3 µm) unicellular forms, possibly *Synechococcus* have recently been found to be widespread [155], and are probably important in oligotrophic tropical oceans. Further details of cyanobacterial biology can be found in Carr and Whitton [27] while the taxonomy of culturable strains is discussed by Rippka et al. [156].

All eukaryotic microbial phototrophs combine oxygenic photosynthesis with O_2-dependent respiration and are thus confined to oxygenated waters and sediments. Diatoms and dinoflagellates seem the most prominent members of the marine phytoplankton, perhaps because their size and rigid walls facilitate sampling and recognition. Small-celled fragile flagellate groups important in oceanic waters are the Haptophyceae represented by coccolithophores and the colonial, non-motile *Phaeocystis*, the Prasinophyceae, Chrysophyceae, and Cryptophyceae. Freshwater habitats contain a large contingent of planktonic and benthic Chlorophyceae (green algae).

Chemolithotrophic sulphur-oxidizing and nitrifying bacteria occur in the water column, on surfaces and within sediments in fresh and salt waters. Recently, interest has been aroused by the discovery of remarkable hydrogen-oxidizing lithotrophs capable of growth at up to 105 °C in hot solfataric springs in the sea floor [56]. Lower temperature hydrothermal vents, first investigated in the Galapagos Rift, also show chemolithotrophic primary production by H_2S-oxidizing bacteria [157] and are surrounded by dense and unusually fast growing animal communities [158]. Among these is an exceptionally large gutless vestiniferid worm, *Riftia pachyptila* which depends on symbiotic H_2S-oxidizing bacteria

Table 11. Bacterial biomass and surface area compared to other oceanic planktonic organisms in an experimental marine enclosure[a]

Organisms	Biomass		Surface area	
	μg dry weight l^{-1}	% total	$cm^2\ l^{-1}$	% total
Bacteria	26	4.6	24.6	69
Protozoa	9.2	1.7	0.27	0.7
Algae	310	56	10.7	30
Zooplankton (animals)	206	37	0.32	0.9
Total	551		35.9	

[a] Modified from [164]

within its tissues (see Sect. on S oxidation). This worm possesses a novel sulphide-binding blood protein [159] which probably both transports the reductant to the bacteria and prevents toxic accumulations of free H_2S. Shallower water marine sediments can support significant chemolithotrophic production at the oxic/anoxic zone boundary [160] as well as symbiotic associations between S-oxidizing bacteria and the bivalve mollusc *Solenmya* and other marine animals [161–163].

Chemoorganotrophic bacteria are ubiquitous in aquatic regions of the biosphere forming an appreciable part of the biomass and a major part of the organism surface area in plankton communities (Table 11) [164]. Peak numbers in the water column are spatially and seasonally associated with phytoplankton maxima in oceans and lakes. Other regions of high organotrophic activity near the sediment surface in lakes [165] and shallow coastal waters [166, 167] include contributions from aerobes above the oxic/anoxic boundary and from NO_3^-, SO_4^{2-}-respiring anaerobes (see Sect. on Nitrate Reduction and S-Reduction) below it. Sedimentary chemoorganotrophs are ultimately dependent on the downward fluxes of phytoplankton or macrophyte-derived particles such as faecal pellets. In shallow lakes and inshore waters these fluxes may be rapid and heavy so sediment microbes may account for a large proportion of organic decomposition. This can profoundly alter the chemical composition of the overlying water as demonstrated by the inception of the anoxic hypolimnion in lakes. In very deep lakes and in oceanic regions sediment microbes probably account for 10% or less of all organotrophic activity [168, 169].

Identification of aquatic chemoorganotrophs by numerical taxonomy [170–174] has confirmed the predominance of Gram-negative types, which can account for 90% of isolates on conventional media. As in soils, isolates do not correspond to accepted reference strains. Where identification is possible, *Achromobacter, Alcaligenes, Alteromonas, Flavobacterium, Pseudomonas,* and *Vibrio* are frequently encountered aerobic Gram-negative genera. In one marine study [173] budding bacteria, mostly *Hyphomicrobium,* formed 10% of the isolates, while various morphologically distinct prosthecate bacteria can be detected in freshwaters [170]. Gram-positive representatives are commonest in sediments, rivers, and streams; they include *Bacillus* spp., various micrococci and anaerobic *Clostridium* spp.

Characterization of aquatic bacteria is far from complete. The low number of specifically marine chemoorganotrophic genera [175] reflects the inadequacy of current identification methods rather than the implied failure of bacteria to thrive in marine habitats. True marine bacteria have an absolute Na^+ requirement [176, 177]. As identification of strains and species improves, many are likely to show habitat specificity. The numerical taxonomic approach has already distinguished between oligotrophic and copiotrophic lake bacteria [178] and between bacteria of different localities in Arctic coastal waters [174]. Substrate uptake studies hint at physiological differences between free-floating and particle-attached chemoorganotrophs [179] as well as enhanced activity by the latter [180].

Aquatic prokaryotes demonstrate a range of structural and physiological adaptions to their environment. Proteinaceous gas vesicles which act as pressure sensors and buoyancy devices have been found in cyanobacteria, purple S bacteria, chemoorganotrophs, and arachaebacterial halobacteria [181]. These structures maintain planktonic microbes at appropriate depths in the water column and could be an alternative to flagella. Magnetotactic bacteria are found worldwide in sediments [182] and contain magnetite inclusions. This response enables the bacteria to return to the sediment if they become suspended, although it is inoperative at the magnetic equator. Pressure-dependant barophilic bacteria are only likely to occur in oceans below 1,000 m [67]. The search for these has probably been hampered because pressure-retaining samplers may fail to protect against rising temperatures during recovery, killing sensitive psychrophiles [67]. At present, barophiles have mainly originated from the deep sea amphipod guts [66, 183] and have not been found suspended in abyssal waters. The high temperature hydrothermal vent communities [55] depend on pressure maintaining superheated water in a liquid state and are certainly barotolerant if not barophilic.

A high proportion of marine planktonic bacteria are less than 0.3 μm in diameter on collection and have been termed *ultramicrobacteria* [136]. This dwarfing may be a phenotypic response to starvation but it is possible that some small cells are obligate oligotrophs corresponding to the model proposed by Poindexter [130]. Coastal ultramicrobacteria have reverted to phenotype resembling *Vibrio, Aeromonas, Pseudomonas,* and *Alcaligenes* [137].

Aquatic fungi have been discussed in [184], and marine types are dealt with by [185]. Fungi are much less prominent in waters than in soils and there is no quantitative information in their activities. The zoospore-producing types, principally Chytridiomycetes and Oomycetes are primarily aquatic and are characteristic decomposers and parasites in freshwaters, some existing as anaerobes [186] while relatively few are marine [184]. Septate higher fungi may have reinvaded watery environments after evolving as parasites or decomposers of land plants. Most aquatic types are ascomycetes or imperfect states mainly found on leaves and other land plant debris in lakes and streams, in coastal mangrove swamps and on submerged wood in the open sea. Some are parasitic or symbiotic on marine macro-algae and very few have been found below 500 m. Spores of aquatic fungi are frequently highly branched (tetraradiate), sigmoid or spiny as aids to floatation or impaction on suitable substrates. Yeasts are widely distributed in the sea, usually $100\ l^{-1}$ but more in coastal areas. The slime-mold-like Labyrinthulales and Thraustochytridiales are exclusively marine, causing damage to mollusc shells and eel grasses.

Phagotrophic flagellate, ciliate, and amoeboid protozoa are characteristic of aquatic environments [187]. Many are bacterivorous but quantitative studies of their grazing on bacteria in nature are at an early stage [188–190]. It is possible that they promote bacterial growth by releasing DOC as well as regulating numbers by grazing [189].

Interactions with Other Organisms

General Aspects

Certain microbial interactions have particular biogeochemical relevance. The first category centres on mixed communities of interdependent microbes which rely on metabolic co-operation (*synergism*) between their members. The communities in this category can be important in the C cycle as decomposers of persistent natural or man-made (*xenobiotic*) organic residues. Members of these essentially mutualistic associations are readily capable of independent existence and do not need direct contact with their partners for successful metabolic exchanges. By contrast the second category consists of highly evolved, closely linked symbioses such as nitrogen-fixing nodules, the mycorrhizae of higher plants, lichens, and theoretically the eukaryotic cell itself (see Sect. on Prokaryotes, Eukaryotes, and Archaebacteria). In these symbioses one partner is the main provider of carbon and energy sources and membership of the symbiosis is frequently obligatory on the recipient organism. Intimate contact occurs between the participants often involving formation of symbiosis-specific structures.

Successful participation in interactions with other organisms as well as survival in a changing environment, may depend on the expression of genetically determined properties additional to those needed for the basic functioning of the cell. Encoding the information specifying such peripheral though potentially advantageous properties, requires the maintenance of DNA sequences that are rarely expressed. Such "luxury" information can be accommodated in eukaryotic chromosomal DNA, but its inclusion within much smaller prokaryotic chromosomes might harm the survival of the population as a whole. An alternative strategy for prokaryotes is to carry these additional genes in a floating library of extrachromosomal DNA [191]. These extrachromosomal elements, represented by plasmids or transposons, are possessed by a few members of the population but are capable of being transferred to other receptive cells. They can thus spread to the rest of the population at time when their presence and expression may aid survival. Plasmid-specified properties of this type include degradation of aromatic compounds by *Pseudomonas*, dehalogenase production by *Pseudomonas* and other Gram-negative bacteria, antibiotic resistance in many bacteria, pathogenicity plasmids in *Agrobacterium* and symbiotic nitrogen-fixing plasmids in some rhizobia. Further details of the distribution of these extrachromosomal elements in natural environments and discussion of factors limiting their spread are provided by Reanney et al. [191, 192].

Synergism

Synergistic associations are capable of effecting chemical transformations which are beyond the separate abilities of each species in the community or *consortium*. Furthermore, the member species are metabolically interdependent, with the possible exception of those initiating the sequence of transformations. Most described synergistic associations are involved in the degradation of complex or persistent organic molecules. Synergistic *co-metabolism* occurs when an organism degrades a compound which cannot be used for growth while growing at the expense of a compound produced by another organism (see Sect. on Co-Metabolism). Identification of synergistic communities is laborious as it depends on knowledge of the relevant biochemical properties of all the species involved. The most readily demonstrable examples occur where environmental parameters or the nature of the substrate restrict the diversity of the community. This situation often occurs in thermophilic habitats or other "stressed" habitats and in the presence of unusual man-made (*xenobiotic*) organic molecules. The oxic decomposition of plant residues (see Sect. on Decomposition of Cellulose and Lignin) represents a natural example of synergism, and intuition suggests that synergism is widespread in microbial communities.

Some of the best characterized natural cases of synergism occur during degradation of plant residues in anoxic environments. Here, lignin remains undegraded while mixed microbial populations are needed to decompose polysaccharides to CO_2 and H_2O. Methane (CH_4) production is an ancillary activity resulting from interspecies H_2 transfer which is a special case of synergism. The general sequences shown in Fig. 8 are liable to disruption in the presence of excess

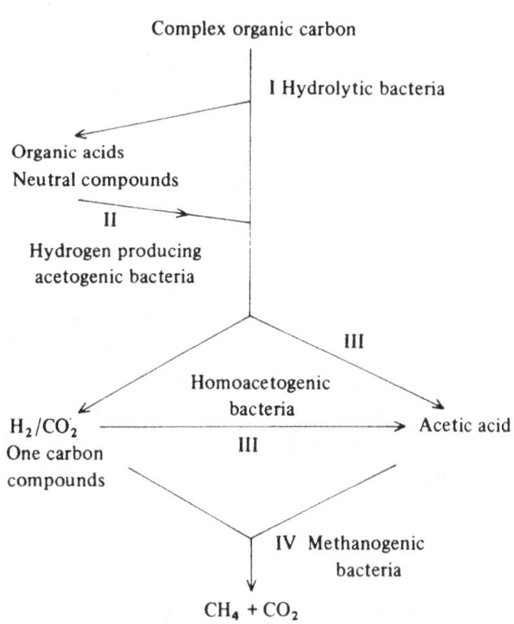

Fig. 8. Metabolic groups (I–IV) participating in anoxic biodegradation. From [193], with permission

Table 12. Potential for synergism in an anaerobic thermophilic community. Methanogenic and sulphidogenic bacteria transform products of the hydrolytic species [a]

Organisms and physiological group	Energy sources	Fermentation products
A. Hydrolytic bacteria		
Thermoanaerobium brockii	Hexose, starch, pentose Saccharides $+S_2O_3^{2-}$	Lactate, ethanol, acetate H_2/CO_2 H_2S + acetate + CO_2
Clostridium thermohydrosulfuricum	Starch, pentose, hexoses Saccharides $+S_2O_3^{2-}$	H_2/CO_2, ethanol, lactate, acetate H_2S, acetate + CO_2
Clostridium thermosulfurogenes	Pectin, starch, pentoses, hexoses	Ethanol, lactate, H_2/CO_2, methanol, acetate
Thermobacteroides acetoethylicus	Pentose, hexose, starch Saccharides $+S_2O_3^{2-}$	Ethanol, H_2/CO_2, acetate H_2S + acetate + CO_2
B. Methanogenic bacteria		
Methanobacterium thermoautotrophicum	H_2/CO_2	CH_4+CO_2
C. Sulphidogenic bacteria		
Thermodesulfotobacterium commune	H_2, lactate, pyruvate + $S_2O_3^{2-}$ or SO_4^{2-}	Acetate + H_2S+CO_2
	Pyruvate	H_2+CO_2 + acetate

[a] Simplified from [193]

inorganic electron acceptors (i.e. NO_3^-, SO_4^{2-}). Zeikus [193] cites specific examples from a volcanic hot spring (Table 12) and lakes, while interesting, mainly archaebacterial communities occur in Icelandic hot springs [57–60]. A similar, more complex series of interactions is found in the rumen [40, 42, 43].

Synergistic degradations of persistent xenobiotic molecules such as herbicides, pesticides, and plasticizers are largely known from laboratory enrichments [79, 194]; extension of these observations to natural environments remains conjectural. Numerous permutations of mutualistic interactions are possible in such communities (Fig. 9) and some members may not even have direct involvement in transformations of the main substrate. Bacteria, particularly *Pseudomonas,* feature prominently but yeasts (*Candida*) and fungi (*Cylindrocarpon, Penicillium*) have also been found [195]. Examples of synergistic communities are given in Table 13; a more complete listing is provided by Slater & Bull [194]. The development of such communities is not automatic. Where the breakdown products are toxic there may be selection for mutants deficient in the pathway(s) needed to catabolize the xenobiotic compound, enhancing its recalcitrance [194].

Interspecies Hydrogen and Sulphur Transfer

Hydrogen transfer links closely coupled bacterial associations like *"Methanobacillus omelianskii"* (ethanol-utilizing, hydrogen-evolving "S" organism + hydrogen-oxidizing *Methanobacterium bryantii*) as well as looser microbial communities in the rumen, aquatic sediments and sludge digesters [196]. It plays a key

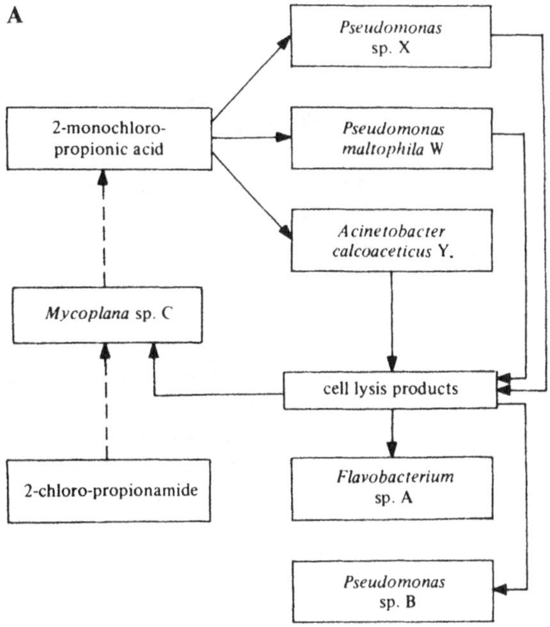

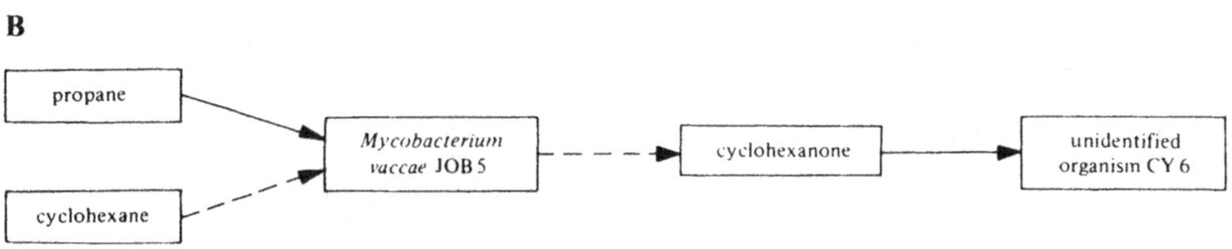

Fig. 9 A, B. Synergistic microbial communities observed in the laboratory. **A** Mutualistic catabolism of 2-chloropropionamide (2CPA). **B** Commensal mineralization of cycloalkanes by two bacteria. Co-metabolic steps are indicated by broken arrows. From [194] with permission

role in methanogenesis [197] (See Sect. on Methanogenesis and Methylotrophy). H_2 is a product of anaerobic fermentations in which the oxidation of organic matter is coupled to proton reduction by *hydrogenogenic* organisms, although H_2 frequently fails to accumulate in anoxic environments since it is subsequently oxidized by *hydrogenotrophic* bacteria to obtain energy for growth coupled to the reduction of inorganic or organic electron acceptors (Table 14). There are numerous H_2-evolving bacteria including *Clostridium, Selenomonas, Ruminococcus, Veillonella,* and many enteric forms. Methanogens depend on H_2 transfer for their existence [197] and are the most prominent hydrogenotrophs. Confusingly, *Acetobacterium woodii* and *Desulfovibrio* can behave as either hydrogenogens or hydrogenotrophs [196]. Other examples of this versatility among anaerobes are discussed by Zeikus [193].

Table 13. Synergistic communities[a]

Substance	Organisms
Diazinon	*Arthrobacter* *Streptomyces*
Linear alkyl benzene sulphonates (LAS)	Multimember community with *Pseudomonas, Alcaligenes* dominant
3,4-Dichloropropionanilide	*Penicillium piscarium* *Geotrichum candidum*
3-Methyl heptane	*Pseudomonas* sp. *Nocardia* sp.
n-Hexadecane	Seven-membered community including species of *Aeromonas, Pseudomonas* and *Vibrio*
Polychlorinated biphenyls	Soil or sediment communities dominated by *Alcaligenes odorans, A. denitrificans* and unidentified bacterium
2-(2-methyl-4 chloro)-phenoxypropionic acid	Undefined, includes 10–12 bacteria with fluorescent pseudomonads

[a] Compiled from [195]

Table 14. Hydrogen-consuming reactions performed by hydrogenotrophic bacteria[a]

H_2-consuming reaction	Hydrogenotrophic organism
$4H_2+CO_2 \rightarrow CH_4+CO_2$	Methanogenic bacteria
$4H_2+CO_2 \rightarrow CH_3COOH+2H_2O$	*Acetobacterium woodii*
$4H_2+SO_4^{2-} \rightarrow S^{2-}+4H_2O$	*Desulfovibrio desulfuricans*
$5H_2+2NO_3+2H^+ \rightarrow N_2+6H_2O$	*Paracoccus denitrificans*
$2H_2+O_2 \rightarrow 2H_2O$	*Paracoccus denitrificans*
$H_2+NO_3^- \rightarrow NO_2^-+H_2O$	*Vibrio succinogenes*
$4H_2+NO_2^- \rightarrow N_2+H_2O$	*Vibrio succinogenes*
$4H_2+NO_3^-+2H^+ \rightarrow NH_4+2H_2O$	*Vibrio succinogenes*
H_2+fumaric acid→succinic acid.	*Vibrio succinogenes*

[a] Modified from [196]

Transfer of H_2 between species is not essential for the anaerobic decomposition of organic residues but it strikingly modifies the end products (Fig. 10). The reasons for this are incompletely known [196]. However, hydrogenotrophic H_2 consumption seems to increase H_2 output in their hydrogenogenic partners by redirecting electron flow away from ethanol, lactate and other electron sinks, thus benefiting hydrogenogens in these partnerships as extra pyruvate becomes available for ATP synthesis.

Interspecies sulphur and sulphide transfer occurs in consortial relationships involving phototrophic bacteria and S-reducing bacteria [198]. Much higher cell densities are achieved by mixed cultures of this type than by either of the partners growing under similar conditions, presumably by minimizing loss of electron donors and acceptors from the system. It is possible that similar transfers of S occur between archaebacterial types resident in geothermal springs [56].

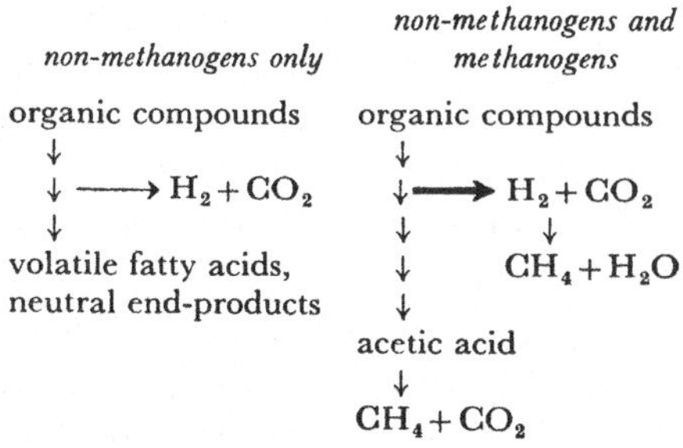

Fig. 10. The effect of interspecies H_2 transfer to methanogens on fermentation end products. From [197] with permission

Nitrogen-fixing Symbioses

The ability to reductively assimilate N_2 (*diazotrophy*) is confined to prokaryotes [81, 82]; eukaryotes and most prokaryotes depend on NO_3^-, NH_3 or organic N as N sources. In many natural undisturbed ecosystems these N sources are not growth-limiting, but ready access to biologically fixed N_2 has had sufficient survival value to encourage the evolution of diverse symbioses and less defined *biocoenoses* between diazotrophic prokaryotes and phototrophic eukaryotes. Quantitatively the symbioses are the dominant diazotrophic systems (see Sect. on Nitrogen Fixation) with the legume root nodules making an overwhelming contribution. Other diazotrophic symbioses of ecological or agronomic importance include the actinomycete induced root nodules of *Alnus* and other pioneering trees and shrubs and the association between *Anabaena* (cyanobacteria) and *Azolla*, an aquatic fern prominent in rice paddies. Identification of such symbioses has been greatly aided by the exploitation of the acetylene reduction assay for nitrogenase. The recent discovery of intracellular diazotrophic bacteria in oceanic diatoms [199] illustrates the versatility of the assay. General accounts of diazotrophic associations are provided by Sprent [81] and Postgate [82]; these include references to more specialized treatments.

Angiosperm root nodule symbioses are the most intimate, complex and successful diazotrophic associations and will serve as examples of mutualism between plants and chemoorganotrophic microorganisms. They are divided into the *Rhizobium*-legume and *Frankia*-non-legume symbioses. There are some 18,000 species of legumes, mostly tropical, with two-thirds in the sub-family Papilionoideae which contains most cultivated and temperate forms. Between 80% and 90% of the Papilionoideae are thought to be nodulated. The readily cultured bacterial partner, *Rhizobium*, is solely identified by its ability to nodulate legumes and in reality includes at least two disparate groups [200]. One group contains fast-growing, acid-producing rhizobia closely related to the plant tumour inducing genus *Agrobacterium*. Members of this group nodulate temperate crop and forage legumes including clovers (*Trifolium*), field peas (*Pisum*), and beans (*Phaseolus*) with considerable host specificity, but also occur in the tropics. The second group are slow-growing rhizobia, nodulating soybean (*Glycine max*), many tropical and sub-tropical legumes and temperate forms including lupins and some *Phaseolus* beans. Slow-growing rhizobia are relatively unspecific in their choice of hosts, some "cow-pea group" rhizobia even nodulate the non-legume *Parasponia cannabina* [201]. The division of rhizobia into cross-inoculation groups on the basis of host preference has been used to distinguish species. This has practical value but limited taxonomic validity as symbiotic properties are sometimes determined by transferable extrachromosomal plasmids [202].

Nitrogen-fixing root nodules have been identified in more than 120 non-leguminous plant species among 17 largely unrelated genera including *Alnus* (alders), *Myrica* (bog-myrtle), *Ceanothus*, and the tropical *Coriaria* and *Casuarina*. Many are trees or shrubs which colonize newly exposed, developing or wet soils. Their actinomycete partners were not axenically cultured before 1978 [203], but were earlier named *Frankia* after ultrastructural characterization. Strains differ in host range, nitrogen fixing ability, and serological characters and do not conform completely as host-specific species [204].

Nodule formation involves integration of plant and microbial activities to produce unique organs where photosynthate-driven prokaryotic N_2 fixation is coupled to efficient transfer of combined N to the plant. Mutual recognition is a key feature of nodulation as the diazotroph has to induce nodulation and achieve intracellular placement while circumventing the plant's normal defences against invasion. This recognition must remain operational as nodulation progresses from multiplication of prospective symbionts around young roots, through initial contact and prokaryotic invasion via root hairs or wounds, to induction of plant cell division and the proliferation, differentiation, and envelopment of the diazotrophs within the nodule cells. Mature legume nodules are novel, relatively short-lived organs with a central core of symbiotically-infected cells. Non-legume nodules are more persistent, modified lateral roots with peripheral symbiotic tissues around a central vascular axis. The genetic control of nodulation and the molecular basis for recognition is only partially known for a few crop legumes. Both plant and bacterial genomes determine the success of the process. Some fast-growing rhizobia carry the genes for nitrogen fixation (*nif*) and bacterial involvement in nodulation on plasmids but this is not universal [205]. Early recognition in clover (*Trifolium*) may be through root-hair surface *lectin* (polysaccharide-recognizing proteins) binding to compatible rhizobia [206]. Nodulation is not obligatory on legumes, being suppressed by NO_3^- additions, by pH extremes and other soil factors. Even when formed, nodules can be ineffective in nitrogen fixation.

Nitrogenase (see Sect. on Nitrogen Fixation) activity within the symbiotically infected tissues is confined to differentiated diazotroph cells, the fragile enlarged bacteroids of *Rhizobium* or globose vesicles on hyphae of *Frankia*. Both structures have been induced in vitro by succinate [207, 208], a likely photosynthate derivative in nodule cells. Rhizobial nitrogenase closely resembles that of free-living diazotrophs and there is no evidence for marked differences in *Frankia*. Once N_2 reduction is complete, rhizobia, like symbiotic cyanobacteria, rapidly export NH_4^+. Incorporation of NH_4^+, largely via the glutamine synthetase-glutamate synthetase (CS-GOGAT) pathway, is performed by the plant. Legume nodules additionally assimilate NH_4^+ into carriers with high N:C ratios such as asparagine, allantoin, and allantoic acid to facilitate transport to growing regions and avoid toxic levels of NH_3. Non-legume nodules similarly produce citrulline.

Nitrogen fixation in nodules is totally dependent on host photosynthate, which may be supplied to bacteroids as organic acids rather than sucrose or its monomers. The overall reaction for nitrogen reduction

$$N_2 + 12\,ATP + 6\,e^- + 8\,H^+ \rightarrow 2\,NH_4^+ + 12\,ADP + P_i$$

consumes the equivalent of 0.75 mol glucose for each mol N_2 reduced. NH_4^+ assimilation requires an equal amount of glucose mainly for carbon skeletons and the cost is further increased by H_2 efflux and other inefficiencies. Legume nodules receive up to about 30% of the plant's photosynthetically fixed carbon, use about a third of this for nitrogen reduction and return perhaps half as N carrier carbon skeletons; the residue is used for nodule growth and maintenance. As ATP generation is via oxidative phosphorylation the symbiosis has to provide an adequate O_2 supply while protecting nitrogenase from free O_2. Legumes co-operate with rhizobia to produce haemoglobins with high O_2 affinity. These permit rapid fluxes of bound O_2 to a suitable bacteroid terminal oxidase. Non-legume mechanisms are largely unknown apart from compartmentalization of nitrogenase in the vesicles, although haemoglobins have recently been reported [204].

The nodulated plant obviously benefits from the nitrogen provided by the symbiosis, regulating the diurnal and long term activity of the nodules to suit its needs. *Frankia* produces spore-like bodies within the nodules and presumably owes its survival to its non-legume hosts. The gains to rhizobia from their association with legumes remain enigmatic. It is unlikely that the osmotically sensitive bacteroids or residual rhizobia in the infection threads contribute much to the long term survival of populations. Possibly the true gainers are the rhizosphere rhizobia that should flourish around a healthy activity growing root system [202].

Mycorrhiza

Plant roots are invaded or invested by a diversity of fungi to produce the composite mycorrhiza (= "fungus-root"). *Ectomycorrhizae* generally have a sheath of fungal tissue around a modified feeder root and its apex; *endomycorrhizae* have largely unmodified primary roots penetrated in their outer cortical zone by intercellular and intracellular hyphae. Both major mycorrhizal forms often have hyphae extending far into the surrounding soil. These external hyphae apparently function as substitute root hairs, although grasses can have both root hairs and endomycorrhizae, while some ectomycorrhizae lack external hyphae.

Three mycorrhizal types have major ecological significance. In order of importance these are vesicular-arbuscular mycorrhizae, ectomycorrhizae (sheathing mycorrhizae) and ericoid mycorrhizae [210]. The first and last are endomycorrhizae. Vesicular-arbuscular mycorrhizae, named after their swollen or much branched intracellular hyphae, occur on the underground parts of mosses and ferns, the roots of grasses, tropical forest trees, temperate fruit trees, and many other angiosperms. Unusual and so far non-cultivable zygomycetes belonging to the Endogonaceae are the obligately biotrophic fungal partner in this ancient symbiosis, including the genera *Glomus* and *Acaulospora*. Ectomycorrhizae, the first recognized mycorrhizal symbiosis, form mainly on temperate or boreal conifers and deciduous trees. Participating fungi include basidiomycetes (*Amanita, Boletus, Suillus, Lactarius, Russula, Pisolithus, Rhizopogon,* etc.), ascomycetes (*Tuber, Terfezia,* etc.), and imperfect fungi (*Cenococcum*) [210, 211]. These symbionts vary considerably in host specificity and capacity for axenic growth. Ericoid mycorrhizae, as the name implies, are formed on the fine hair roots of heathers and other members of the Ericaceae by imperfect fungi or ascomycetes (*Pezizella*), and are important on heathlands, moors, and in the tundra.

Experimental and some field observations demonstrate that mycorrhizal plants are larger and better developed than non-mycorrhizal plants of the same species. The reasons for this are diverse and the picture is complicated as not all susceptible plants become mycorrhizal while infection is not always beneficial [212]. Mycorrhizal plants can have enhanced P, N, K contents [210, 213] and may acquire water [214], other minerals, growth factors and protection against pathogens from their fungal partners [210, 215]. The wide distribution of mycorrhizal symbioses [211] implies that they are thus responsible for a large, but at present unmeasurable, part of mineral exchanges between plants and soils.

Studies on mycorrhizal function have largely followed P uptake [212]. Vesicular-arbuscular and ectomycorrhizal fungi accumulate P_i from soil, storing, and translocating P in hyphae as polyphosphate [216, 217] and actively transfer P_i from their internal pools to the plant. Whether mycorrhizal fungi mobilize non-labile P sources normally unavailable to plants remains an open question [212]. The increased P flux into mycorrhizal roots may relate to the large soil volume contacted by the external hyphae. Ericoid mycorrhizae are more important as suppliers of N which the fungi scavenge as NH_4^+ in the acidic soils typically colonized by the Ericaceae [218]. Host photosynthate usually supplies the growth and energy requirements of mycorrhizal fungi which act as photosynthate sinks by converting it into forms unavailable to the plant [219]. Ectomycorrhizal fungi convert soluble carbohydrates from the plant into trehalose and mannitol; vesicular-arbuscular mycorrhizae probably employ lipids as their photosynthate sink [220].

Lichens

Lichens represent ecologically obligate symbioses between their chemoorganotrophic mycobionts (fungi) and the phototrophic phycobionts (cyanobacteria and eukaryotic algae) [221, 222]. The fungus is the morphologically dominant partner. Phototrophs may occupy $<10\%$ of the thallus volume although present at more than 10^8 cells g^{-1} dry weight [223]. Lichens are principally identified by mycobiont characters and chemotaxonomy. As about half of all known ascomycetes are lichenised they are the dominant mycobionts among the 18,000 lichen species. There are a few genera of imperfect fungi and about 20 species of basidiomycetes. The majority of the phycobionts are green algae (Chlorophyceae) including *Trebouxia* (largely confined to lichens), *Coccomyxa, Myrmecium,* and *Trentepholia.* About 8% of lichens contain cyanobacteria, mainly *Nostoc* or *Scytonema,* either exclusively or in parts of the thallus separate from the green algae. Most lichens have an ordered arrangement of fungal tissues with an upper compact cortex, a medulla of loosely interwoven hyphae and a lower cortex. The algae or cyanobacteria are usually confined to a narrow zone between the upper cortex and the medulla. Lichen morphologies range from flat crustose forms, which may be endolithic (*Lecidea*) or more superficial (*Perfusaria, Rhizocarpon, Lecanora*), through leaf-like foliose types (*Peltigera, Lobaria, Parmelia, Xanthoria*) to fruticose filamentous, strap-shaped or finger-like forms (*Usnea, Ramalina, Thamnolia*). Superficially they appear to be plants and are usually considered primary producers although this may not be an adequate reflection of their physiological behaviour.

Phycobiont photosynthesis is undisputably a key feature of the symbioses. When in physical contact with the fungus the phycobiont C assimilation processes are modified so that as much as 90% of their photosynthate [224] is exported to the mycobiont as single symbiosis-specific carbohydrates. Cyanobacteria export glucose, green algae release ribitol or sorbitol or erythritol [225]. As in mycorrhizae the fungus acts as a photosynthate sink, but the proportion utilized for growth is remarkably low, perhaps $<10\%$ [224, 226], with the bulk entering fungal polyol pools [227]. The reasons for this massive diversion of a potential carbon source lie in the physiological ecology of lichens. Most are extremely slow growing (radial expansion rates of c 1 mm y^{-1} or less) and characteristically inhabit rock, soil, tree bark, and other surfaces subject to cycles of wetting and drying. Lichens lack an impervious cuticle and readily gain or lose water, subjecting their cells to periodic exposure to low a_w. They accordingly need the protection provided by the high solute concentration of their internal mannitol or arabitol pools which may be near saturation [227]. Rewetting also imposes a demand to replace losses caused by metabolite leakage through damaged membranes and enhanced respiration during the process [228]. Thus, most of the phycobiont photosynthate is employed to provide "physiological buffering" [224] that permits lichen survival in harsh environments. Lichens may be viewed as fungi with inherently slow growth rates whose symbiosis with algae provides the latter with an ecological niche and enables both organisms to survive even in environments that are too severe for most living things.

Despite their slow growth lichens probably usefully contribute to primary production in tundra, hot and cold deserts and other stressful environments. Those containing cyanobacteria obtain a supply of biologically fixed N (see Sect. on Nitrogen Fixation). Other facets of lichen biology of biogeochemical relevance are their propensity to accumulate metal cations [229], and their putative role in soil formation [230]. Lichen thalli have been found to contain up to 93,400 ppm Zn, 78,000 ppm Fe, 3,000 ppm Cu, and 3,000 ppm Ni [229] when growing on mine or metallurgical wastes or exposed to smelter emissions. Arctic lichens achieved notoriety necause their acquisition of ^{90}Sr, ^{137}Cs, and other radionuclides from atmospheric nuclear weapons tests exposed man and other predators feeding on caribou to abnormal radiation levels [229]. Thus both underlying soils or minerals and the atmosphere can act as cation sources. The chief metal cation uptake mechanism seems to be ion-exchange reactions modified by metal complexing [231], involving extracellular sites in the fungal walls or the surrounding matrix [232]. Little of the metals enters the cells. Lichen substances, the largely symbiosis-specific depsides, depsidones, and other phenolics do not seem to participate [233], despite being potential chelating agents. This might be due to their low solubility in water, a property that might also restrict their role in mineral breakdown during soil formation.

Linnaeus was one of the first to suggest that lichens aided soil formation [230]. This relates to the ability of lichens to cover rock or soil-less surfaces in areas where plant growth is severely handicapped. Endolithic and other rock-colonizing lichens undoubtedly make a slow contribution to weathering, perhaps through the agencies of lichen substances, oxalic acid or water retention. Lichen crusts in the tundra and deserts serve as foci for the accumulation of organic debris. Lichens are not, however, the first colonizers of newly exposed volcanic soils; this and their slow distribution and growth casts doubt on their pioneering role in pedogenesis.

Lichens have attracted attention as monitors of environmental pollution [234] since many are extremely sensitive to SO_2. Metal accumulation by lichens can also be used to assess specific atmospheric pollution by smelters and refineries [233].

Associations with Animals

Detailed discussion of the majority of microbial associations with animals is beyond the scope of this chapter as many are not biogeochemically significant. The interested reader can discover further details of microbial invasion of man and animals in Isenberg & Balows [235] and Linton [236]. Persistent microbial associations with other organisms which are important environmentally are located in invertebrate and vertebrate alimentary tracts. Gut microbiota play a significant role in degradation of plant residues. Ecologically important associations include the rumens of herbivores such as cattle, sheep, goats, antelopes, deer, and camels [40–42] and the hindguts of lower termites such as *Reticulotermes* [237]. Rumen environments are predominantly anoxic and the breakdown of ingested plant residues to organic acids, CO_2, and CH_4 is dependent on consortia of strictly anaerobic cellulolytic, hemicellulolytic, pectolytic acetogenic, and hydrogenogenic bacteria along with methanogens. Other rumen inhabitants include an experimentally dispensable although large population of ciliates [41, 42], and unusual anaerobic chytrid-like fungi [238, 239]. Lower termites depend on hindgut anaerobic protozoa such as *Trichomitopsis* and *Trichonympha* to digest cellulose and produce acetate that they absorb [237]. By-products include CO_2 and H_2, and, as in the rumen, H_2 may be converted via interspecies H_2 transfer through methanogens to CH_4. The hindgut bacteria, unlike those in the rumen, do not seem to be directly implicated in cellulolysis but may be important to termite survival as sources of biologically fixed N. The rumen is clearly the most ecologically significant association between animals and microorganisms but the remarkable associations between chemolithotrophic sulphur-oxidizing bacteria and marine animals (see Sect. on S-Oxidation) may well prove to be widespread and of much greater significance than presently recognized.

Microorganisms as Geochemical Determinants

The Carbon Cycle

General Aspects and Primary Productivity

Global aspects of the carbon cycle have been reviewed extensively in recent years [240–242] including elsewhere in this series [243] and these reviews should be consulted for detailed information or quantitative aspects. There are two parts to the carbon cycle; the *geological* carbon cycle is concerned with the transfer of CO_2 between the atmosphere, the oceans and the enormous reservoirs bound as carbonates in limestones and dolomites [243]. Microorganisms are, of course, indirectly concerned with the precipitation and solubilization of carbonates (See Sect. on Transformation of other elements), but here we concentrate on the *biological* cycling of carbon between CO_2 and biomass, concentrating on the degradative aspects.

Organisms deriving carbon from the reductive assimilation of CO_2 are the primary producers upon which all other trophic levels depend [83]. Global primary production has been estimated at about 7.6×10^4 Tg y^{-1}, the contribution from terrestrial sources being twice that of the marine environment [243]. Vascular land plants undoubtedly make the major contribution to global primary production but microorganisms also play a significant part since planktonic microalgae contribute about 35% of global productivity [175]. The part played by prokaryotic productivity is difficult to quantify, in particular the contribution from cyanobacteria remains unknown [244]. It is possible that the cyanobacterial contribution is more significant than supposed in both terrestrial and marine environments. Chemolithotrophic primary production is probably globally insignificant, although locally it can be important. Similarly, the contribution of anoxygenic phototrophic bacteria is likely to be only of localised significance, confined to certain anoxic stratified lakes [245].

The bulk of primary production becomes partitioned between herbivores and *decomposers*. The latter group consists of chemoorganotrophic microorganisms and invertebrate animals that utilize dead organisms and their residues (*detritus*) as carbon and energy sources. On land only 5%–10% of net primary production is consumed by herbivores, the rest being transformed by decomposers [246]. The relatively small proportion of terrestrial primary production consumed by herbivores is a consequence of the recalcitrance of the tough lignocellulose cell walls of multicellular land plants. In marine environments grazing accounts for the bulk of the far less recalcitrant primary production [164, 168].

Decomposers frequently encounter their potential metabolic substrates in the form of whole or fragmented animal corpses, plant roots, leaves, fruits, branches, and microbial cells. Ecologists studying decomposition prefer to call such complex structured entities *resources,* using substrate only in the enzymic sense [246, 247]. As most intracellular macromolecules are relatively labile, the rate limiting step in the biological carbon cycle is the depolymerization of the structural polymers of plant and microbial cell walls and arthropod exoskeletons.

Under certain incompletely understood circumstances, the decomposition of organic carbon is arrested and carbon escapes from the biological carbon cycle as the humic fraction of soils, kerogen in sediments, and as the precursors of fossil fuels. An important contributory factor in this process is the development of anoxic conditions. Such persistent residues exceed the living biomass carbon by many hundred-fold. On re-exposure to oxygen these residues may again become accessible to microbial attack. However, the long-term accumulation of reduced carbon has contributed towards the maintenance of an oxygen atmosphere which would rapidly be consumed if all fossil carbon was reconverted to CO_2.

Decomposition of Cellulose and Lignin

Cellulose and lignin are probably the first and second most abundant forms of cellular C in the biosphere [248], although the prevalence of chitin, especially in the oceans, should not be overlooked. The decomposition of the lignocellulose complex in plant cell walls is thus critical to the operation of the present day biological C cycle. Lignin, and to some extent cellulose, also make major contributions to the soil organic fraction and to peat and its derived fossil fuels. The mechanisms and organisms involved in the degradation and transformation of the lignocellulose complex have increasingly attracted the attention of biotechnologists in view of the ready availability of such a potentially valuable carbon and energy source. There is little doubt that the fungi make the largest contribution to lignocellulose decomposition. Detailed accounts of cellulose breakdown are provided by Eriksson [249] and Burns [250] while lignin degradation has been extensively dealt with by Kirk et al. [251], Crawford [252], and Zeikus [248].

The structure of individual cellulose molecules (Fig. 11) consisting of D-glucose monomers and regular β 1–4 glycosidic linkages does not suggest particular resistance to depolymerization despite molecular weights $>1.5 \times 10^6$. The recalcitrance of cellulose largely arises from the linear assemblage of the molecules into microfibrils and the microfibrils into fibrils with tightly packed crystalline regions. The resultant structure restricts the access of both the cellulolytic organisms and their depolymerases. Lignin (Fig. 12) is markedly different in structure, consisting of cinnamyl alcohol-derived p-hydroxyphenylpropane monomers joined by heterogeneous C–C and C–O–C linkages. As the final random assemblage of monomers is achieved by non-enzymic dehydrogenative polymerization, lignin is refractory to most hydrolytic depolymerases active on enzymically produced polymers [248, 252].

Fungal mechanisms for cellulose breakdown are better characterized than those of bacteria and probably differ in detail. In both the imperfect fungal genus *Trichoderma* and in the basidiomycete *Phanerochaete chrysosporium* (= *Sporotrichum pulverulentum*) cellulolysis is effected by synergistic batteries of highly specific enzymes which carry out chain breakage in the intact fibrils, exposing reduc-

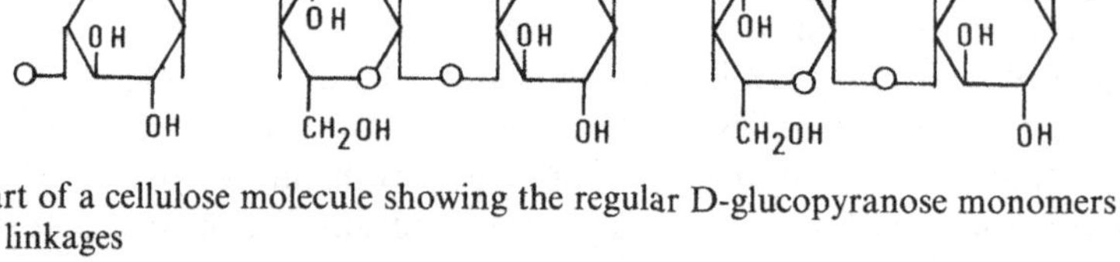

Fig. 11. Part of a cellulose molecule showing the regular D-glucopyranose monomers and β1–4 glucosidic linkages

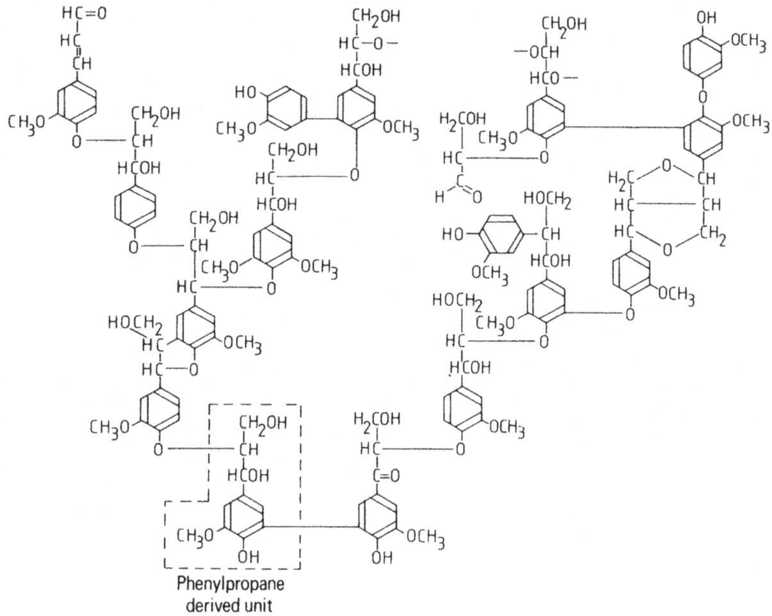

Fig. 12. Part of a lignin molecule illustrating the variety of linkages between phenylpropane units. From [4] with permission

ing and non-reducing ends of individual cellulose molecules, releasing oligomers and eventually glucose monomers by hydrolytic cleavage of the β 1–4 glycosidic linkages. These cellulolytic systems are induced by dimers such as cellobiose and are subject to end-product inhibition. Their complexity is indicated by the considerable number of different glucanases detected in cellulolytic *P. chrysosporium* and *Trichoderma* spp. [250]. Bacterial systems probably have fewer enzyme species [250]. Brown-rot basidiomycetes may initiate cellulose breakdown by non-enzymic attack using free radicals generated from H_2O_2 in the presence of Fe^{2+} [249, 253, 254]. This mechanism is O_2-dependant but most cellulolytic systems are not and can be possessed by both aerobic and anaerobic microorganisms.

The study of lignin degradation has largely depended on the use of synthetic lignins (dehydrogenative polymerizates = DHP's) or uniformly-labelled plant lignins as enzyme substrates together with the careful chemical analyses of naturally degraded lignins. By 1980, studies with the thermotrophic white-rot basidiomycete *Phanerochaete chrysosporium* had demonstrated that the lignolytic system was O_2 dependant, appeared in response to N depletion and was dependant on the presence of other carbon and energy sources, features associated with fungal secondary metabolism. Early models suggested *in situ* ring cleavage as part of the mechanism [255]. However, later studies with *P. chrysosporium* have shown the breakage of inter-monomer C–C bonds and these studies of Tien and Kirk [256] indicate the presence of a single, oxidative, non-specific, H_2O_2-requiring lignolytic enzyme (MW ~ 42,000) capable of degrading at least two representative model lignin structures (Fig. 13) as well as spruce and birch lignins. Since this sys-

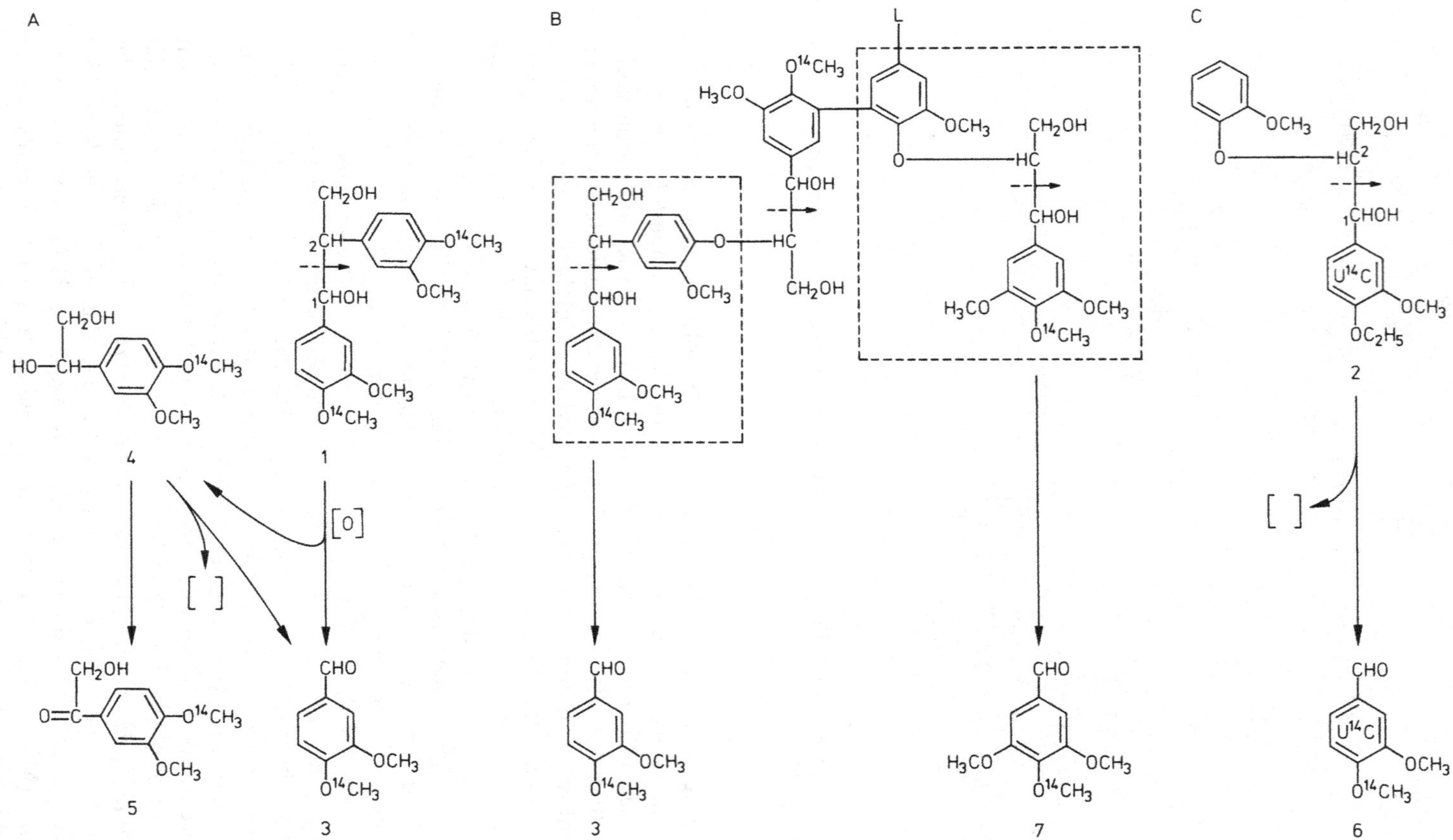

Fig. 13. Lignin biodegradation. Cleavage of C-C intermonomeric bonds at the arrowed positions by H_2O_2 and a lignolytic enzyme from *Phanerochaete chrysosporium*. The breakdown of ^{14}C-labelled natural lignin is shown in B where the dashed boxes indicate the substructures present in model compounds. The left hand box and compound [1] represent 1,2 diarylpropane whose breakdown is shown in A. The right hand box and compound [2] represent arylglycerol-β-aryl ether whose breakdown is shown in C. End products include vanillin methyl ether [3], vanillin ethyl ether [6], and syringaldehyde methyl ether [7]; the ketol [5] is a minor product, while unidentified products are shown by empty brackets. From [256] with permission

tem is presumably confined to aerobes, this does much to explain the persistence of lignin residues in the biosphere.

The white-rot fungi, mainly basidiomycetes, are pre-eminent in wood decay [257] because they possess both cellulolytic and lignolytic systems, *Coriolus versicolor* being able to cause an almost total loss of dry weight in test wood blocks. Brown-rot basidiomycetes produce a dispersed attack on cellulose but modify rather than depolymerize lignin. Many ascomycetes, imperfect fungi, oomycetes, and chytrids are cellulolytic. The chytrids include interesting obligately anaerobic fungi found in the rumens or caeca of herbivorous mammals [238, 258]. Cellulolytic aerobic bacteria include members of the genera *Bacillus, Cellulomonas, Cytophaga, Erwinia, Pseudomonas,* and *Sporocytophaga* as well as several genera of actinomycetes. Anaerobic cellulose decomposers are found among the genera *Clostridium, Ruminococcus*, and *Bacteroides*. The question of whether there are truly lignolytic bacteria has only been recently settled by use of ^{14}C labelled lignins [252]. Isolates of lignolytic bacteria include strains of *Streptomyces, Nocardia, Bacillus,* and *Pseudomonas,* but none of these bacterial isolates are as active as white-rot fungi in lignolysis.

Co-Metabolism

Co-metabolic transformations may help determine the fates of many relatively recalcitrant organic compounds within the biosphere. The term *co-metabolism* has recently been redefined [259, 260] as "Co-metabolism – the transformation of a *non-growth substrate* in the obligate presence of a growth substrate or another transformable compound." Non-growth substrates are characterized as substances that do *not* support cell division (as opposed to an increase in biomass). The key to an organism's ability to participate in co-metabolic reactions lies in the possession of enzymes with dual or multiple specificity.

Stirling and Dalton's [259] redefinition of co-metabolism was largely founded on studies with the methylotroph. *Methylococcus capsulatus* which progressively oxidizes methane (CH_4) to an assimilable C_1 form, formaldehyde (HCHO). The initial oxidation of methane to methanol (CH_3OH) is mediated by a NADH-linked *methane monooxygenase* (Fig. 14) which in cell-free extracts is capable of oxidizing 31 non-growth substances. This remarkably broad substrate specificity permits whole cells to co-metabolically oxidize (*co-oxidize*) at least 7 non-growth substances, showing the greatest activity on carbon monoxide (CO), considered a competitive inhibitor of bacterial methane oxidation. Thus, when methane is limited considerable oxidation of CO to CO_2 is possible. This co-metabolic activity of methylotrophic bacteria may account for the bulk of atmospheric CO oxidation [261] in view of the unknown but believed small contribution of the CO-consuming carboxydobacteria [262].

Reports of co-oxidative transformations have dominated the literature ever since Leadbetter and Foster's [263] original discovery of ethane oxidation by an obligate methane-oxidizer. Industrial and environmental considerations have concentrated attention on the importance of microbial degradation of hydrocarbons and their derivatives [264]. The most detailed information about co-oxidation has been obtained from pure culture studies in the laboratory (Table 15). Cir-

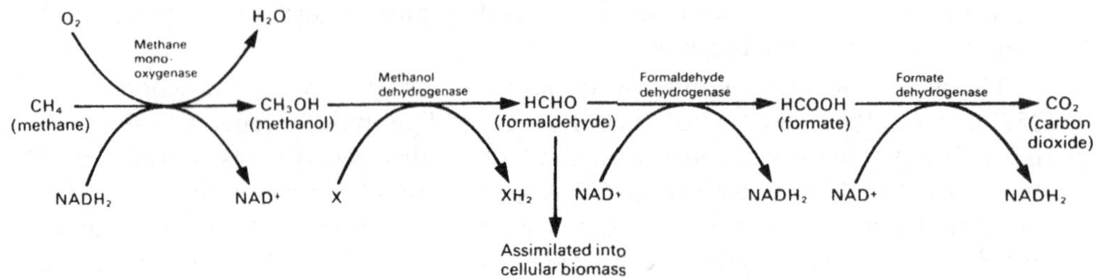

Fig. 14. Microbial oxidation of methane. The initial enzyme, methane mono-oxygenase, can participate in co-metabolic transformations. From [260] with permission

Table 15. Examples of microbial co-metabolism[a]

Organism	Growth Substrate	Co-metabolized Substrate	Co-metabolic products
Methylomonas methanica	Methane	Ethane	Ethanol, acetaldehyde, acetate
Nocardia salmonicolor	n-Hexadecane	Toluene	2,3-Di-hydroxybenzoic acid and α-Methyl muconic acid
Nocardia salmonicolor	n-Hexadecane	Ethylbenzene	Phenylacetic acid
Nocardia corallina	n-Hexadecane	Benzene	Muconic acid
Nocardia corallina	n-Hexadecane	p-Chlorotoluene	p-Chlorotoluic acid
Nocardia corallina	n-Hexadecane	2,7-dimethyl-napthalene	7-Methyl-2-napthoic acid
Corynebacterium sp.	n-hexadecane	Napthalene	Salicyclic acid
Corynebacterium sp.	Glucose	Anthracene	2-Hydroxy-3-napthoic acid
Pseudomonas sp.	Glucose	Limonene	Perillic acid, perillyl alcohol
Mycobacterium sp.	Propane	Cyclohexane	Cyclohexanone
Candida cloacae	Acetate	n-Dodecane	1,12-Dodecanoic acid
Candida cloacae	Acetate	n-Tridecane	1,13-Tridecanoic acid
Candida cloacae	Acetate	n-Hexadecane	1,16-Hexadecanoic acid

[a] Compiled from [260], [264]

cumstantial evidence [265, 266] suggests similar processes in nature and a significance role for methane oxidizers has been postulated [267]. Other bacteria performing co-oxidations include members of the genera *Mycobacterium, Nocardia, Pseudomonas, Corynebacterium, Bacillus,* and *Micrococcus* [260, 264]. Yeasts (*Candida*) have also been implicated and there is no reason to exclude filamentous fungi from this activity.

It is likely that microbial co-metabolic modifications of recalcitrant natural molecules do occur although our knowledge of this area is fragmentary. It can be argued that microbial generation of peroxide (H_2O_2) radicals in cellulose and lignin degradation could constitute a co-metabolic phenomenon.

The evolutionary advantages of co-metabolic capabilities are uncertain [260], although this activity might aid its possessor's survival as part of a synergistic community.

Escapes from the Biological Carbon Cycle

Incompletely oxidized organic C accumulates in the biosphere as soil *humus* (*heterocondensate*), humus analogs in aquatic environments and as peats. Some of these pools of undecomposed reduced C are matured by diagenetic and catagenetic processes to form fossil fuels such as crude oils, natural gases, and coal. Even though the global reserves of fossil organic C are vast, their production only demands a small fraction ($<1\%$) of annual net primary production. Microbiologists face the question "why has decomposition been arrested?" The answer lies partly in the structures of the long-lived organic residues and partly in the nature of the environments that permit their accumulation.

Soils and sediments contain appreciable quantities of proteins, carbohydrates, lipids, and other cell constituents which have been protected from rapid decomposition by bonding to clays, metals or pre-existing humic substances [113, 268]. The humic substances are the longest-surviving soil organic fraction, being up to 7×10^3 y old in some temperate and tropical soils [269]. Both soil humus and the humus-like components of the water column and sediments can originate under oxic conditions where depolymerization and oxidation are possible and accumulation may occur due to microbial failure to metabolize depolymerization products. This seems particularly true of the soil humus complex which is largely based on phenolic units derived from lignin [269]. Some of the aromatic residues in humus may also be of microbial origin and its structure is further complicated by physical or chemical bonding to proteins, carbohydrates, or other biological molecules. Humus recalcitrance thus derives from its aromatic content, variable cross-linkages and bonding to clay minerals and metals [104]. Sedimentary humus-like materials have a similarly bewildering multiplicity of units and cross-linkages, but differ from soil humus in being largely aliphatic [268].

Sedimentary organic complexes may be one of the sources of *kerogens,* the insoluble, amorphous, ill-defined organic residues in rocks of all ages [270]. Kerogens are of interest as possible sources of catagenically derived hydrocarbons. Recent analogs of kerogen have been described under cyanobacterial mats in Baja California [271]. Kerogens and more mobile precursors of crude oils need to be present at 0.5%–5% of sediment weight, to be potential sources of petroleum [272]. Low concentrations of kerogen can accumulate in anoxic sediments beneath oxygenated waters, but significant concentrations additionally require anoxic conditions in the overlying water column. Appropriate anoxic conditions are relatively rare in present day marine systems but do occur in the Black Sea, the Cariaco Trench and some fjords with shallow sills. One postulated scheme for the production of some oil-producing sediments supposes the injection of hypersaline water from adjacent aquifers into the lower parts of lakes or enclosed marine basins [273]. This would produce marked chemical stratification with a lower salt-rich anoxic region where lipids, waxes, and other hydrocarbon precursors would be conserved. Debris from anaerobic bacteria below the chemocline would contribute significantly to the crude oil precursors in the underlying sediments.

At present the extent to which water column or sediment microbes modify or add to sediment hydrocarbons is unknown [274].

Peat formation is another process dependent on the development of anoxic conditions. The detailed causes of peat formation remain obscure [275], although a general requirement is the input of terrestrial or coastal plant debris to waterlogged environments. Developing peats have active populations of microbes at their surfaces but there is a rapid decline in microbial activity, particularly of fungi and other aerobes, with depth [275]. This may reflect the decline in E_h with depth but other contributory factors include inorganic nutrient limitation and chemical inhibition. Whatever the causes, the consequences are that plant tissues, structures, and biopolymers are conserved, often in macroscopically recognizable forms. Peats can have, on a dry weight basis, up to 20% lignin, 30% cellulose, and 36% hemicellulose [275]. An obvious consequence of O_2 deficiency is the cessation of lignin decomposition and the conservation of other associated plant cell wall polymers. Other plant phenolics may also be conserved, notably tannins, and their presence may account for the preservation of 2,000 year old human corpses in Danish bogs. Decomposition inhibitors, if present, must be relatively labile as peats re-exposed to air are rapidly decomposed [276].

Methanogenesis and Methylotrophy

Methane (CH_4) production (*methanogenesis*) and growth on methane or other R–CH_3 compounds lacking carbon-carbon bonds (*methylotrophy*) are distinctive and important microbial contributions to the biological carbon cycle. About 5% of the annual global total of photosynthetically-fixed carbon is eventually converted to methane as an end product of anaerobic decomposition by methanogenic bacteria [277]. Conversely, much of the 10^3 Tg methane produced annually [277] is oxidized and assimilated by methylotrophic bacteria on its release to oxic environments. Recent books by Anthony [278] and Large [279] should be consulted for detailed information on the biology and biochemistry of these processes. Zeikus [280] also includes a very useful treatment of methanogenesis in a broad review of anaerobic C_1 metabolism. Carbon monoxide (CO) constitutes another potentially oxidizable C_1 compound present in significant amounts in the environment. Microbial oxidation of CO includes both direct oxidation by certain hydrogen-oxidizing bacteria [281] and co-metabolic oxidation by methane-oxidizing bacteria [261, 162].

Methanogenesis is the exclusive property of a taxonomically diverse group of strictly anaerobic archaebacteria, the *methanogens*. In natural environments these unusual prokaryotes obtain energy largely by reducing CO_2 at the expense of hydrogen, thus:

$$CO_2 + H_2 \rightarrow CH_4 + H_2O; \ \Delta G_\theta = -131 \text{ kJ mol } CH_4^{-1}$$

or by a unique aceticlastic reaction:

$$CH_3\,COO\,H \rightarrow CH_4 + CO_2; \ \Delta G_\theta = -31 \text{ kJ mol } CH_4^{-1}.$$

Reduction of CO_2 at the expense of the oxidation of H_2 is the predominant methanogenic route in the rumen and other regions of alimentary tracts, yielding $\simeq$ 1 ATP for each CH_4. A possible pathway [278] for this process is shown in

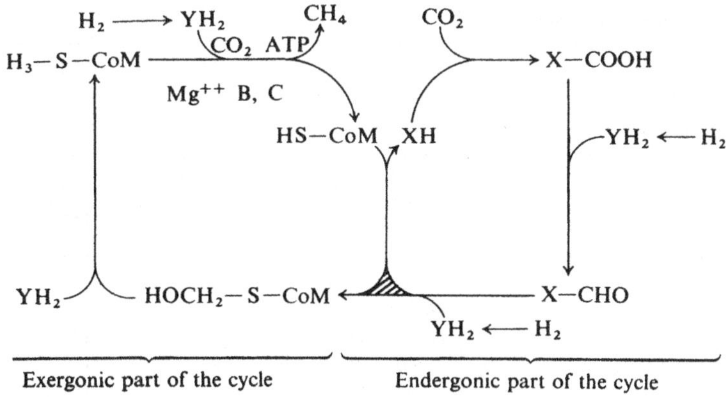

Fig. 15. A scheme for the reduction of CO_2 to methane by methanogenic bacteria. CO_2 is bound to an unknown C_1 carrier X on which it is reduced to an aldehyde. The C_1 unit is then transferred to the methyl carrier coenzyme-M (CoM = 2-mercaptoethane sulphate) on which reduction to methane is completed. The final reduction involves a methyl-reductase protein (C) which is activated by CO_2, ATP, Mg^{++}, and compound B. For thermodynamic reasons the reductant YH_2 cannot be the same in all cases. Further details are in [278] from which this figure is reproduced, with permission

Fig. 15. Since the E_0' values of redox couples involving CO_2 are very electronegative (Fig. 3), the initial stages of methanogenesis with CO_2 as an electron acceptor must presumably involve an energy input powered by energy-generating terminal stages [44]. Evidence is now beginning to emerge for CO as an enzyme-bound intermediate in the reductive process [282].

Acetate is the main source of methane in sediments, swamps, sludge digesters, and most waterlogged environments. The details of the aceticlastic mechanism and ATP yield are unknown.

The methanogens obtain their cellular carbon largely by reductive CO_2 fixation and, hence, can be regarded as autotrophs. Carbon assimilation has only been studied in detail in *Methanobacterium autotrophicum* and *Methanosarcina barkeri,* and apparently follows novel incorporation routes distinct from any known in other prokaryotes, probably involving condensation into acetyl CoA [278, 280]. Assimilative CO_2 reduction is thought to occur on C_1 carriers, possibly corrinoids [283] which occur in high concentrations in methanogens, and again perhaps with bound carbon monoxide (CO) as an intermediate [280, 282].

Currently recognized genera of methanogens are listed in Table 16 [284]. It is likely that current taxonomic criteria do not adequately separate physiologically distinct isolates [280].

Methanogens are active only at redox potentials more negative than −330 mV. Their main habitats include aquatic sediments, swamps, waterlogged soils, and the anoxic regions of intestinal tracts such as the rumen. It is likely that novel forms are to be discovered in surface and submarine geothermal springs [55, 285]. Other methanogens occur as natural or introduced inhabitants of ground

Table 16. Taxonomy of methanogenic bacteria.

Order	Genus	Number of species
Methanobacteriales	*Methanobacterium*	(3)
	Methanobrevibacter	(4)
	Methanothermus	(1)
Methanococcales	*Methanococcus*	(5)
Methanomicrobiales	*Methanomicrobium*	(2)
	Methanogenium	(4)
	Methanospirillum	(1)
	Methanoplanus	(1)
	Methanosarcina	(3)
	Methanococcoides	(1)
	Methanothrix	(1)
Order unassigned	*Methanolobus*	(1)

[a] Modified from [284]

waters associated with oil or gas deposits [286]. Through their participation in acetate utilization and interspecies H_2 transfer [196, 197], methanogens play a key role in anaerobic decomposition both through regulation of proton and electron availability and because they, along with some of the sulphate-reducing bacteria are the only anaerobes capable of the complete breakdown of acetate under anoxic conditions.

The hydrogenogenic partners of methanogens in natural environments include many different anaerobes able to use protons as an electron sink when the partial pressure of H_2 is kept low by methanogenic activity, but also include the more specific obligate proton-reducing acetate-producing forms such as *Syntrophomonas wolfei* and *Syntrophobacter wolinii* [197]. Sulphate-reducing bacteria can also act as proton-reducing partners under certain conditions, but more commonly compete with methanogens for H_2 and acetate. The latter activity does not necessarily mean that sulphide and methane production are mutually exclusive [280], but in the main, methanogenesis occurs below the sulphate-reducing zone in sediments and is more significant in freshwaters than in sulphate-rich marine environments.

Methylotrophs constitute a less cohesive physiological group and include both prokaryotes and eukaryotes [278]. The latter are mainly yeasts capable of growing on methanol. Methylotrophic bacteria are most important in natural environments, participating in all or some of the sequence of oxidations shown in Fig. 14. There are two major groups defined by the ability or inability to grow on methane. The *methanotrophs* possess methane monooxygenase, unlike methylotrophs which can only grow on methanol or other R–CH_3 substrates (Table 17). Most members from both groups assimilate one-carbon fragments as formaldehyde (HCHO) via either the ribulose-monophosphate (RuMP) or serine pathways, although a few exceptions oxidize methanol to CO_2 and then assimilate it using ribulose diphosphate carboxylase [278].

Table 17. Growth substrates for methylotrophs [a]

One carbon substrates		Substrates without C–C bonds	
CH_4	Methane	$(CH_3)_2O$	Dimethylether
CH_3OH	Methanol	$(CH_3)_2NH$	Dimethylamine
CH_2NH_2	Methylamine	$(CH_3)_3N$	Trimethylamine
HCHO	Formaldehyde	$(CH_3)_4N^+$	Tetramethylammonium
HCOOH	Formate	$(CH_3)_3NO$	Trimethylamine N-oxide
$HCONH_2$	Formamide	$(CH_3)_3S^+$	Trimethylsulphonium
CO	Carbon monoxide		

[a] After [278]

The current taxonomy of the methanotrophs is in a state of flux [287] with only the genera *Methylococcus* and *Methylomonas* formally recognized although the genera *Methylosinus* and *Methylovibrio* have been described. Methylotrophs include obligate methanol utilizers such as the industrial strain "*Methylophilus*" grown for animal feed protein. There are also facultative methylotrophs, including atypical pseudomonads and members of the stalked, budding genus *Hyphomicrobium*. The hyphomicrobia are a widespread and unusual aquatic group growing only on C_1 compounds and other carbon sources with less than 3 carbon atoms. Since hyphomicrobia can oxidize methanol using NO_3^- as an electron acceptor they can grow in anoxic environments closed to other methylotrophs.

Methylotrophic bacteria are probably widely distributed in nature but information about their occurrence is largely confined to methane oxidizers [288]. These are found in soil, natural waters, and sediments, even when methane is not detectable. Various reports describe anaerobic production of CO_2 from methane [288] but this may merely reflect exchange reactions involving the C_1 carriers of methanogens [280]; overall, methane is conserved in anoxic environments.

Microbial metabolism plays a significant part in the oxidation of carbon monoxide (CO), a potentially toxic atmosphere pollutant. The atmosphere contains some 4.9×10^2 Tg CO from anthropogenic and natural sources, with a residence time of 0.1 y^{-1} [243]. Soils alone may consume 5×10^2 Tg CO y^{-1} through microbial activities [289]. Biological CO oxidation in natural environments is brought about by *carboxydobacteria* and by co-metabolic oxidations by methanotrophs and other microorganisms with broad specificity oxygenases [262]. The carboxydobacteria are capable of chemolithotrophic growth with CO as the sole carbon and energy source, oxidizing CO to CO_2 and fixing 4%–16% of the CO_2 into cell carbon via the Calvin-Benson cycle [281]. They are, in fact, aerobic hydrogen-oxidizing bacteria, growing better on H_2/CO_2 than on CO, but distinguished from other members of the group by the possession of an inducible carbon monoxide oxido-reductase [281]. So far six species have been isolated, four belonging to the genus *Pseudomonas* [281]. It is unlikely that these versatile organisms account for much CO oxidation in soils since kinetic studies on pure cultures [262] have led to the conclusion that they would have to exist in populations of 10^8–10^9 g^{-1} soil in order to match observed CO consumption rates. This population size is extremely unlikely, and the co-metabolic route may thus be the dominant biological means of CO disposal. Carbon monoxide consumption is also possible in anoxic environments through exchanges with the C_1-incorporating systems of methanogens [282].

Hydrocarbons

Hydrocarbons enter the biosphere as components of microbial and plant cells [290], as methane from swamps and sediments and as oil and gas from fossil sources with or without the intervention of man. The capacity to grow on and degrade hydrocarbons has thus long been part of the microbial metabolic repertoire. The investigation of microbial growth on hydrocarbons has been spurred by concern over oil pollution, by the potential of hydrocarbons as carbon sources for single-celled protein and by the economic consequences of microbial spoilage of refined oil products. The article by Gibson [291] in this series and the review by Higgins and Gilbert [292] should be consulted for biochemical details of hydrocarbon metabolism. Atlas [293] discusses environmental aspects of hydrocarbon degradation.

The diverse components of crude oils fall into three broad classes, the alkanes (saturates), the aromatic fraction and the complex condensed asphaltenes (polar fraction). Microbial degradation of the fractions in the environment is essentially oxidative. Anaerobic transformation are possible but are thought to be environmentally insignificant [293]. The most readily decomposed hydrocarbons are the liquid straight chain C11–C19 n-alkanes. Monooxygenases initially attack the C_1 methyl groups, converting the n-alkanes to primary alcohols, aldehydes, and finally carboxylic acids that can be cleaved into acetate units by β-oxidation [291, 192]. Branched and cyclic alkanes are increasingly resistant to microbial attack, particularly complex alicyclics like the environmentally persistent hopanes [293]. Aromatic hydrocarbons can be metabolized via oxidative ring scission [291], but the presence of sulphur or halogen substitutions or the formation of multiple ring structures increases recalcitrance to microbial decomposition. Thus, the highly condensed asphaltenes are largely resistant to biological attack. Hydrocarbons are also subject to modification by co-metabolic reactions [264].

Most hydrocarbons are water-insoluble and since microbial growth is dependent on the availability of water, hydrocarbon utilizers are restricted to oil-water interfaces. There they can either absorb hydrocarbons by direct contact, or they produce surfactant-like molecules [294] which can dispense hydrocarbons in water as fine emulsions (*pseudosolubilization*) suitable for transport into the cell [295]. Gaseous hydrocarbons present less of a problem being significantly water-soluble. Where large amounts of hydrocarbons are present, as after oil spillages, microbial degradation may be slowed by nitrogen or phosphorus limitation [293]. The application of inorganic fertilizers can thus accelerate the microbial dispersal of oil.

Organisms capable of growth on non-gaseous hydrocarbons include numerous bacteria, yeasts, and filamentous fungi, while phototrophic cyanobacteria and eukaryotic algae may oxidize aromatic hydrocarbons [293]. Among the bacteria, strains of *Pseudomonas* are particularly prominent and widespread, but at least 22 genera are implicated in aquatic environments [296] and some 28 genera in soils [297] including *Achromobacter, Alcaligenes, Arthrobacter, Bacillus, Flavobacterium,* and *Mycobacterium.* The pseudomonads are particularly interesting as their degradative abilities may be determined by extrachromosomal plasmids [298]. These degradative plasmids, found particularly in *Pseudomonas putida* but tranferable to other *Pseudomonas* species, regulate all or part of the pathways for the degradative metabolism of n-alkanes, aromatic, and polynuclear aromatic hydrocarbons, terpenes, and chlorinated hydrocarbons [299]. Such plasmids are good examples of the widespread genetic elements that can confer metabolic adaptability on microbial communities [194].

Candida is the most notable of the 15 genera of yeasts capable of utilizing hydrocarbons [293, 297], some strains having been grown industrially on oil products to provide single-celled protein. Members of nearly 40 genera of filamentous fungi may grow on hydrocarbons [293, 297]. Of these, *Amorphotheca* (*Cladosporium*) *resinae* is noteworthy because of its widespread and potentially disastrous accurrence in water-contaminated jet fuel.

The distribution of these hydrocarbon-utilizing microbes in natural environments is widespread in habitats from the poles to the tropics [293]. Where oil spillages are frequent the ability to degrade hydrocarbons may be shown by practically all the organotrophic bacterial population, but in most soil, marine, and freshwater habitats less than 10% of the population have this ability [293].

Growth on gaseous hydrocarbons has been most studied in methylotrophs although ethane and propane-utilizing bacteria are known, including lipid-rich strains of *Pseudomonas, Mycobacterium*, and *Nocardia* [300, 301]. These, with fungi, may form waxy deposits of "paraffin dirt" around natural gas seeps and surveys for ethane utilizers have been tested as a prospecting aid [300].

The Nitrogen Cycle

General Aspects

The inorganic salts of N are extremely water soluble and therefore N does not generally occur in concentrated smount in geological deposits, but rather as an evenly distributed component of sedimentary deposits and rocks. A few exceptions are to be found where exceptionally arid climatic conditions have prevented the leaching of nitrogenous material, as in the guano deposits on islands off the coast of South America.

N has a crucial role to play in primary productivity on earth since it is the plant nutrient required in greatest quantity. In the biosphere it undergoes what is essentially an eight electron shuttle between the most oxidized form at valence +5 (NO_3^-) and the most reduced form at valence −3 (NH_3 or NH_4^+). A number of stable intermediate valence states including NO_2^-(+3), NO(+2), N_2O(+1), N_2(O), and NH_2OH(−1) have a significant impact on the biosphere. The main features of the N cycle are shown in Fig. 16. The chemistry of the various valence states, and the global inventories of N for the terrestrial, aquatic, and atmospheric systems have been recently reviewed [302, 303]. Here we concentrate on the importance of microbial N transformations in the continuing turnover of the N cycle.

N in biomass is in its most reduced form at valence −3 and may be released unchanged in valence as NH_3 by the process of *ammonification*. Ammonification, which is largely a microbially mediated decomposition process, probably consti-

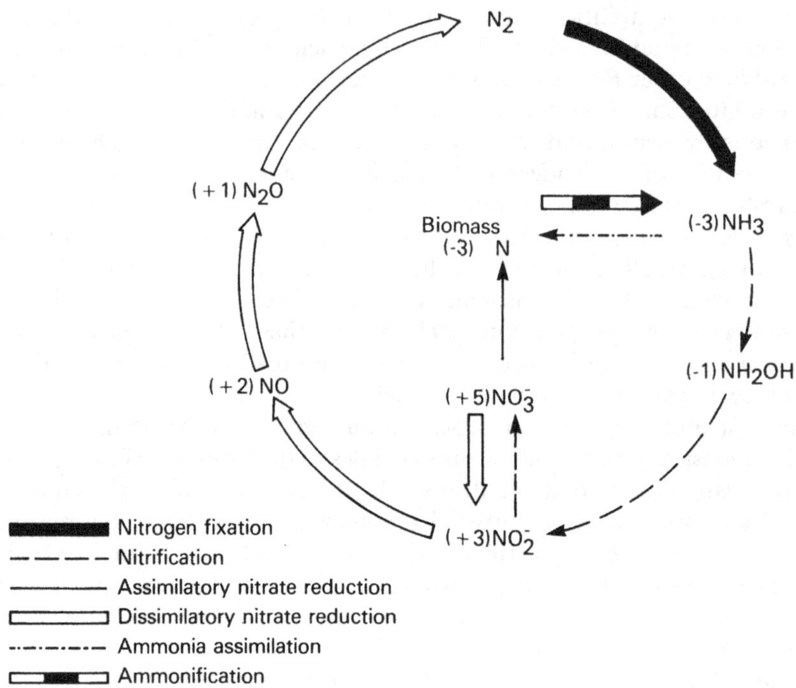

Fig. 16. The nitrogen cycle. Modified from [4] with permission

tutes the major flux in the N cycle, producing a form of N that is readily assimilable by most primary producers. However, under oxic conditions, *nitrification* occurs when a ubiquitous group of chemolithotrophic prokaryotes, the nitrifying bacteria, oxidize NH_4^+ to NO_3^- (Fig. 16). Accordingly, excepting gaseous nitrogen (N_2), NO_3^- is the most widespread inorganic form of N in oxic habitats. Plants and microorganisms that use NO_3^- as an N source must therefore expend reducing equivalents to generate a form of N that is suitable for incorporation into biomass (*assimilatory nitrate reduction*). One of the consequences of nitrification is that it produces oxidized N compounds which can be used as terminal electron acceptors (*dissimilatory nitrate reduction*) by other microorganisms under anoxic conditions. If this nitrate reduction should progress as far as the reduced gaseous forms of N (namely N_2 and N_2O), then this can constitute a significant loss of plant nutrient from the immediate environment (*denitrification*). Fortunately, globally denitrification is roughly balanced by an opposing utilization of N_2 (*nitrogen fixation*) by certain specialized prokaryotes sometimes called *diazotrophs.* Accordingly the global N cycle is presently in overall balance, although in local areas balance may not occur due to excessive losses of N by denitrification or leaching.

Nitrification

In quantitative terms the most significant nitrification is carried out by the chemolithotrophic nitrifying eubacteria. A few organotrophic microorganisms

Table 18. Nitrifying bacteria and their habitats

Substrate	Species	Habitat
NO_4^+ oxidized to NO_2^-	*Nitrosomonas europaea*	Soil, fresh water
	Nitrosospira briensis	Soil
	Nitrosolobus multiformis	Soil
	Nitrosovibrio tenuis	Soil
	Nitrosococcus nitrosus	Soil
	Nitrosococcus oceanus	Marine
	Nitrosococcus mobilis	Marine
NO_2^- oxidized to NO_3^-	*Nitrobacter winogradskyi*	Soil, fresh water
	Nitrospina gracilis	Marine
	Nitrococcus mobilis	Marine

including soil bacteria of the genus *Arthrobacter* and common soil fungi such as *Aspergillus* spp. appear to be capable of carrying out a non-energy yielding nitrification, but the amount of N transferred is several orders of magnitude less than that by chemolithotrophic nitrifying bacteria [304–306]. Inhibition of chemolithotrophic nitrification, by inhibitors such as N-Serve [2-chloro-6-(trichloromethyl) pyridine] and AN(2-amino-4-chloro-6-methyl pyrimidine) can be used to determine the proportion of organotrophic nitrification occurring in the field [307]. These compounds have some commercial use as inhibitors of nitrification since the products of nitrification are more readily leached than NH_4^+ and may constitute a serious loss of N from the system.

Two distinct groups of obligately aerobic chemolithotrophic bacteria are responsible for the major part of nitrification in soils and water. Table 18 summarizes the currently recognised species and habitats. Members of the group with the generic prefix "Nitroso" carry out a six electron oxidation of NH_4^+ to NO_2^-, whereas the second "Nitro" group carry out a two electron oxidation of NO_2^- to NO_2^-. Complete oxidation of NH_4^+ to NO_2^- is then a sequential consortial process, and in heterogeneous environments like soil, parameters like water activity (a_w) have a major impact on the process both by affecting redox potential and the motility of the organisms.

Nitrifying bacteria are found in any oxic environment where mineralization of organic material is occurring. Most nitrifiers are found in the upper layers of soils and either at the sediment water interface in its aquatic environment, or in the photic zone. High concentrations may be found in sewage treatment systems where very rapid controlled mineralization occurs. Most isolates grow optimally at neutral pH and 25°–30 °C. The numbers of nitrifiers in fertile loam soils may be from 10^3 to 10^5 g^{-1}, both groups being equally represented. Comparable numbers are found in aquatic sediments. With the exception of a few strains of *Nitrobacter winogradskyi* which can be grown organotrophically, all of the isolates are strict chemolithotrophs, and isolation and culture are both difficult and time consuming. The state of the taxonomy of these bacteria and the methods employed for isolation and culture has been recently reviewed [308] as has the ecology and biochemistry of the group [309].

Considerable overall free energy change is available from the oxidation of NH_4^+ to NO_2^- (-276 kJ mol^{-1}), rather less from the oxidation of NO_2^- to NO_3^- (-73 kJ mol^{-1}). The oxidation of NH_4^+ to NO_2^- occurs via the formation of hydroxylamine $NH_2OH(-1)$ and probably nitroxyl HNO(+1). The E_0' of the NH_2OH/NH_4^+ couple (+899 mV) is too high to allow the oxidation of NH_4^+ via the respiratory chain (Fig. 3) and it is likely that the formation of NH_2OH is accomplished via a reductive oxygenation reaction similar to that employed by methane-oxidizing bacteria [310]. The NO_2^-/NH_2OH couple then has a redox appropriate to enter the electron transport chain (Fig. 3). However, reducing power is lost by the primary oxygenation reaction and the NO_2^-/NH_2OH couple has a redox inappropriate for the direct reduction of NAD, hence energy has to be expanded to mobilise "reverse electron flow" in order to produce reducing equivalents for CO_2 fixation. Accordingly, these bacteria have their environmental impact because they transfer very large amounts of substrate (50×10^6 µg N day^{-1} g^{-1} cells) in order to sustain growth [305]. The oxidation of NO_2^- to NO_3^- is via a normal respiratory mechanism since the E_0' value of the NO_3^-/NO_2^- couple (+420 mV) is appropriate. However, the limited redox span between NO_3^-/NO_2^- and ½O_2/H_2O (Fig. 3) dictates that large amounts of substrate are consumed. There is no evidence that any of the organotrophic nitrifiers are able to conserve energy via the process [305], and its function in these organisms remains obscure. Certain of the chemoorganotrophs appear to be capable of the complete sequence of oxidation from NH_4^+ to NO_3^- [306, 308].

Important factors affecting nitrification include redox and temperature. Nitrification is considered not to occur in significant amount at redox values more negative than +200 mV, but paradoxically, sewage treated systems that are extremely low in dissolved oxygen support large numbers of nitrifiers. NO_2^- oxidation is more sensitive to low levels of oxygen and NO_2^- also tends to accumulate at low temperatures [308, 309]. The importance of nitrification is that it is virtually inevitable in any oxic environment, producing an oxidized form of N that can participate in denitrification, permitting a localised loss of N from the system. However, conversely, it can be argued that nitrification permits mobilization of N, allowing a major transfer of N from the terrestrial to the aquatic environment.

Whether one believes nitrification is advantageous or disadvantageous, it is clear that the leaching of NO_3' into inland waterways due to the nitrification of N fertilizers is in part responsible for major increases in the nutrient status of lakes and rivers in many parts of the world, evidenced by algal blooms [311]. High NO_3^- levels in drinking water also indirectly are responsible for the condition of methaemoglobinaemia seen in infants. In this condition NO_3^- is reduced to NO_2^- in the less acid stomachs of these infants, and NO_2^- combines with haemoglobin producing a "blue baby" syndrome. Concern has also been voiced over the formation of carcinogenic nitrosamines in the environment by the condensation of NO_2^- with secondary amines. This reaction is also possible in less acid stomach contents [302, 303].

A further question arises over the environmental effects due to the production of nitrous oxide (N_2O) and possibly nitric oxide (NO) as a by-product of nitrification. The biochemistry of nitrification in both organotrophs and lithotrophs is relatively obscure. The process probably proceeds in the following manner:

$$NO_4^+ \rightarrow NH_2OH \rightarrow (HNO) \rightarrow NO_2^- \rightarrow NO_3^- .$$

The "Nitroso" group of organisms evolve some N_2O and NO probably as a decomposition product of the postulated intermediate nitroxyl (HNO), since nitrifiers do not appear to use these compounds. N_2O omissions from nitrification probably contribute around 1% of the total atmospheric N_2O [302, 303]. The photodissociation of N_2O depletes stratospheric ozone (O_3) through a series of photochemical and chemical reactions via singlet oxygen ($O_2^{\cdot\cdot}$) thus

$$N_2O \xrightarrow{\text{light}} N_2 + O_2^{\cdot\cdot}$$

$$N_2O + O_2^{\cdot\cdot} \rightarrow 2\,NO$$
$$NO + O_3 \rightarrow NO_2 + O_2 ,$$

N_2O is also produced by denitrification in roughly equal amounts [302, 303]. As to the extent of ozone depletion by the above mechanisms, calculations vary widely [302, 303]. Significant quantities of N oxide are also produced by the internal combustion engine, but as yet the quantity of oxides produced by biological N cycling probably exceeds man-made pollution.

Nitrate Reduction

In recent years a number of review articles on the microbiological, biochemical, physiological, and ecological aspects of nitrate reduction have appeared and these should be consulted for detailed information [305, 312–317].

The capacity for assimilatory nitrate reduction is widespread amongst prokaryotes and eukaryotes. The end product is NH_4^+ and the pathway is subject to feedback inhibition by NH_4^+ [312, 313]. The pathway is probably the reverse of that occurring in nitrification with NH_2OH as an intermediate [313, 314]. A dissimilatory non-NH_4^+-inhibited version of this pathway seems to operate under certain conditions amongst certain prokaryotes including members of the genera *Bacillus, Escherichia,* and *Vibrio*. This kind of *nitrate ammonification* contributes NH_4^+ to the deeper regions of some lakes [318].

The dissimilatory reduction of NO_3^- to NO_2 and ultimately to gaseous products is by far the most significant nitrate reduction process in the biosphere. A number of eukaryotes have been shown to reduce NO_3^- to NO_2^- under anoxic conditions [319], but to date there is only one recorded example of true *nitrate respiration,* i.e. dissimilatory nitrate reduction coupled to energy conservation [22]. There is no doubt that prokaryotic nitrate respiration contributes a major flux in the N cycle. Dissimilatory nitrate reduction is clearly distinct from assimilatory nitrate reduction in that the former is inhibited by O_2 and not by the final reaction products of the reaction. The physiological substrates for nitrate respiration are generally the same as for aerobic respiration, but in view of the reduced redox span between donor and acceptor (Fig. 3) the switch from aerobic to nitrate respiration (all nitrate respirers are facultative), is accompanied by a reduction in molar growth yield of 50% [320]; accordingly large amounts of N compounds are transferred and released to the environment.

Dissimilatory nitrate reduction probably occurs by the following reactions, not all prokaryotes being capable of the complete reductive process: $NO_3^- \rightarrow NO_2^- \rightarrow [NO] \rightarrow N_2O \rightarrow N_2$: whether free NO is an intermediate is in dispute, but evidence is emerging for a bound form [321]. In theory, the use of N_2O/NO and N_2O/N_2 redox couples should be energetically very favourable (Fig. 3), but there is no known energy-conserving mechanism for use with these couples [44]. The production of gaseous products (mainly N_2O and N_2) clearly has the most impact on the environment, both in terms of N loss from the immediate environment, and in the formation of nitrous oxides with possible detrimental effects on the ozone layer (see Sect. on Nitrification).

Organisms capable of reducing NO_3^- only as far as NO_2^- are by far the most numerous, some 70 genera being recorded as containing representatives capable of this transformation [322], including common soil and marine isolates. Organisms capable of the further reductive steps resulting in denitrification are less common. A compendium of species able to carry out denitrification has recently been published [316]. Genera from 9 out of the 19 form groups in Bergey's 8th Manual are represented, with examples from the genera *Pseudomonas* and *Bacillus* probably most important in numerical terms in the environment. In those organisms capable of both nitrate assimilation and nitrate dissimilation, the process can be separated by exposure to oxygen [323].

The key factor in determining whether denitrification will take place is the redox of the environment. At redox levels of c + 200 mV dissimilatory nitrate reduction is activated. Clearly, waterlogged soils and sediments at depth are sites where active dissimilation of nitrate occurs. Soil NO_3^- fulfills a useful function as a redox "buffer", maintaining redox potentials at around +200 mV until all NO_3^- is exhausted. The extent of denitrification in a particular environment has usually been estimated to be the discrepancy between total nitrogen present and that remaining after a given period of time, taking account of leaching losses [307]. Alternatively, since acetylene is a specific inhibitor of N_2O reductase, estimates can be made by infusion of the habitat with C_2H_2 [307]. In non-waterlogged soils it is well established that denitrification activity is stimulated in the rhizosphere area surrounding plant roots due to the presence of high concentrations of substrates and lower redox levels as a consequence of plant and microbial respiration [305, 324].

Denitrification clearly constitutes an important local loss of fixed N from the N cycle. In fertilized waterlogged fields such as rice paddies the losses can be as much as 50% of the applied N fertilizer [307]. Even in non-fertilized fields up to 35% of N accretion can be lost [307]. Fortunately, on a global scale denitrification is largely balanced by biological and chemical N_2 fixation. Denitrification is sometimes considered to be a desirable aim of sewage treatment processes since N_2 is the least troublesome of the N compounds released in large amounts by mineralization [325]. Denitrification also presumably has an important function in transferring N from the world's oceans back to the terrestrial environment via N_2 fixation.

Nitrogen Fixation

Biological N_2 fixation transforms smaller amounts of N than ammonification or the direct assimilation of NH_4^+, but plays a key role in offsetting denitrification. Estimates of global biological N_2 fixation are generally around 2×10^2 Tg yr^{-1}, terrestrial grassland, forest, and the marine environment contributing roughly equal amounts [302, 303]. Chemical fixation by the expensive and very energy-demanding Haber process probably accounts for an additional 50 Tg yr^{-1} [303]. The expense of chemically-fixed N fertilizer has largely restricted its use to the developed countries of the world. Many soils are limited in their ability to produce crops due to severe deficiency of available soil N. The key to maximizing food production in under-developed areas lies in part in exploiting biological N_2 fixation. The realisation that N_2 fixation is crucial to world agriculture has provided the impetus for much research over the last 2–3 decades. A vast literature exists with many original papers and review articles, recently splendidly collated and discussed by Postgate [82]. This work should be consulted for detailed information on the chemistry, biochemistry, genetics, and microbiology of N_2 fixation. Ecological aspects have also been recently reviewed in detail [81].

The capacity to reductively assimilate N_2 has not been shown by any eukaryotes so far investigated (there is however a report of an alga of uncertain taxonomic position awaiting clarification [326]). The ability to fix N_2 is, however, widespread amongst prokaryotes, including cyanobacteria, but as yet no archaebacterial example has been reported. Diazotrophs from nearly every major "form group" of prokaryotes have been reported (Table 19). A number of these form associations with higher plants ranging from true symbiosis as in the case of the legume-*Rhizobium* symbiosis to rather loose associations so-called *biocoenoses* sometimes called *associative symbioses* exemplified by the association of *Azospirillum* and *Azotobacter* spp. with the rhizosphere of certain plants.

The enzyme responsible for N_2 fixation in called *nitrogenase*. Nitrogenase is composed of an association between two proteins, an iron sulphur protein (Protein 2) and a larger iron, molybdenum sulphur protein (Protein 1). If complementary proteins from various sources are mixed they will often yield active nitrogenase, indicating that the system is highly conserved. Cell-free studies have indicated that the system is highly oxygen-sensitive. Protein 2 accepts electrons from an appropriate electron donor, usually the iron sulphur protein ferredoxin, and participates in a cyclical ATP and Mg^{2+} stimulated reduction of protein 2, which binds N_2 probably through the molybdenum atom (Fig. 17). Electrons are added sequentially to N_2 and the end product NH_3 is presumably generated via a series of dinitrogen hydride intermediates [82]. The binding of N_2 involves the displacement of an initial dihydride group, evolving H_2, constituting a loss of reducing power from the system. So-called "efficient" diazotrophs minimise the loss through the operation of a unidirectional *hydrogenase* which "captures" H_2 evolved by nitrogenase activity [82, 327]. A number of cycles are required to achieve the six electron reduction of N_2 to two molecules of NH_3. NH_3 as NH_4^+ can be assimilated into tissues via condensation reactions with α-keto-glutaric acid and glutamic acid. Nitrogenase has the property of reducing a variety of triple bonded compounds in addition to N_2, notably acetylene (C_2H_2). The ability to qualitatively reduce C_2H_2 to ethylene (C_2H_4) is the basis of a rapid and sen-

Table 19. N_2-fixing systems [a]

A. Free-living genera and groups

Aerobes	Microaerophilic	Facultative anaerobes	Anaerobes
Azotobacter	*Corynebacterium*	*Klebsiella*	*Clostridium*
Beijerinckia	*Azospirillum*	*Bacillus*	Anoxygenic phototrophic bacteria
Azomonas	*Frankia*	*Citrobacter*	
Azotococcus	*Rhizobium*	*Erwinia*	*Desulfovibrio*
Derxia	*Thiobacillus*	*Enterobacter*	*Desulfotomaculum*
Xanthobacter	*Arthrobacter*		
Cyanobacteria	Methane oxidizers		
	Cyanobacteria		

B. Symbiotic associations

Nodule	Rhizosphere		Phyllosphere	Cyanobacteria
	Temperate	Tropical		
Legume + *Rhizobium*	*Bacillus* (non-specific)	Azospirillum (non-specific)	*Beijerinckia*	Lichens
Alder etc. + *Frankia*	*Klebsiella* (non-specific)	*Beijerinckia* (non-specific)	*Azotobacter*	*Azolla* + *Anabaena azollae*
		Azotobacter paspali + *Paspalum notatum*		Cycads + *Anabaena* or *Nostoc*
				Gunnera + *Nostoc*
				Protozoa?
				Sponges?
				Corals?

[a] Compiled from [4, 82]

sitive assay procedure for nitrogenase [82, 328], and since the advent of the acetylene reduction test the list of prokaryotes considered to be diazotrophs has grown enormously.

The oxygen sensitivity to nitrogenase is reflected to some extent in the kinds of prokaryotes listed in Table 19. Anaerobes, facultative anaerobes, and microaerophiles clearly avoid the problem of O_2-mediated damage when fixing N_2. Aerobes have evolved a variety of O_2 protection mechanisms, including O_2-binding proteins or functional changes in nitrogenase proteins, high affinity terminal oxidases which "scavenge" traces of O_2, and in the case of certain of the filamentous oxygenic cyanobacteria, separate anoxygenic cells known as heterocysts [82]. Structures that morphologically resemble heterocysts are known from the fossil record to pre-date the appearance of eukaryotes, and if one accepts the ancient endosymbiont theory for the origins of eukaryotes, it is odd that eukaryotes have not managed to acquire such a useful attribute. Postgate [329], however, believes

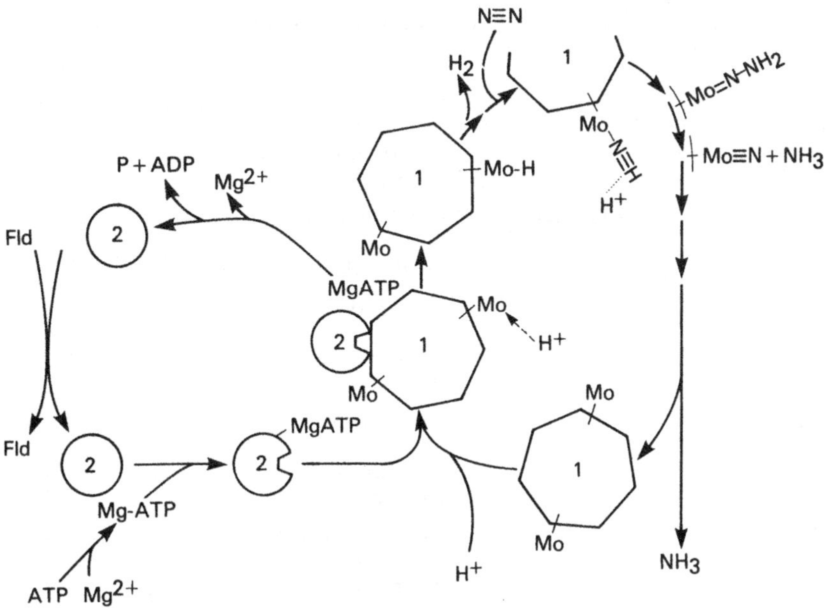

Fig. 17. The mechanism of nitrogenase. The small subunit [2] is reduced by an electron donor (Fld) and forms an ATP-Mg complex. This shuttles reductant to the large subunit [1] which in turn binds H^+ and N_2. Modified from [82]

that N_2 fixation capability is a relatively late evolutionary acquisition, and questions the interpretation of pre-eukaryotic fossils as heterocysts.

A consideration of the postulated biochemical mechanism for N_2 reduction indicates that large amounts of reducing power and energy are required. Phototrophs are the only free-living diazotrophs with unlimited access to energy and reducing power, and the other free-living diazotrophs make a relatively insignificant contribution to global N_2 fixation. The most efficient N_2 fixation occurs in various symbiotic associations between prokaryotes and plants. Table 20 illustrates some of the fixation rates measured for various systems. Agronomically, the most significant N_2 fixation occurs in the legume-*Rhizobium* root nodule symbiosis. A significant part of the increase in crop yields this century is attributable to the introduction of new or improved legumes into areas previously devoid of such crops. Much research continues into selecting effective competitive and persistent strains of rhizobia, as well as into the selection, breeding, and physiology of host legumes. The potential of tropical legumes is as yet not fully exploited, and this area is receiving much attention [330].

The potential of the non-legume root nodule symbioses is also under intense scrutiny. About 175 plant species covering 17 genera, 8 families and 7 orders have been reported to produce N_2-fixing nodules containing actinomycete-like endosymbionts currently classified in the genus *Frankia*. These generally woody plants often play a prominent role in plant succession following soil erosion, volcanic activity or human activities like road building or forest clearance. No agricultural crops of the group have yet been identified although some browsing of

Table 20. Relative agronomic efficiencies of some N_2-fixing systems [a]

		Organism or System	N fixed ($kg\ ha^{-1}\ yr^{-1}$)
Nodules	Legumes:	Tick clover (tropical)	897
		lucerne (temperature)	45–673
	Non-legumes	*Alnus* (temperature)	139
		Alnus (arctic)	62
		Casuarina (tropical)	52
Rhizosphere associations		Rye grass (temperate)	61
		Paspalum (tropical)	15–93
		grassland (temperate)	39
Cyanobacterial associations		Lichens (tropical)	10–100
		Azolla (tropical)	83–125
		lichens (arctic)	1–2
Free-living		Cyanobacteria (temperate)	15–20
		Cyanobacteria in rice paddies (tropical)	10–80
		Azotobacter	<1
		Clostridium	<1

[a] After [4]

Purshia occurs. However, significant N transfer presumably occurs in the early stage of land reclamation, leading to conditions more appropriate for later climax vegetation. It is only in recent years that the endosymbionts have been isolated, shown to produce nitrogenase in the laboratory and produce effective nodules on reinfection [331, 332].

Cyanobacteria contribute significantly to N_2 fixation in the biosphere, being very widely distributed and presumably responsible for the bulk of marine fixation [333], although chemoorganotrophs in mats of eukaryotic algal growth may have a significant role to play [333, 334]. Cyanobacteria are important primary colonizers of stressed environments [27]. Cyanobacterial N_2 fixation rates are greater in tropical habitats although significant rates are to be found in temperate regions (Table 20). Rice paddies constitute a major habitat. The economic significance of rice culture has provided impetus to continuing research into the exploitation of cyanobacteria as green "manure" [27, 335]. Cyanobacteria also form associations with a variety of higher plants, and in particular the association between *Anabaena* spp. and the water fern *Azolla* is of great significance for rice culture. The *Azolla* symbiosis is more efficient at N_2 fixation than cyanobacteria alone and is probably the most efficient recorded to date (Table 20). *Azolla* is now systematically grown and distributed to farmers' cooperatives in the Far East. The significance of cyanobacteria and *Azolla* in rice culture has recently been reviewed [336] – there is no doubt as to the contribution of paddy fields to global N_2 fixation (Table 21).

Table 21. Global N_2-fixation rates ($Tg\ yr^{-1}$)[a]

			Area under cultivation ($ha \times 10^6$)
Agriculture	Legumes	35	250
	Rice	4	135
	Remaining terrestrial	100	12,000
Marine		60	36,000
Industrial		30	

[a] After [4]

In recent years considerable attention has been focussed on rhizosphere associations between diazotrophs and the roots of certain tropical and semi-tropical grasses and cereal crops. These *biocoenoses* are intermediate between root nodule symbiosis and the normal non-specific colonization of the rhizosphere since some tissue penetration seems to occur [81]. Certain cultivars of the forage grass *Paspalum notatum* are specifically colonized by *Azotobacter paspali,* and the roots of maize and sorghum are colonized less specifically by *Azospirillum* spp. Early reports suggested high N_2 fixation rates and considerable N transfer to the plants [82] but more recent reports have indicated much lower N_2 fixation rates and rather less benefit to the plants than had once been supposed [337]. The picture is complicated since these bacteria also produce plant growth-regulating substances, and there are those who believe that any improvement in yield could be due to these substances alone [338]. Similar less specific temperate associations are known, and the agronomic potential remains to be explored fully.

Some 17 genes are known to be involved in the expression of nitrogenous activity, some coding for structural elements, some concerned with the regulation of activity [82, 339]. Other genes involved in NH_4^+ assimilation and hydrogenase function are also necessary for an efficient system. In recent years the techniques of restriction and ligation [340] have enabled mobilizable genetic elements containing nitrogenase genes to be manufactured and these have been used to produce new diazotrophs [82]. There is now a real possibility of transfer to eukaryotes with the eventual goal of a diazotrophic cereal crop. However, a successful system has to have adequate supplies of reducing power, energy, and an O_2 protection mechanism as well as the large number of genes involved in expression and control of nitrogenase itself. Accordingly, solving the world's food supplies problem in this way is some way off.

The Sulphur Cycle

General Aspects

The S cycle bears some similarity to the N cycle since it also is essentially an eight electron shuttle between the most oxidized valence +6 form (SO_4^{2-}) and the most reduced valence −2 form (S^{2-}). The main reservoirs of SO_4^{2-} are seas and evaporites, whereas those of S^{2-} are sediments, shales, metamorphic, and igneous

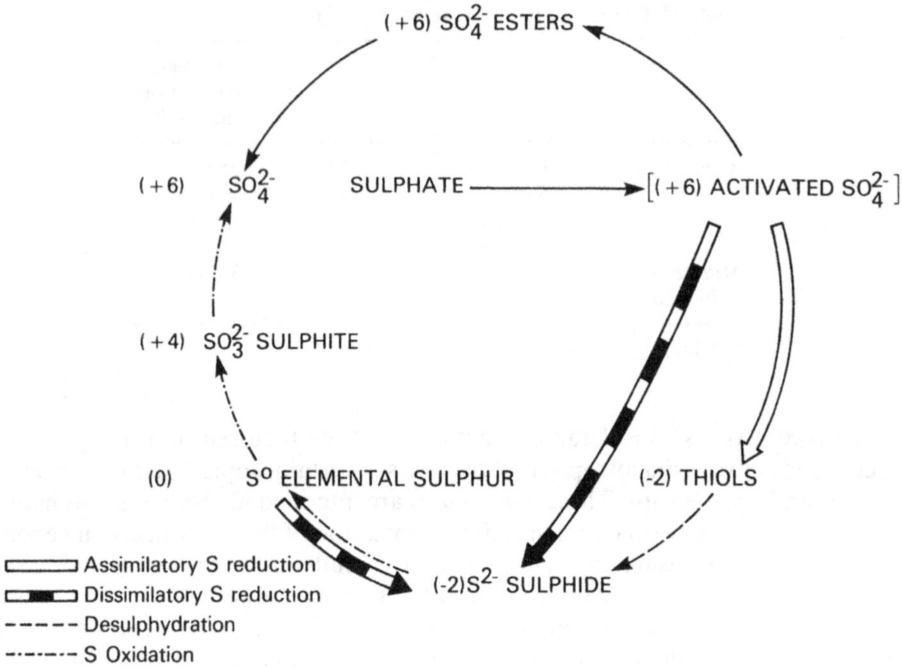

Fig. 18. The sulphur cycle. Modified from [4] with permission

rocks. Sulphur is continually cycled between oxidized and reduced forms in oceans, sediments, evaporites, and biomass, and although the terrestrial reservoir of S in biomass is small (5.6×10^3 Tg) compared with the other reservoirs [341, 342], microbial activity is responsible for major fluxes in the S cycle. Unlike the N cycle, the atmosphere is insignificant as a reservoir, the main transfers occurring between the dissolved SO_4^{2-} of the world's oceans and evaporites, and S^{2-} in sediments and shales. Quantitative aspects of the global S cycle have been reviewed elsewhere in this series [342]. Here we concentrate on microbially-mediated aspects of the S cycle outlined in Fig. 18.

S is an essential element since it is contained in several amino acids in the form of sulphydril thiol groups (R–SH). S in biomass is mainly in this -2 valence, but sulphate esters and aromatic sulphates also occur [341–343]. Under oxic and anoxic conditions thiol groups can be *desulphydrated* releasing S unchanged in valence state. There is an equilibrium between the equivalent valence states S^{2-}, HS^-, and H_2S depending on the pH of the environment and the capacity for volatilisation of the gaseous H_2S. H_2S is the major species under more acid conditions whereas S^{2-} predominates under alkaline conditions. However, some of each species are always present, and for convenience S^{2-} will be used as a general term for the -2 valence state. The capacity to desulphydrate is possessed by many microorganisms, both eukaryotes and prokaryotes, and this process probably accounts for a significant source of S^{2-} in fresh waters. The degradation of methionine may also generate the gaseous product demethylsulphide and related compounds [341, 342] and it is possible more atmospheric S than suspected may be a consequence of volatile organic S production [341, 342, 344].

S^{2-} is oxidized back to SO_4^{2-} through a series of intermediate valence states including elemental S (S^0), partly chemically and partly through the activities of certain specialised prokaryotes under both oxic and anoxic conditions. The microbial contribution to global S oxidation is difficult to quantify, due to difficulties in distinguishing chemical and microbial oxidation under oxic conditions, but it is likely to be 50%–60% of the total S oxidation observed [342].

SO_4^{2-} is the most appropriate form of S for higher plants. All organisms that utilize SO_4^{2-} have a requirement for a reductive assimilation back to the -2 valence state (*assimilatory sulphate reduction*), a mechanism directly analogous to assimilatory nitrate reduction. However, one of the most important features of the S cycle is the generation of vast amounts of S^{2-} from SO_4^{2-} by specialised prokaryotes, by the different process known as *dissimilatory sulphate reduction*. This particular type of anaerobic respiration directly analogous to nitrate respiration, is particularly important in the marine environment that has more than 500 times more oxidation equivalents in the form of SO_4^{2-} than in the form of O_2. This process may account for up to 50% of the carbon mineralized in certain marine sediments [345, 346]. There is no known chemical reaction for the reduction of SO_4^{2-} to S^{2-} at normal temperature and pressures, so there is little doubt that these prokaryotes are responsible for the bulk of the transfer of SO_4^{2-} to S^{2-}. In certain specialised environments S (S^0) can be used as an electron acceptor, but this particular flux in the S cycle is insignificant in quantitative terms being restricted to situations where active S^0 deposition is occurring [346] or where volcanic S^0 is present.

S Reduction

Dissimilatory sulphate reduction contributes a major flux in the S cycle at between 1–2 × 10^3 Tg y^{-1} from both the marine environment and the terrestrial environment [341, 342]. Swamplands and waterlogged soils presumably account for most of the S^{2-} in the atmosphere, since S^{2-} originating in marine sediments would be oxidized before reaching the surface [343]. Sulphate-reducing bacteria utilize organic electron donors similar to those used for aerobic respiration, and the SO_4^{2-}/SO_3^{2-} redox couple is therefore at an inappropriate potential to function directly as an electron-accepting couple (Fig. 3). Accordingly, there is a requirement for SO_4^{2-} to be "activated" as adenosine 5′ phosphosulphate (APS) (Fig. 19). APS is the real electron acceptor for dissimilatory sulphate reduction, and the APS/AMP + SO_3^{2-} couple has an appropriate redox potential to allow utilization of organic electron donors (Fig. 3). The mechanism of dissimilatory sulphate reduction is probably via a cyclical generation of SO_3^{2-} as outlined in Fig. 20, and the process is a cytochrome-limited energy-conserving respiration [347]. This reductive process is quite different from the necessary assimilatory sulphate reduction process in organisms that use SO_4^{2-} as sole S source. Here the substrate is 3′-phosphoadenosine-5′-phosphosulphate (PAPS) and the process is both enzymically different and energy-demanding [348].

A few prokaryotes can reduce elemental S (S^0) in a dissimilatory fashion. This property, first described for *Desulfuromonas acetoxidans* [349] has now been

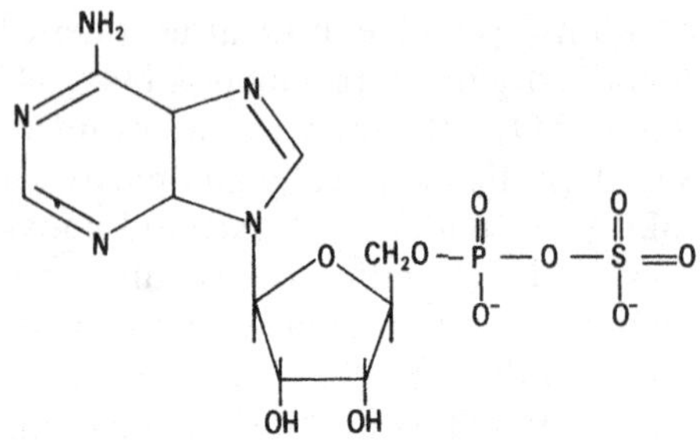

Fig. 19. Adenosine-5′-phosphosulphate (APS)

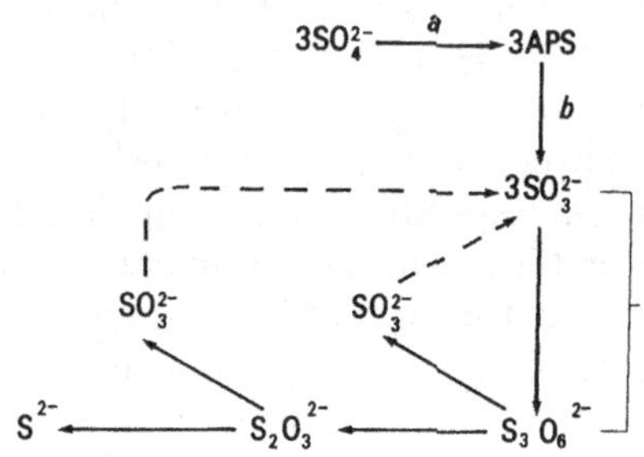

Fig. 20. Dissimilatory sulphate reduction. *a* ATP sulphurylase; *b* APS sulphotransferase; *c* sulphide reductase. From [4] with permission

shown to be a feature of certain archaebacteria occupying geothermal environments [350, 56], and curiously, all of the archaebacterial methanogens tested so far can carry out this reduction [351]. S^0 reducers are only found in consortial relationships where active S^0 deposition occurs due to geothermal activities or the activities of S-oxidizing bacteria [346, 350]. Table 22 lists the genera and habitats of S-reducing bacteria described to date.

If electron donors and acceptors are present in sufficient amount, then the redox of the environment determines whether dissimilatory sulphate reduction occurs (Fig. 4) since sulphate reducers are ubiquitous. In normal soils sulphate reduction only occurs in significant amounts under waterlogged conditions [342]. However, the anoxic condition with negative redox potential is a constant feature of many sediments at depth [354]. High levels of S^{2-} can have serious consequences, since S^{2-} is toxic to many plants, and has caused human fatalities in closed environments, such as sewerage conduits [355]. The problem of S^{2-}-induced wilt of rice plants in paddy fields is well recognized and of economic concern [335, 356]. The major geochemical consequence of SO_4^{2-} reduction is the formation of iron sulphides since Fe^{2+} is stable under anoxic conditions and combines with S^{2-} to form the black precipitate which is a characteristic of most anoxic environments. It is now believed that the initial iron sulphide precipitate is poorly crystallized mackinawite ($FeS_{0.94}$) and that eventual geochemical transformation to framboidal pyrite (FeS_2) occurs via the following sequence [357]:

$$FeS_{0.94} \xrightarrow{\text{limited } O_2} FeS_{1.1} \longrightarrow Fe_3S_4 \longrightarrow FeS_2$$

mackinawite	pyrrhotite	greigite	pyrite
(tetragonal)	(hexagonal)	(cubic, spheres)	(spheres, framboids)

Table 22. Genera and habitats of S-reducing bacteria [a]

	Genus or Group	Habitat	Comments
(1) $SO_4^{2-} \rightarrow S^{2-}$			
	Desulfomonas	Animal intestine	lactate incompletely oxidized
	Desulfotomaculum	Soils, water	thermophile, one species oxidizes acetate
	Desulfovibrio	Soils, water, marine	lactate incompletely oxidized
	Desulfobulbus	Water, marine	fatty acids incompletely oxidized
	Desulfobacter	Marine	fatty acids completely oxidized
	Desulfococcus	Water	fatty acids completely oxidized
	Desulfonema	Marine	fatty acids completely oxidized
	Desulfosarcina	Marine	fatty acids completely oxidized
	Thermodesulfotobacterium	Geothermal spring	thermophile
(2) $S^{\circ} \rightarrow S^{2-}$			
	Desulfuromonas	Water, marine	Forms consortia with S°-precipitating bacteria
	Thermoproteus	Geothermal springs	thermophilic arachaebacteria
	Thermodiscus	Marine, geothermal vents	thermophilic archaebacteria
	Pyrodictium	Marine, geothermal vents	thermophilic archaebacteria
	Thermofilum	Geothermal springs	thermophilic archaebacteria
	Desulfurococcus	Geothermal springs	thermophilic archaebacteria
	Thermococcus	Marine, geothermal springs	thermophilic archaebacteria
	Methanogens	Intestines, soils sediments, geothermal springs	Facultative S°-reducers, archaebacteria

[a] Compiled from [352, 353, 193]

Pyrite is by far the most abundant mineral in sedimentary rocks, and biogenic S^{2-} is probably the main precursor [358].

Sulphate-reducing bacteria are also of significance in amplifying the electrochemical corrosion of ferrous metals under anoxic conditions. When ferrous metals are immersed in a solution, metal ions pass into solution producing an anodic site, thus:

$$Fe \rightarrow Fe^{2+} + 2e .$$

Electrons move to less negatively charged areas constituting a cathode. Under anoxic conditions the following cathodic reactions occurs:

$$2H^{+} + 2e \rightarrow H_2 .$$

Under sterile conditions H_2 eventually blankets the cathode, preventing further electron flow and stopping the corrosive anodic reaction. Utilization of H_2 by a lithotrophic or mixotrophic organism will clearly allow the process to continue, and sulphate-reducing bacteria are particularly implicated in this kind of effect. A further amplification of the process also occurs since iron sulphides are formed on the surfaces of the metal and these are cathodic to ferrous metals. This kind of corrosion is obviously important in environments unlimited in SO_4^{2-} and electron donors. Off-shore oil installations are particularly vulnerable, and indeed any ferrous metal in contact with high sulphate sediments is at risk. Metal corrosion is however a complex process and many factors including other kinds of microbial activity have profound effects on the outcome. The electrochemistry and microbiology of corrosion has recently been extensively reviewed [359].

In the anoxic environment many different kinds of microorganisms ferment organic material. The end products of the fermentation of complex organic matter are H_2, CO_2, and acetate in the proportions: 4:2:2 respectively [360]. As a result, not more than 33% of organic matter can be degraded completely if acetate cannot be further decomposed under anoxic conditions [361]. Certain of the sulphate-reducing bacteria are capable both of the sulphate-linked oxidation of acetate, probably via the citric acid cycle [362] and of the complete oxidation of fatty acids, and are thus of major importance as terminal oxidizers of organic material under anoxic conditions (Table 22). Methanogenic bacteria fulfill a similar role since they also utilize acetate and H_2 under anoxic conditions. Methanogenic bacteria are more important than sulphate-reducing bacteria under conditions of sulphate limitation in fresh water environments. Methane formation from acetate in natural habitats is inhibited by sulphate apparently because methanogenic bacteria exhibit a higher K_S for acetate than sulphate-reducing bacteria [362].

S Oxidation

The problem of differentiating the enzymic and chemical oxidation of S has prevented a full understanding of the mechanism of biological S oxidation to date. Two groups of prokaryotes are considered to have a major role in S oxidation, the aerobic chemolithotroph bacteria (the colourless S bacteria) and the anaerobic anoxygenic photolithotrophic and photomixotrophic bacteria (the purple and green S bacteria). In recent years a number of cyanobacteria have been shown to carry out a limited oxidation of S under anoxic conditions [363]. Table 23 lists some of the genera known to be involved in S oxidation. The prokaryotes listed are to be found in oxic soils and waters, the marine environment, anoxic aquatic environments, and geothermal habitats including deep sea hydrothermal vents [157, 365, 366]. In quantitative terms the aerobic chemolithotrophs probably make the greatest contribution to S oxidation globally. In the most remarkable habitats of all, water from hydrothermal vents at a depth of 2,500 m shows numerous bacteria including chemolithotrophic *Thiomicrospira* spp. [366] flourishing in the sulphide-containing waters. Near the vents, in complete darkness, dense populations of clams, mussels, and worms thrive, presumably dependent on chemolithotrophic bacteria for their food supply. Bacteria have also been found *inside* a number of the larger animals including the mouthless large worm *Riftia pachyptila*. Sulphur inclusions [161] indicate that these bacteria are likely to be S-

Table 23. Sulphur-oxidizing bacteria [a]

	Genus or group	Habitat	Comments
Chemotrophs	*Sulfobacillus*	Mine tips	Lithotroph, spore, Fe(II) oxidizer, thermophilic
	Thiobacillus	Soil, water, marine	Mostly lithotrophs, one Fe(II) oxidizer, some denitrifiers, some thermophiles: deposit S^0 outside cells
	Thiosphaera	High sulphide effluent	Lithotroph, denitrifier
	Thiomicrospira	Marine	Lithotroph
	Sulfolobus	Geothermal springs	Lithotroph, thermophile, archaebacterium, Fe(II) oxidizer
	Thiobacterium	Water	Not grown in pure culture, deposit S^0 inside cell, energy status not known
	Macromonas	Water	
	Thiovulum	Water	
	Thiospira	Water	
	Beggiatoa	Water, soil, marine	Gliding bacteria; deposit S^0 inside cells, difficult to isolate and energy metabolism unknown
	Thioploca	Water, soil	
	Thiothrix	Water, soil, marine	
	Achromatium	Water	
	Thiodendron	Water, soil	Appendaged, energy status unknown
Phototrophs	Chromatiaceae (purple S bacteria)	Water, marine	Mixotrophs, deposit S^0 inside cell except for *Ectothiorhodospira* spp. which deposit S^0 outside cell
	Chlorobiaceae (green S bacteria)	Water, marine	Lithotrophs, S^0 deposited outside cell
	Chloroflexaceae	Geothermal springs	Mixotrophs, S^0 deposited outside cell; thermophiles
	Cyanobacteria	Water, soils	S^0 deposited outside cell under anoxic conditions

[a] Compiled from [4, 364]

oxidizing chemolithotrophs fixing CO_2 and thus producing organic carbon for the worm. Since that remarkable discovery a number of other symbiotic chemolithotrophs have been found in worms and molluscs in more normal high-sulphide habitats [161, 367], and this kind of non-light dependent CO_2-fixing symbioses may be of considerable importance throughout the world. Chemolithotrophic S oxidation appears to be confined to prokaryotes, but a number of microorganisms, including eukaryotes are known to be capable of oxidizing S when growing organotrophically [368, 369]. The benefit to the microorganisms, and their overall contribution to the S cycle, remain unknown.

The enzymic mechanisms involved in S oxidation have been largely gleaned from the study of chemolithotrophic *Thiobacillus* spp. and photomixotrophic *Chromatium* spp. A generalised scheme for S oxidation is shown in Fig. 21. Not all the interconversions outlined occur in all S oxidizers, but a consensus is begin-

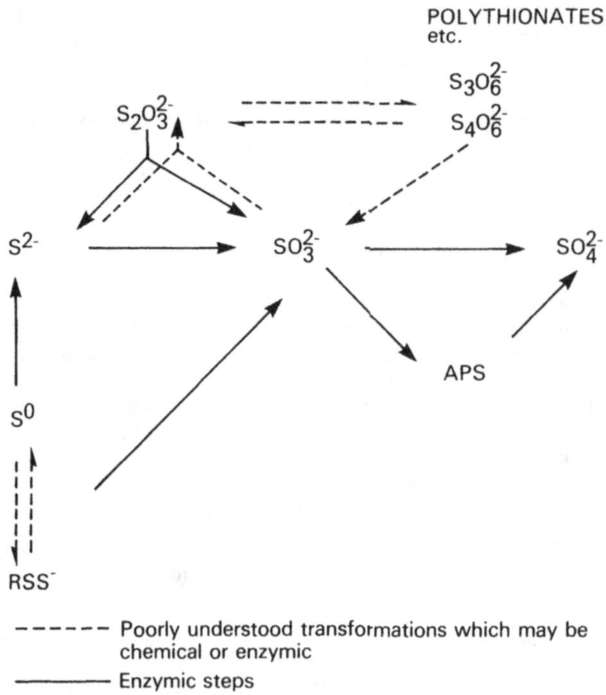

Fig. 21. Oxidative transformations of sulphur compounds. From (4] with permission

ning to emerge over the importance of SO_3^{2-} as a key intermediate, and the process has considerable similarity to dissimilatory SO_4^{2-} reduction in reverse. The biochemistry of S oxidation has been recently reviewed for both phototrophs [370] and chemotrophs [371]. Like nitrifying bacteria, S oxidizers often show low growth yields and thus transform large amounts of substrate, because certain of the electron donors used have redox potentials unsuitable for the direct generation of reducing equivalents for CO_2 fixation (Fig. 3).

Elemental S (S^0) is usually deposited by cells growing on S^{2-} under oxic or anoxic conditions. Some cells deposit S^0 *inside* the cell, others deposit S^0 *outside* the cell (Table 23). The nature of the S^0 deposition has been used as a cardinal taxonomic feature amongst the phototrophic bacteria, and recent reviews have discussed the taxonomy and ecology of these bacteria in relation to S metabolism [153, 372]. The geological consequences of biogenic S^0 deposition are beyond dispute [373]. Many of the prokaryotes listed in Table 23 further oxidize S^0 to SO_4^{2-} despite the hydrophobic nature of the substrate. It is now believed that there is an equilibrium between the crystalline form of S^0, which is an eightmembered ring structure, and a more soluble straight chain polysulphane form that is the actual substrate for the S^0-oxidizing enzyme [374].

A number of the bacteria listed in Table 23 are capable of oxidizing metal sulphides, ubiquitous *Thiobacillus* spp. being particularly implicated. One of the consequences of the complete oxidation of S compounds is the generation of H_2SO_4. Fe S_2 is one of the most abundant minerals and exposure to attack by

S oxidizing bacteria can have catastrophic effects on the surrounding environment due to acid run off thus:

$$Fe\ S_2 + H_2O + 3\tfrac{1}{2}O_2 \rightarrow Fe\ SO_4 + 2\ H_2\ SO_4.$$

Acidification is inimical to plant life, promotes metal corrosion, and stone deterioration, solubilizes metals obviously including iron which later precipitates as unsightly iron complexes or "yellowboy."

Acidification is also caused by "acid rain," some 40 Tg yr^{-1} of acid-generating SO_2 being released by industrial processes [341, 342]. SO_2 also has a direct corrosive effect on marble and limestone, thus:

$$SO_2 + Ca\ CO_3 + \tfrac{1}{2}O_2 \rightarrow Ca\ SO_4 + CO_2.$$

Microorganisms are not involved in these kinds of effects directly, but transform the S in the terrestrial and marine environments when it leaves the atmosphere.

The acidification promoted by S-oxidizing bacteria plays an important role in the extraction of certain low grade ores. Hydrometallurgy consists of the dissolution of metal from minerals by the percolation of a leach solution through beds of ore. Copper in particular is extracted from low grade ore ($<1\%$ Cu by weight) by this process, although the process has some commercial use in the extraction of uranium and possibly other metals. Ores from sedimentary strata contain substantial amounts of $Fe\ S_2$ which can be oxidized by thiobacilli. One particular thiobacillus, *Th. ferrooxidans* also lithotrophically oxidizes Fe(II) to Fe(III), and the solubilisation of copper (or other metals) is the consequence of several reactions.

a) $Fe(II)S_2 + 3\tfrac{1}{2}O_2 \rightarrow Fe(II)SO_4 + H_2SO_4$

b) $4Fe(II)SO_4 + O_2 + 2H_2SO_4 \rightarrow 2Fe_2(III)(SO_4)_3 + H_2O.$

Acidification clearly promotes solubilisation of metals, but the generation of $Fe_2(SO_4)_3$ is crucial to the process since this is a powerful chemical oxidizing agent which catalyses the chemical oxidation of minerals. Thus chalcopyrite (Cu Fe S_2) can be oxidized rapidly:

c) $Cu\ Fe\ S_2 + Fe_2(SO_4)_3 + H_2O + 3\tfrac{1}{2}O_2$

$\rightarrow Cu\ SO_4 + 3FeSO_4 + H_2SO_4.$

$Fe(II)SO_4$ generated by reaction c) can then be recycled lithotrophically to $Fe_2(III)(SO_4)_3$ (reaction b)). The process is cheap and efficient and widely used. *Th. ferrooxidans* has a central role in the process, but a complex microbial population of interracting lithotrophs and organotrophs develops in leach piles and recent reviews of the microbiology should be consulted for detailed information [375–377].

The S Cycle and Ore Deposition

The observation that present day S bacteria deposit elemental S and generate metal sulphides as a consequence of sulphate reduction, has led to speculation that S ores may be a consequence of microbial activity in the past.

Much of our understanding of the biogeochemistry of S in the geological record is based on isotope data from metal sulphides, elemental S and $CaSO_4$ (an-

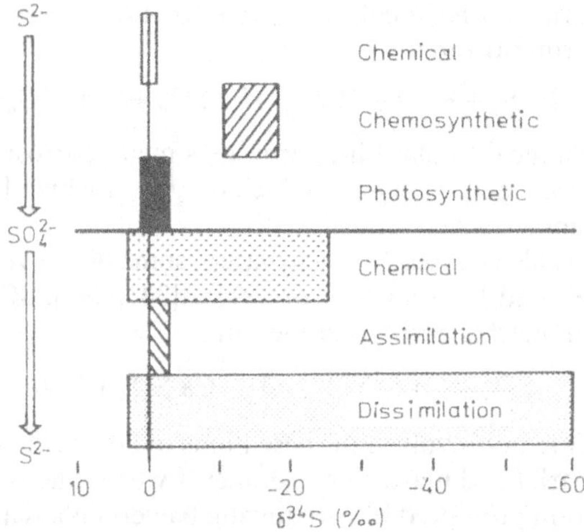

Fig. 22. Kinetic sulphur isotope effects associated with chemical and biological sulphur oxidoreductions: $\delta^{34}S$ values are expressed relative to the initial substrate. From [373] with permission

hydride, gypsum) deposits. Isotope compositions are expressed by comparing the ratio ($\delta^{34}S$) of the two most abundant stable S isotopes ^{32}S and ^{34}S in a sample and a standard. Meteoric troilite is the accepted geological standard. Thus:

$$\delta^{34}S = \frac{(^{34}S/^{32}S)\ \text{sample}}{(^{34}S/^{32}S)\ \text{standard}} - 1 \times 1\,000\,,$$

Meteoric S has an isotopic composition similar to volcanic S and this is believed to represent the isotopic composition of primordial terrestrial S.

Present day S bacteria fractionate S isotopes due to kinetic effects. Fractionation is particularly pronounced during dissimilatory SO_4^{2-} reduction (Fig. 22). If such fractionation has occurred in the past it would be predicted that biogenically produced sulphide deposits would have a tendency to be enriched in ^{32}S compared to source SO_4^{2-}, as is seen in present day environments like the Black Sea [378]. Examination of Pre-Cambrian sedimentary strata indicates that the isotope effects expected from bacterial sulphate reduction begin at around 2.8 Ma, and there is little dispute that pyrite in sedimentary rocks is largely a consequence of sulphide biogenesis [358]. Present day marine sediments are however limited in metal sulphide generation by lack of reactable metals apart from Fe(II) and would at best produce low grade ores. High grade non-ferrous metal ore genesis does occur however at a few tectonically active sites supplying reduced metals and chemically produced sulphide [357, 358]. Some isotopic indication of biological genesis is to be found in a few copper deposits such as the White Pine site in the USA [379] and there is similar evidence for sedimentary zinc and lead deposits which even harbour bacterial fossils [380], although in both cases there is an indication of a dual source of reduced S. The general consensus of opinion is that bacterial S reduction has made an indirect contribution to the formation of economically significant sulphide ores [357, 358, 373].

A number of present day environments have been described where microbial activity leads to the deposition of significant amounts of S^0. These are usually stratified lakes where sulphide reduction occurs in the anoxic light-free depths, and where S^{2-} diffuses to either oxic or photic zones lithotrophic oxidation occurs producing elemental S. Examples include Lake Eyre [381], the the Cyrenaecian lakes [381, 382] and coastal areas of the Bay of Bengal where phototrophic bacteria are particularly important. A consideration of the isotope fractionation occurring during the oxidation of S^{2-} to S^0 indicates that $\delta^{34}S$ produced by sulphate reduction is largely preserved by biogenic S^0 deposition (Fig. 22) and data from present day sites where S^0 deposition occurs are commensurate with a biogenic origin [358]. Sedimentary S^0 forms *syngenetic* deposits, and a biosyngenetic origin has been ascribed to some economic sulphur deposits such as those of Sicily [383], Cis-Carpathia [384], and Krasnovodsk [385]. *Epigenetic* deposits, which are formed subsequent to the host rock by replacement of $CaSO_4$ by S^0 and calcite ($CaCO_3$), are of much greater economic importance, accounting for some 90% of mined elemental S [386]. Epigenetic deposits are associated with evaporite beds of $CaSO_4$ in association with oil deposits (and are thus the remains of ancient seas). It is supposed that SO_4^{2-} reduction occurs in groundwater percolating through the evaporite (oil probably acts as a source of electron donors), and that where oxygenation occurs near the surface S^0 is deposited. It would be expected that biological involvement would deplete anhydrite $CaSO_4$ in ^{32}S and enrich S^0 in ^{32}S. $^{34}S/^{32}S$ abundances in a number of epigenetic deposits including the Texas and Louisiana salt domes [387, 388] the Shorsu deposits [385] and the Lower Fars Formation [389] are consistent with microbial involvement. Present day epigenetic deposition occurs at certain of these sites [358], thiobacilli being implicated.

Biological systems also fractionate the stable isotopes of carbon, and in particular the reductive fixation of CO_2 produces characteristic $^{13}C/^{12}C$ abundance in biomass. In epigenetic S^0 deposits there is a replacement of $CaSO_4$ by $CaCO_3$ due to Ca^{2+} mobilised by SO_4^{2-} reduction reacting with CO_2 produced by biological activity. The source of CO_2 is believed to be the oil deposits which, being largely the remains of marine phytoplankton, are enriched in ^{12}C. Accordingly, the calcite ($CaCO_3$) which replaces $CaSO_4$ is isotopically similar to the oil in adjacent strata. Table 24 indicates S and C isotopes in a number of bioepigenetic S^0 deposits. In general, taking into account the effects of the differential diffusion of different isotopes, the fact that these are not idealised closed systems, the evidence is compelling that global deposits of elementary S other than the 5% or so believed to be of volcanic or hydrothermal origin, were produced by oxidation of S^{2-} generated by bacterial SO_4^{2-} reduction [358, 373].

Transformations of Iron and Manganese

Fe and Mn can be conveniently considered together since they have properties in common, and the simultaneous occurrence of these elements is a regular feature of sedimentary deposits. Fe occurs in the divalent Fe(II) form and the trivalent Fe(III), both forms occurring in sedimentary rocks with the Fe(III) form being more common. Mn occurs in a number of valency states dependent on both redox

Table 24. Sulphur and carbon isotopes in bioepigenetic sulphur deposits[a]

Deposit	$\delta^{34}S$[b] (‰)		$\delta^{13}C$[c] (‰)	
	Anhydrite	Sulphur	Calcite	Oil
Cap rock				
Challenger Knoll (Gulf of Mexico)	+30	+14.2	−21.7 to −25.3	−26.6 to −26.7
Boling (Gulf Coast)	+12.2 to +39.2	0 to +4.0		
Moss Bluff (Gulf Coast)	+14.0 to +52.7	−10.8 to −5.2	−23.8 to −51.1	−24.9 to −27.1
Spindletop (Gulf Coast)	+17.6 to +61.7	+ 0.9 to +15.3		
Stratabound				
Castile (Texas)	+26.6	+ 6.7	−24.1 to −38.0	−26.1

[a] After (373)
[b] Relative to meteoritic sulphur
[c] $\delta^{13}C$ is analogous to $\delta^{34}S$: the reference standard (PDB) is belemnite from the fossil cephalopod, *Belemnitella americana* , from the Peedee Formation, South Carolina

and pH, and complex interconversions of higher valence states occur [390]. However, the main valence states in the environment are considered to be the Mn(II) and Mn(IV) forms. An important feature of the various valence states of both Fe and Mn is their ability to form complexes with organic material. Such complexes are often more soluble than the original form of the metal, and consequently valence states that would normally form insoluble precipitates survive in solution for long periods of time [390–394].

Iron is the fourth most abundant element in the earth's crust and large amounts are evenly distributed in sediments as pyrite (FeS_2) [358], as a direct consequence of microbial activity. However, most of the world's iron and manganese ore resources are concentrated in the banded iron formations (BIF) of Precambrian age found in many areas of the world. BIF's consist of layers of chert (SiO_2) and iron minerals such as siderite ($FeCO_3$) and haematite (Fe_2O_3). Workable Mn deposits (MnO_2, $MnCO_3$) occur in, or are closely associated with, these iron formations. Phanerozoic deposits also occur and may be locally important, but in global terms insignificant in quantity. The time span encompassed by iron and manganese deposits thus is very great – from early Archaean to Devonian, but the most important ores representing at least a 5–7 fold increase over crustal abundance are of Precambrian age [391, 395]. In present day environments Fe and Mn accumulate in freshwater "bog ores" and marine ferromanganese nodules. These accumulations are small in comparison to the deposits in BIF's. The distribution in time of BIF's and their paleoenvironmental significance has been recently extensively reviewed [395].

In the normal pH range of oxygenated water (pH 6–9), Mn(II) is soluble over a wide range of redox potential, either as Mn^{2+} or $MnOH^+$, whereas Mn(IV) is essentially insoluble in the form of MnO_2. Under the same conditions, Fe(II) is readily autoxidised to Fe(III) at redox values $> +200$ mV, resulting in the precipitation of insoluble Fe hydroxides and oxides. Accordingly, although oceans and

Table 25. Organisms involved in Fe and Mn transformations [a]

Genus, species or group	Habitat	Comments
Thiobacillus ferrooxidans *Leptospirillum ferrooxidans* *Sulfolobus* *Sulfobacillus thermosulfoxidans*	Mine waste minerals	Lithotrophs, Fe(II) only, S oxidizers and acidophiles
Sheathed bacteria including *Leptothrix, Sphaerotilus* etc.	Water	Fe(II) and Mn(II) oxidizers, some not grown in pure culture, probably organotrophs
Appendaged bacteria including *Pedomicrobium, Metallogenium Gallionella*	Soil, Water	Fe(II) and Mn(II) oxidation; probably organotrophic. *Gallionella* and some *Metallogenium* spp. may be lithotrophic
Siderocapsaceae	Soil, water, marine	Taxonomy dubious; Fe(II) and Mn(II) oxidizers, probably common soil microorganisms e.g. *Arthrobacter, Bacillus*
Arthrobacter *Bacillus*	Soil, water	Organotrophic
Fungi (*Cephalosporium, Cladosporium*, etc.)	Soil	Organotrophic

[a] Compiled from [4, 391, 392)

freshwaters contain roughly equal amounts of soluble Fe and Mn (3 $\mu g\ l^{-1}$ $-1\ mg\ l^{-1}$ depending on position in the water column), much of the Fe is probably in the form of organometal complexes [391, 393].

There is little doubt that under present day conditions, major Fe and Mn transformations depend on the influence of microorganisms. When Mn and Fe are weathered largely out of sedimentary rocks, they are not in general retained in solution for long periods but deposited as oxides and hydroxides. Many microorganisms are known to deposit Fe(III) or Mn(IV) under certain conditions and are therefore by implication supposed to be involved in Fe(II) and/or Mn(II) oxidation. Prokaryotes are considered to be most significant and Table 25 lists some of the major groups and habitats. However, the direct enzymic oxidation of Fe(II) and Mn(II) is confined to a restricted range of organisms and the majority of the bacteria listed in Table 25 cause precipitation of Mn and/or Fe by indirect means by altering E_h or pH which in turn leads to chemical oxidation and precipitation. Alternatively, the organic portion of organometal complexes may be utilized, with similar results. Chemolithotrophic oxidation of Fe(II) is confirmed for *Thiobacillus ferrooxidans* [396, 397], *Sulfolobus* spp. [397, 398], and probably *Gallionella* spp. [399]. As yet there is no convincing evidence for chemolithotrophic Mn(II) oxidation, although there may be non-energy-conserving enzymic oxidation in the *Leptothrix* group and some *Arthrobacter* spp. [390, 394], despite the theoretically higher energy yields for Mn(II) oxidation (-148 k J mol^{-1}) compared with Fe(II) oxidation (-46 k J mol^{-1}). The importance of the S and Fe-

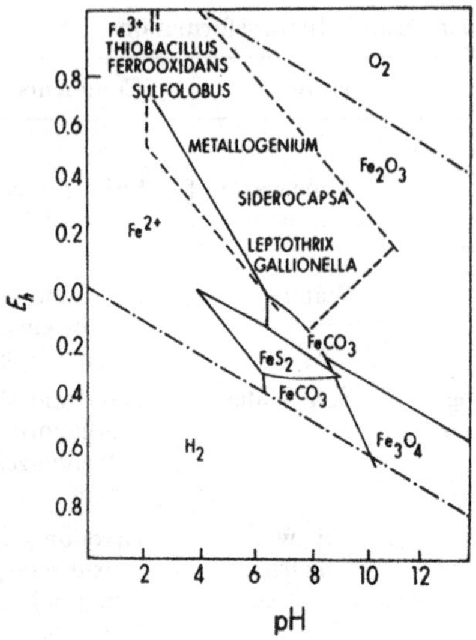

Fig. 23. E_h/pH iron stability diagram showing the natural domains of the main groups of Fe bacteria. From [391] with permission

oxidizing *Th. ferrooxidans* in ore leaching has already been discussed. Chemolithotrophic Fe(II) oxidizers clearly transform very large amounts of substrate since little energy is available from the oxidation. However, these bacteria are confined to oxic acid environments since O_2 is the only appropriate electron acceptor and Fe(II) is unstable under oxic conditions unless the pH is less than 4. Iron sulphides and carbonates which are stable at neutrality under oxic conditions are substrates for *Gallionella* and *Leptothrix* spp., and Fig. 23 illustrates the E_h/pH iron stability diagram showing the natural domains of iron bacteria. Since Mn(II) is stable over a much wider range of E_h and pH, these kinds of clearly-defined microbial Mn oxidizer domains are less apparent [394].

The economic effects of Fe and Mn precipitation include the clogging of pipes [390, 391, 400] by both Mn and Fe oxides and hydroxides, usually produced by appendaged and sheathed bacteria, and some contribution to corrosion by Fe(II) oxidation at anodes. However, most speculation has been concerned with the involvement of Mn and Fe-precipitating bacteria in the formation of ore deposits. Much of the discussion has centred on two areas: [1] the evidence that microorganisms are involved in the formation of present day "bog ores" and ferromanganese nodules; [2] the identification of fossil microorganisms in ore deposits.

Ferromanganese nodules or concretions were first collected by the Challenger expedition in the 1870's and are presently considered to have economic potential as ores, since in addition up to 50% (w/w) Fe and Mn, they often contain significant amounts of Co, Ni, and Cu probably due to chemical uptake mechanisms [390, 391]. Such nodules also occur in freshwater environments ("bog ores"). Some investigators believe that the freshwater and marine nodules have different origins largely based on the different proportions of Fe and Mn in the two types

(2 : 1 Fe Mn in freshwater nodules; 1 : 2 in marine nodules). However, there is considerable overlap in composition between the two types [390, 391] and the usual higher concentration of trace metals in marine nodules can be explained by the fact that these on average grow 3 orders of magnitude more slowly, allowing more extensive absorption of metals [391]. Appendaged and sheathed bacteria are implicated in bog ore formation [401] and indeed these contribute to Fe and Mn nodules and horizons in certain soil types [390, 391, 402]. These particular bacteria have not often been seen or isolated from marine nodules, but other bacteria, generally *Arthrobacter* and *Bacillus* spp. have been implicated [403, 404]. Nodules and soil profiles also contain a variety of organotrophic Mn(IV) and Fe(III)-reducing bacteria [391–393] which mobilize the metals, thus completing the cycle. Whether these metals are specifically utilized as electron acceptors for anaerobic respiration remains to be established.

Nodules often exhibit a layered appearance reminiscent of stromatolite structures [390]. Freshwater nodules are formed as Fe and Mn, probably largely as organometal complexes, are leached from soils into lakes, deposition of the metals occurring in the oxic littoral regions [390, 391]. The ocean floors with the highest concentrations of nodules are in areas with dense phytoplankton concentrations [405]. It is assumed that plankton concentrates metals from sea-water, and that organometal complexes are a major source for the marine ferromanganese nodules. A number of recent reviews have discussed the origins and significance of ferromanganese nodules [406, 407].

Well preserved fossils are to be found in BIF's, and stromatolithic structures are common [390, 391]. Some of our best-preserved Precambrian fossils are from BIF cherts, and assemblages of fossil prokaryotes resembling present day appendaged or sheathed Fe and Mn-precipitating bacteria are well documented. The precise mechanism of formation for particular ores such as magnetic (Fe_3O_4) and haematite (Fe_2O_3) remains uncertain – these ores are found in almost all BIF's and appear very early in the geological record [395]. Present day Fe-precipitating bacteria precipitate $Fe(OH)_3$ initially, and the chemistry of further absorption/transformation including the possible chemical absorption of Mn(II) is not well understood. However, we should not forget that the recently described magnetotactic bacteria apparently initially accumulate perfect crystals of magnetite [182].

Present day Fe and Mn-precipitating bacteria are aerobes and microaerophiles, and the age of many of the BIF's are of an era when anoxic conditions prevailed on the earth [395]. However, small amounts of O_2 may have been present and stromatolites up to 3.5 Ma are known, suggesting possible oxygenic activity [1–3]. Large amounts of Fe(II) would have of themselves kept O_2 levels very low by chemical oxidation and precipitation. Extensive BIF's terminate around 2 Ma with the appearance of chemical evidence for oxygenic photosynthesis, and the appearance of Fe(III)-rich sedimentary deposits (red beds). Phanerozoic deposits are largely of this type.

It is not certain that microorganisms played causative roles in Fe and Mn ore formation, but most workers have been unable to escape the conclusion that some kind of microbial activity like that presently seen in ferromanganese nodule formation was involved [390, 391, 393, 394].

Transformation of Other Elements

Microorganisms are involved in the transformation and mobilisation of a number of different elements. Calcium and phosphorus are largely mobilised and immobilised without change in valence state, whereas certain metal or metalloid elements may be changed in valence state and/or methylated. These transformations and mobilisations are in general carried out by a heterogeneous group of unrelated prokaryotes and eukaryotes, and there are, with one exception, no microorganisms specific to a particular element transformation in the way it is possible to define S bacteria or Fe bacteria. The one exception appears to be *Stibiobacter senarmontii,* an inhabitant of antimony ores, which appears to lithotrophically oxidize Sb(III) to Sb(V) [408]. The remarkable Fe bacterium *Th. ferrooxidans* is also capable of lithotrophically oxidizing Cu(I) to Cu(II) [371].

Microbially-mediated changes in valence state of metals and metalloids, particularly heavy metals, may be a detoxification mechanism related to the solubility of various forms of the element [409, 410]. The capacity to oxidize and/or reduce particular metals is so widespread amongst prokaryotes and eukaryotes that it is unlikely that specific enzymic systems are often involved. Many of the changes are likely to be redox or pH mediated, or alternatively reflect less than specific existing oxidases and reductases [409, 410]. Microbial transformations of this type have been extensively catalogues and discussed [409, 410] and need no general amplification here. Selenium is of special interest since the element is similar to sulphur, existing in the same valence states in nature, the inorganic forms structurally very similar to S analogues. A Se cycle similar to the S cycle is to be found thus:

$$Se^{2-} \rightleftharpoons Se^0 \rightleftharpoons Se_2O_3^{-2} \rightleftharpoons Se_2O_4^{2-},$$

Like S interconversions, some chemical interconversion occurs, but S bacteria have been shown to catalyse these interconversions and there is some evidence of lithotrophic oxidation. Se also replaces S in a number of essential organic compounds, and selenomethionine, and selenocystine are to be found in plants growing in high Se areas. Degradation of these compounds produces volatile Se analogues of S compounds such as dimethylselenide. The microbial aspects of Se biogeochemistry have been recently reviewed [411].

Metal methylation appears to involve the transfer of either the methylcarbonium ion (CH_3^+) from S-adenosyl methionine or perhaps methyl homocysteine, or the transfer of methyl carbanion ion (CH_3^-) from methyl cobalamin. Prokaryotes including archaebacterial methanogens, as well as fungi, have been implicated in methylations of this kind. The transformation is again probably a detoxification process since the methylated metals are volatile and thus escape from the immediate vicinity of the microorganisms. Heavy metal methylation, particularly of lead and tin is of particular interest in view of the toxicity to humans of the end products. Methylation of the metalloid arsenic has also attracted attention, again because of a number of well publicised human poisonings [412]. In many cases microbial involvement in methylation may be indirect since these methyl donors have been shown to donate methyl groups in the absence of viable microorganisms. The comprehensive review of Craig [413] elsewhere in this series

Table 26. Microbial transformations of metals and metalloids excluding Fe and Mn [a]

Transformation	Metal	Microorganisms
Oxidation	As(III)	*Pseudomonas, Actinobacter, Alcaligenes*
	Sb(III)	*Stibiobacter* (lithotrophic?)
	Cu(I)	*Thiobacillus (ferroxidans)* (lithotrophic?)
Reduction	As(V)	*Chlorella*
	Hg(II)	*Pseudomonas, Escherichia, Staphylococcus, Aspergillus*
	Se(IV)	*Corynebacterium, Streptococcus*
	Te(IV)	*Salmonella, Shigella, Pseudomonas*
Methylation	As(V)	*Aspergillus, Mucor, Fusarium, Paecilomyces*, methanogens [reduction to As(III) also occurs]
	Cd(II)	*Pseudomonas*
	Te(IV)	*Pseudomonas*
	Se(IV)	*Pseudomonas, Aspergillus, Candida, Cephalosporium, Penicillium*
	Sn(II)	*Pseudomonas*
	Hg(II)	*Bacillus, Clostridium*, methanogens, *Aspergillus, Neurospora*
	Pb(IV)	*Pseudomonas, Aeromonas*

[a] After [4]

should be consulted for detailed information on the chemistry, biochemistry, and microbiology of metal methylation, or for information on concentration effects and toxicity levels.

Table 26 summarizes the elements presently known to undergo microbially-mediated transformations of the types discussed here.

The elements calcium and phosphorus are rather different, in that transformations occur between different chemical forms at the same valence state. In theory P has several possible oxidation states ranging from $PH_3(-3)$ to $PO_4^{3-}(+5)$ and there is some evidence of interconversions of these valence states in soils [414]. However, these transformations are quantitatively insignificant compared to the mobilisation and transfer of P as PO_4^{3-} in the environment. Ca occurs almost exclusively as Ca(II) in the environment. Ca and P are often considered together since the two elements are associated in metal deposits, particularly in minerals of the apatite series. Most authors will subscribe to the view that microbial activity has a major impact on the mobilisation and transfer of Ca and P from the terrestrial to the marine environment and vice versa, but our knowledge of what happens is deductive and incomplete.

The main reservoir of Ca and P is in apatite rocks [$Ca_5(PO_4)_3$ OH,F] and large numbers of different microorganisms are capable of solubilising apatite by changing the pH or redox or by producing organic acids which chelate divalent cations [415, 416]. Soils in general are P-limited containing 10^{-5}–10^{-8} *M* soluble P [414], far less than required to sustain normal plant growth for the season. Presumably P is constantly produced from apatite minerals and organic P [414] via microbial activity.

Ca has a dominant effect on the Earth's surface since it locks up three vital non-metal elements, carbon, sulphur, and phosphorus as insoluble $CaCO_3$ (calcite), $CaSO_4$ $2H_2O$ (gypsum) and minerals of the apatite series respectively. Carbonate carbon mostly as $CaCO_3$ constitutes the greatest reservoir of carbon on earth [417, 242].

The availability of soluble Ca in the environment is correlated with the partial pressure of CO_2 thus:

$$Ca(HCO_3)_2 \rightleftharpoons CaCO_3 + H_2O + CO_2.$$

This equilibrium has a major global impact as a pH buffer in the marine environment. The partial pressure of CO_2 is largely regulated by respiration and lithotrophic CO_2 fixation, particularly photolithotrophic CO_2 fixation. The large-scale withdrawal of CO_2 by any biological fixation process, results in the precipitation of $CaCO_3$, and many areas of the world's oceans are saturated with Ca(II) and CO_3^{2-} [417, 419] due to biological activity. Present day stromatolites show extensive calcification due to cyanobacterial CO_2 fixation [420]. Accordingly, large deposits of $CaCO_3$ in various crystalline forms such as aragonite and calcite are assumed to be due to biological activity in the past, both due to perturbation of CO_2 partial pressure and by replacement of evaporite $CaSO_4$ during SO_4^{2-} reduction. Purely inorganic precipitation of carbonate may occur, but biologically-mediated precipitation predominates, as evidenced by the constant ratio of reduced carbon to carbonate rock in sedimentary strata since the Precambrian period [417]. Ca(II) precipitation by various kinds of organisms including microorganisms has been recently reviewed [419, 421, 422].

Ca(II) and PO_4^{3-} crystallize in several forms in the alkaline environment (i.e. sea water) to form phosphorite minerals. Phosphorites occur in Precambrian sedimentary deposits including BIF's but the most striking increase in phosphorite concentration occurs at the Cambrian/Precambrian boundary. It is reasonable to suppose the change is linked to the major explosion of taxa known to occur at this point [423, 424]. Phosphorites slowly revert to insoluble minerals of the apatite series particularly hydroxapatite [$Ca_5(PO_4)_3OH$] and fluoroapatite [$Ca_5(PO_4)_3F$]. The final resting place of Ca and P is thus in marine sediments initially as phosphorite nodules, finally being transformed back to apatites. Present day guano deposits also slowly transfer to apatites [425]. P can eventually replace silicon in clay minerals, and rock phosphates of Mg, Al, and Fe are known although they comprise much less extensive deposits.

Extensive accounts of the geochemistry of carbonate precipitation and turnover [420], dissolution [426], and aspects of calcification by bacteria and algae [422] have recently appeared. The biogeochemistry of phosphate minerals [425] and the global P cycle have also been reviewed [424, 427–429] notably by Emsley [430] elsewhere in this series. These compendia should be consulted for quantitative aspects.

The Silica Cycle in Relation to Soil and Sediment Formation

General Aspects of Soil and Sediment Formation

Fully developed soils and sediments are intricate mixtures of organic and inorganic components, their properties depending on the exchange and interaction between organic and inorganic fractions. Soils and sediments differ in that soils

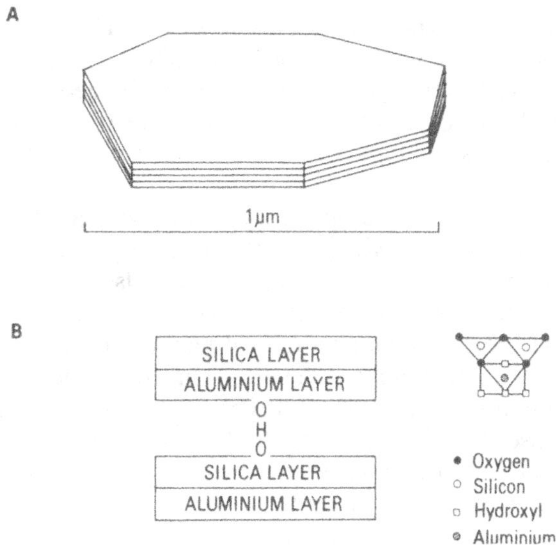

Fig. 24. A A clay platelet. Clay platelets are large, flat, leaf-like structures with silicon, and aluminium atoms arranged alternately in a sandwich-like structure. **B** Diagrammatic representation of kaolinite showing a layer of silicon atoms joined to a layer of aluminium atoms. From [4] with permission

contain a significant proportion of air by volume, whereas a sediment is defined as "solid material that has settled down from a state of suspension in a liquid" [354]. The organic content of both can range from 0.3–0.5% (w/w) to 30% (w/w) depending on topography, climate, and biological activity.

The inorganic portion of soils and sediments is composed of small rock fragments including a wide variety of crystalline and non-crystalline minerals ranging in size from gravels (<2.0 mm) to clays (>0.002 mm). The most abundant elements in rocks are Si, Al, and Fe, and soils and sediments have a mixture of *primary* minerals (those derived from parent igneous and metamorphic rocks), and *secondary* minerals (those derived largely from sedimentary rocks or by chemical modification of igneous or metamorphic rocks). Primary minerals are usually silicates, the most abundant being quartz and feldspars. Quartz is composed of a continuous framework of tetravalent silicon atoms linked by oxygen atoms (silica tetrahedra), whereas feldspars are negatively charged aluminosilicates where tetravalent silicon is partially substituted by trivalent aluminium with varying amounts of cations, most usually Na^+, Mg^{2+}, K^+, and Ca^{2+}. Secondary minerals include *authigenic* minerals (formed *in situ,* in part by biological activity e.g. FeS_2, biologically precipitated silica, $CaCO_3$, etc.) and clays. Clays are layered silicates, also known as phyllosilicates (Fig. 24) and these play a prominent role in determining the physical and chemical properties of soils and sediments by providing a large charged surface area for binding of ions and organic material. The chemical compositions of primary and secondary minerals in soils [104] and sediments [354] have been catalogued.

Soils and sediments develop when parent rock is broken down under a combination of biological, physical, and chemical effects. As the inorganic fraction is being transformed it exchanges material with the biogenic fraction, consisting of organic material and some authigenic minerals, the final product consisting of a layer of material physically and chemically different from the underlying matrix.

In soils, the major organic component is the brown, largely insoluble residue given the generic name *humus*. Most of this material is associated with the secondary mineral fraction [104]. This complex material, also known as *heterocondensate,* is a product of the chemical condensation of microbially-produced phenolics and terpenoids with breakdown products of lignocellulose [104]. The genesis of this material is still incompletely understood, but the structure is considered to be an ether-linked polyphenol [104, 431]. The chemistry of this material has been briefly reviewed elsewhere in this series [104] and both the structure and the genesis of this material have been reviewed many times over the years [432–434]. The overall composition of humus varies from site to site but generally reflects the structure of the lignocellulose in the plant cover. However, the picture is complicated by the invariable association of humus with other organic compounds, such as inositol phosphate [414]. The starting material lignocellulose is only decomposed significantly under oxic conditions. Under anoxic conditions, incomplete transformation results in the genesis of coal and peat. In theory, under oxic conditions complete degradation is possible [435] but in practice hardly ever occurs due to anoxic microenvironments, and binding of material to inaccessible sites.

The organic material brought to the site of decomposition in many freshwater sediments is similar to heterocondensate in structure [354, 436] reflecting run off from surrounding terrestrial areas. In contrast, the organic material derived from primary production by phytoplankton has a much lower C:N ratio [354, 437], and is considered to be almost completely biodegradable under oxic and anoxic conditions. The nature of the organic material in marine sediments is largely unknown but unlikely to be polyphenolic [438, 439]. The amount of organic material in many deep sea sediments is very small (0.3–0.5% w/w) reflecting the long residence time in the water column and the less recalcitrant nature of the low C:N phytoplankton [354, 439]. Some deep sea sediments are oxic since the supply of O_2 exceeds that of organic material [354].

Soils have a more complex structure than sediments, being composed of well-defined soil aggregates which depend on the nature of the parent rock material, the nature of the microbial population, and the heterocondensate. Microbial exopolysaccharides [440], heterocondensates, and clay minerals in particular contribute to the stability of the soil aggregate. Figure 25 depicts an idealized soil aggregate. The condition of a soil, and therefore its agronomic potential reflects the structure of the individual aggregates. Many soils show a series of stratifications with depth known as soil *horizons* which collectively form the *soil profile*. Each soil is characterized by a given sequence of horizons. These horizons are the consequence of weathering, leaching, and precipitation of mineral and organic materials at different depths in the soil profile [104]. Microbial activity is significant both in the solubilization of minerals, the formation of heterocondensate, in pro-

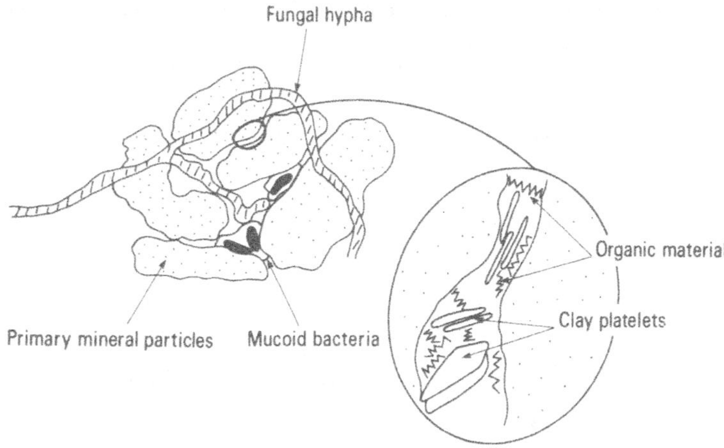

Fig. 25. Diagrammatic representation of a soil aggregate. From [4] with permission

ducing exopolysaccharide "cement" [440], in the production of authigenic minerals and in the mobilization and immobilization of organometal complexes, particularly Mn and Fe-containing complexes [402]. Soils which are constantly waterlogged become anoxic due to iron sulphide production (*gleysols*), as a consequence of microbially mediated SO_4^{2-} and Fe(III) reduction. Details of master soil horizons and overall soil types are given elsewhere in this series [104].

Transformations of Silica

The mineral fraction of soils and sediments develops when rocks are changed by a series of processes collectively known as *weathering*. Silicates are attacked by physical, chemical, and biochemical processes. It is believed that microbial activity has a major part to play in the dissolution and modification of silicates. There are two contrasting aspects of the microbially-mediated decompositon of stones – the weathering of stones is a vital part of soil and sediment fraction, whereas the corrosion of stone as a building material or work of art is clearly detrimental. Nevertheless, the two processes are essentially identical.

In the terrestrial environment soils provide the pioneering inhabitants of freshly exposed silicate rocks. Some rocks are relatively rich in organic material because of the way they have evolved, and organic compounds can penetrate to a depth of several cm under stone [441]. The microflora which colonizes rocks includes both lithotrophic and organotrophic microorganisms, developing in fissures and cracks. Microorganisms are also capable of penetrating and living inside porous rocks [442]. Chemolithotrophic S and N-oxidizing bacteria undoubtedly play an important role in the weathering of silicates by producing H_2SO_4 and $H\ NO_3$. In industrialized areas, the air contains large quantities of acidic chemicals such as SO_2 which contribute to stone decay, but there is little dispute that S-oxidizing bacteria in particular cause extensive deterioration, particularly of limestone, and sandstone [356]. Equations *a* and *b* describe the solubilization of polymeric metal (M)-substituted and non-substituted silica (crystalline or non-crystalline):

a

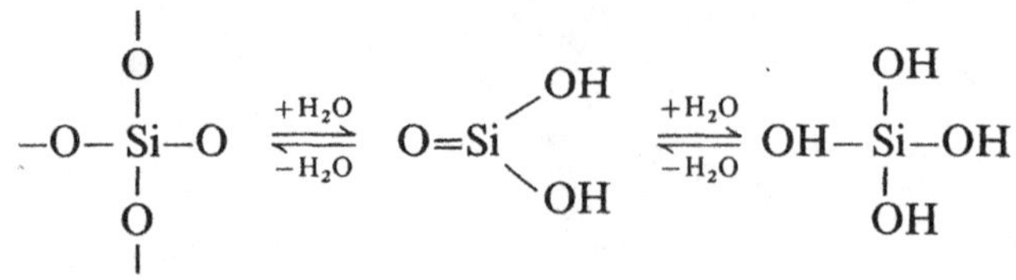

b

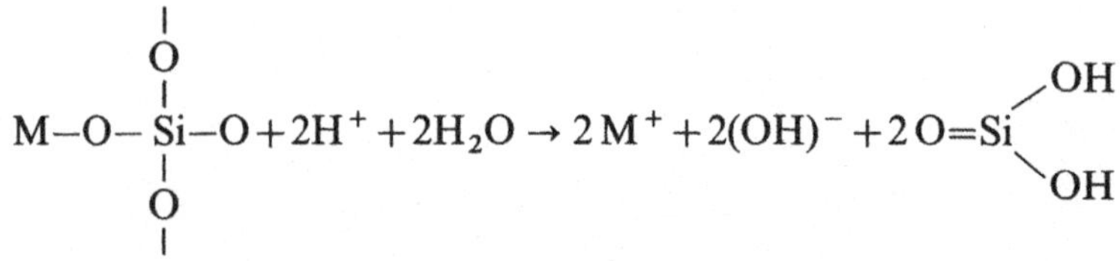

All silicates are to some extent water soluble, and the important role played by H^+ is evident since exchange with metal cations (M) in the lattice occurs, the subsequent rearrangement placing the structure under stress, promoting hydrolytic attack. Silicates harbour extensive populations of organotrophic bacteria in addition to lithotrophs. These microorganisms produce a variety of organic acids which contribute to decay in an additional way through the removal of cations by chelation. 2-keto-gluconic acid is very active in this respect, and eukaryotic microorganisms and in particular fungi are known to produce large amounts of this and other organic chelators [415, 416, 443, 444].

Rocks which contain large amounts of calcite binder are particularly prone to attack by H_2SO_4 generated by S-oxidizing bacteria. $CaCO_3$ is replaced by $CaSO_4$ which hydrates to gypsum, expanding in volume and thus placing the structure under stress. The influence of these largely microbially-mediated processes on different kinds of silicate has been recently reviewed [445–447]. The weathering of silicates has usually been considered to be a long process, but it is clear that it can be reasonably rapid in the presence of microorganisms if the conditions are correct [446].

The main form of dissolved silica in both terrestrial and marine environments is $Si(OH)_4$ (Equation *a*), present at between 5–80 μg l^{-1}, considerably below saturation value of 120 μg l^{-1} [448]. Although equation *a* indicates that polymeric silica can be readily precipitated from aqueous solution, the mechanism of polymerisation of biogenic silica is complex and probably involves the intervention of organic templates or organosilicate intermediates [34]. It is well established that organosilicon esters and ethers can be formed by a wide variety of organisms [34]. Compounds with C - O - Si, C - N - Si, and C - Si linkages have been detected in soils [34]. Figure 26 shows a diagrammatic representation of the process in the formation of siliceous shells.

Silica, fixed and precipitated by marine and terrestrial organisms is essentially comparable in structure. It is a hydrated non-crystalline form given the designation Opal-A, which then transforms eventually to crystalline Opal-CT, and finally becomes quartz [448]. The main sources of biogenic silica in terrestrial sources are

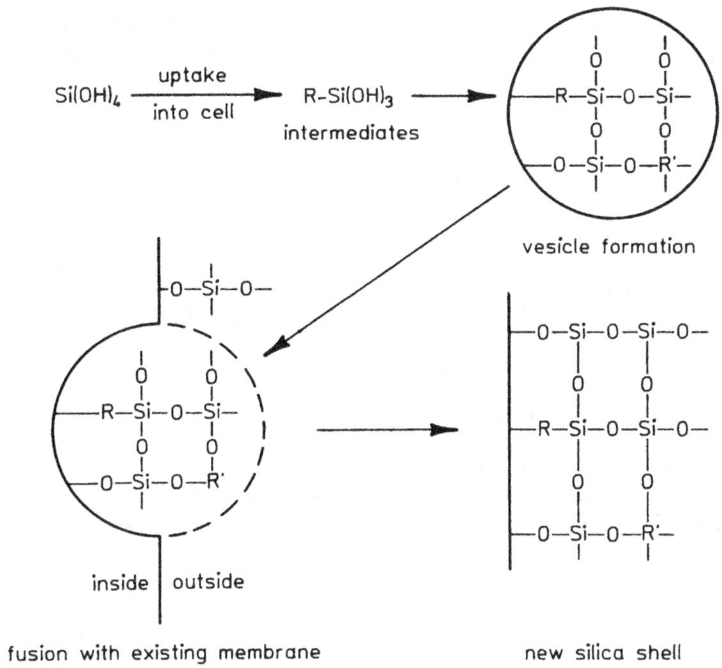

Fig. 26. Diagrammatic representation of processes involved in formation of siliceous walls. From [34] with permission

vascular plants and diatoms [448], whereas in marine environment, radiolarians, silicoflagellates, and sponges are largely responsible, by far the largest contribution being made by marine diatoms [448] which account for 90% of the suspended silica in oceans.

The vast majority of biologically polymerised silica in oceans is redissolved during settling, in part because protective organic coatings are degraded, exposing SiO_2 to $Si(OH)_4$-unsaturated waters, in part because non-crystalline silica is more soluble anyway [448]. The amount of silica annually fixed roughly balances the amount dissolved in the marine environment (25×10^3 Tg) with relatively little being permanently deposited in marine sediments (1×10^3 Tg) [449]. The marine system operates effectively as a closed system and the silica transformations in the marine environment greatly exceed transfers from the terrestrial environment (1×10^3 Tg) which balance deposition to sediments [450]. The silica which does escape dissolution in sediments participates in the formation of authigenic metals such as cherts by replacing calcareous sediment, or reacts with clays to form new silicic minerals. Precipitation of silica in the terrestrial environment to form intergranular cements and overgrowths on primary minerals is considered to be important in the genesis of secondary sedimentary rocks and clay minerals [448].

There are no known records of silica-depositing microorganisms before the Cambrian period. The appearance of silica-depositing organisms from the Cambrian period on has resulted in a roughly ten-fold reduction in the concentration of dissolved silica in the biosphere, providing an excellent example of an alteration in the surface chemistry of the planet brought about by a biological sysem [34].

References

1. Orpen, J.L., Wilson, J.F.: Nature *291*, 218 (1981)
2. Walter, M.R., Buick, R., Dunlop, S.J.R.: Nature *284*, 443 (1980)
3. Lowe, D.R.: Nature *284*, 441 (1980)
4. Grant, W.D., Long, P.E.: Environmental Microbiology. Blackie, Glasgow and London 1981
5. Whittaker, R.H.: Science *163*, 150 (1969)
6. Lewin R.A.: Nature *261*, 697 (1976)
7. Stanier, R.Y., van Niel, C.B.: Arch. Microbiol. *42*, 17 (1962)
8. Brock T.D.: Biology of Microorganisms. 3rd Edition. Prentice/Hall, London 1979
9. Schopf, J.W.: Ann. Rev. Earth Planet. Sci. *3*, 213 (1975)
10. Walter, M.R., Oehler, J.H., Oehler, D.Z.: J. Paleontol. *50*, 872 (1976)
11. Awramik, S.M.: Mineral Deposits and the Evolution of the Biosphere. (Holland, H.D., Schidlowski, M., eds.) Springer-Verlag Berlin Heidelberg New York 1982, p. 67
12. Margulis, L.: Symbiosis in Cell Evolution. W.H. Freeman & Co., San Francisco 1981
13. Küntzel, H., Köchel, H.G.: Nature *293*, 751 (1981)
14. Fox, G.E., Stackebrandt, E., Hespell, R.B., Gibson, J., Maniloff, J., Dyer, T.A., Wolfe, R.S., Balch, W.E., Tanner, R.S., Magrum, L.J., Zablen, L.B., Blakemore, R., Gupta, R., Bonen, L., Lewis, B.J., Stahl, D.A., Luehrsen, K.R., Chen, K.N., Woese, C.R.: Science *209*, 457 (1980)
15. Woese, C.R., Fox, G.E.: Proc. Nat. Acad. Sci. USA *74*, 5088 (1977)
16. Woese, C.R., Fox, G.E.: J. Mol. Evol. *10*, 1 (1977)
17. Stackebrandt, E., Woese, C.R.: Molecular and Cellular Aspects of Evolution (Carlile, M.J., Collins, J.F., Moseley, B.E.B., eds.) Cambridge University Press 1981, p. 1
18. Woese, C.R., Magrum, L.J., Fox, G.E.: J. Mol. Evol. *11*, 245 (1978)
19. Woese, C.R.: Scientific American *245*, 94 (1981)
20. Langworthy, T.A., Tornabene, T.G., Holzer, G.: Zbl. Bakt. Hyg. Orig. *C3*, 228 (1982)
21. Huet, J., Schnabel, R., Sentenac, A., Zillig, W.: EMBO Journal *2*, 129 (1983)
22. Finlay, B.J., Span, A.S.W., Harman, J.M.P.: Nature *303*, 333 (1983)
23. Stanier, R.Y., Pfennig, N., Trüper, H.G.: The Prokaryotes. Volume I (Starr, M.P., Stolp, H., Trüper, H.G., Balows, A., Schlegel, H.G., eds.) Springer-Verlag Berlin Heidelberg New York 1981, p. 197
24. Buchanan, R.E., Gibbons, N.E. (eds.) Bergey's Manual of Determinative Bacteriology 8th Edition. Williams and Wilkins, Baltimore 1974
25. Starr, M.P., Stolp, H., Trüper, H.G., Balows, A., Schlegel, H.G. The Prokaryotes. Volumes I and II. Springer-Verlag Berlin Heidelberg New York 1981
26. Rippka, R., Dereulles, J., Waterbury, J.B., Herdmann, M., Stanier, R.Y.: J. Gen. Microbiol. *111*, 1 (1979)
27. Carr,N.G., Whitton, B.A.: The Biology of the Cyanobacteria. Blackwell, Oxford London Edinburgh Melbourne 1982
28. Deacon, J.W.: Modern Mycology. Blackwell Oxford London Edinburgh Melbourne 1980
29. Webster, J.: Introduction to Fungi. 2nd Edition. Cambridge University Press 1980
30. Burnett, J.H.: Fundamentals of Mycology. Edward Arnold 1976
31. George, G.A.: Brit. Phyc. J. *11*, 49 (1976)
32. Metting, B.: Bot. Rev. *47*, 195 (1981)
33. Round, F.G.: The Biology of the Algae. 2nd Edition. Edward Arnold 1973
34. Heinen, W., Oehler, J.H.: Biogeochemical Cycling of Mineral-Forming Elements (Trudinger, P.A., Swaine, D.J., eds.) Elsevier 1979, p. 431
35. Corliss, J.O.: The Ciliated Protozoa. Pergamon Press New York 1979
36. Kudo, K.: Protozoology 5th Edition. C.C. Thomas Illinois 1971

37. Hall, R.P.: Protozoa Holt, Rinehart and Winston 1964
38. Pike, E.B., Curds, C.R.: Microbial Aspects of Pollution (Sykes, G., Skinner, F.F., eds.). Academic Press London New York 1971, p. 123
39. Curds, C.R.: Ecological Aspects of Used-water Treatment. Volume 1 (Curds, C.R., Hawkes, H.A., eds.). Academic Press London New York 1975, p. 203
40. Wolin, M.J.: Adv. Microb. Ecol. *3*, 49 (1979)
41. Hobson, P.N., Wallace, R.J.: CRC Crit. Rev. Microbiol. *9*, 165 (1982)
42. Hobson, P.N., Wallace, R.J.: CRC Crit. Rev. Microbiol. *9*, 253 (1982)
43. Stout, J.D.: Adv. Microb. Ecol. *4*, 1 (1980)
44. Jones, C.W.: Bacterial Respiration and Photosynthesis. Nelson 1982
45. Knowles, C.J.: Diversity of Bacterial Respiratory Systems. Volume I. CRC Press, Boca Raton Florida 1980
46. Lehninger, A.L.: Biochemistry, 2nd Edition. Worth Publishers New York 1975
47. Gregory, R.F.P.: Biochemistry of Photosynthesis. Wiley - Interscience 1974
48. Kelly, D.P.: Phil. Trans. R. Soc. Lond. *B298*, 499 (1982)
49. Brown, C.M., McDonald-Brown, D.S., Meers, J.L.: Adv. Microb. Physiol. *11*, 1 (1974)
50. Morris, J.G.: J. Appl. Bacteriol. *40*, 229 (1976)
51. Kong, S., Davison, A.J.: Arch. Biochem. Biophys. *204*, 18 (1980)
52. Brock, T.D.: Thermophilic Microorganisms and Life at High Temperatures. Springer-Verlag New York Heidelberg Berlin 1978
53. Kushner, D.J. (ed.): Microbial Life in Extreme Environments. Academic Press New York London 1978
54. Shilo, M. (ed.): Strategies of Life in Extreme Environments. Dahlem Conference Report. Verlag-Chemie Weinheim New York 1979
55. Baross, J.A., Deming, J.W.: Nature *303*, 423 (1983)
56. Escher, F., Zillig, W., Stetter, K.O., Schreiber, G.: Nature *301*, 511 (1983)
57. Zillig, W., Holz, I., Janeković, D., Schäfer, W., Reiter, W.D.: System. Appl. Microbiol. *4*, 88 (1983)
58. Hüber, H., Thomm, M., König, H., Thies, G., Stetter, K.O.: Arch. Microbiol. *132*, 47 (1982)
59. Zillig, W., Holz, I., Schäfer, W., Reiter, W.D.: Syst. Appl. Microbiol. *4*, 88 (1983)
60. Zillig, W., Gierl, A., Schreiber, G., Wunderl, S., Janecović, D., Stetter, K.O., Klenk, H.P.: Syst. Appl. Microbiol. *4*, 79 (1983)
61. Zillig, W., Stetter, K.O., Prangishvilli, D., Schäfer, W., Wunderl, S., Janeković, D., Holz, I., Palm, P.: Zbl. Bakt. Hyg. I, Abt. Orig. *C3*, 304 (1982)
62. Zeikus, J.G., Hegge, P.W., Anderson, M.A.: Arch. Microbiol. *122*, 41 (1979)
63. Heinrich, M.R. (ed.): Extreme Environments. Mechanisms of Microbial Adaption. Academic Press 1976
64. Vishniac, H.S., Baharbeen, S.: Int. J. Syst. Bact. *32*, 437 (1982)
65. Deming, J.W., Colwell, R.R.: App. Env. Microbiol. *44*, 1222 (1982)
66. Yayanos, A.A., Dietz, A.S., van Boxtel, R.: Science *205*, 808 (1979)
67. Yayanos, A.A., Dietz, A.S., van Boxtel, R.: App. Env. Microbiol. *44*, 1356 (1982)
68. Yayanos, A.A., Dietz, A.S.: Science *220*, 497 (1983)
69. Morita, R.Y.: Origins of Life *10*, 215 (1980)
70. Morita, R.Y.: Can. J. Microbiol. *26*, 1375 (1981)
71. Tindall, B.J., Ross, H.N.M., Grant, W.D.: Syst. Appl. Microbiol. *5*, 41 (1984)
72. Imhoff, J.F., Sahl, H.G., Soliman, G.S.H., Trüper, H.G.: Geomicrobiol. J., *1*, 219 (1978)
73. Galinski, E.A., Trüper, H.G.: FEMS Microbiol. Lett. *13*, 357 (1982)
74. Mescher, M.F., Strominger, J.L.: Energetics and Structure of Halophilic Microorganisms (Caplan, S.R., Ginsburg, M., eds.): Elsevier/North-Holland 1978, p. 503
75. Stoeckenius, W., Bogolmi, R.A.: Ann. Rev. Biochem. *51*, 587 (1982)
76. Brock, T.D., Brock, K.W., Belley, R.T., Weiss, R.L.: Arch. Microbiol. *84*, 54 (1972)
77. Krulwich, T.A., Guffanti, A.A.: Adv. Microb. Physiol. *24*, 173 (1983)
78. Bull, A.T., Slater, J.H.: Microbial Interactions and Communities. Volume 1 (Bull, A.T., Slater, J.H., eds.) Academic Press London 1982, p. 1
79. Bull, A.T.: Contemporary Microbial Ecology. (Ellwood, D.C., Hedger, J.N., Latham, M.J., Lynch, J.M., Slater, J.H., eds.) Academic Press London 1980, p. 107

80. Wimpenny, J.W.T.: Biol. Rev. *56,* 295 (1981)
81. Sprent, J.I.: The Biology of Nitrogen-fixing Organisms. McGraw-Hill Book Company (U.K.) London 1979
82. Postgate, J.R.: The Fundamentals of Nitrogen Fixation. Cambridge University Press 1982
83. van der Ploeg, S.W.F.: Handbook of Environmental Chemistry. Volume 1 B, (Hutzinger, O., ed.) Springer-Verlag Berlin Heidelberg New York 1982, p. 1
84. Lewis, D.H.: Evolution in the Microbial World (Carlile, M.J., Skehel, J.J., eds.) Cambridge University Press 1974, p. 367
85. Starr, M.P.: Symbiosis (Jennings, D.H., Lee D.L., eds.) Cambridge University Press 1975, p. 1
86. Bull, A.T., Slater, J.H.: Microbial Interactions and Communities (Bull, A.T., Slater, J.H., eds.) Academic Press London 1982, p. 13
87. Odum, E.P.: Fundamentals of Ecology. Saunders, Philadelphia 1953
88. Cooke, R.: The Biology of Symbiotic Fungi. John Wiley and Sons, London 1977
89. De Bary, A.: Comparative Morphology and Biology of the Fungi, Mycetozoa and Bacteria. Oxford University Press 1887
90. Berkeley, R.C.W., Lynch, J.M., Melling, J., Rutter, P.R., Vincent, B.: Microbial Adhesion to Surfaces. Ellis Horwood Chichester 1980
91. Wardell, J.W., Brown, C.M., Flannigan, B.: Microbes in their Natural Environments (Slater, J.H., Whittenbury, R., Wimpenny, J.W.T., eds.) Cambridge University Press 1983, p. 351
92. Brown, C.M., Ellwood, D.C., Hunter, J.R.: FEMS Microbiol. Lett. *1,* 163 (1977)
93. Costerton, J.W., Geesey, G.G., Cheng, K.J.: Scientific American *238,* 76 (1978)
94. Marshall, K.C.: Interfaces in Microbial Ecology. Harvard University Press Cambridge 1976
95. Costerton, J.W., Irvin, R.T., Cheng, K.J.: Ann. Rev. Microbiol. *35,* 299 (1981)
96. Ellwood, D.C., Keevil, C.W., Marsh, P.D., Brown, C.M., Wardell, J.W.: Phil. Trans. R. Soc. Lond. *B297,* 517 (1982)
97. Wangersky, P.J.: Ann. Rev. Ecol. Syst. *7,* 161 (1976)
98. Norkrans, B.: Adv. Microb. Ecol. *4,* 51 (1980)
99. Wimpenny, J.W.T.: Phil. Trans. R. Soc. Lond. *B297,* 497 (1982)
100. Rowbury, R.J., Armitage, J.P., King, C.: Microbes in their Natural Environments (Slater, J.H., Whittenbury, R., Wimpenny, J.W.T., eds.) Cambridge University Press 1983, p. 299
101. Westall, J., Stumm, W.: Handbook of Environmental Chemistry. Volume 1 A (Hutzinger, O., ed.). Springer-Verlag Berlin Heidelberg New York 1980, p. 17
102. Wangersky, P.J.: Handbook of Environmental Chemistry. Volume 1 A (Hutzinger, O., ed.). Springer-Verlag Berlin Heidelberg New York 1980, p. 51
103. Wimpenny, J.W.T., Coombs, J.P., Lovitt, R.W., Whittaker, S.G.: J. Gen. Microbiol. *127,* 277 (1981)
104. Paul, E.A., Huang, P.M.: Handbook of Environmental Chemistry. Volume 1 A (Hutzinger, O., ed.). Springer-Verlag Berlin Heidelberg New York 1980, p. 69
105. Chapman, H.D.: Ecology of Soil-Borne Plant Pathogens (Baker, K.F., Snyder, W.C., eds.). John Murray London 1966, p. 120
106. Griffin, D.M.: Adv. Microb. Ecol. *5,* 91 (1981)
107. Parkin, T.B., Brock, T.D.: Arch. Microbiol. *125,* 19 (1980)
108. Greensmith, V.T.: Petrology of the Sedimentary Rocks. George Allen and Unwin Ltd., London 1971, p. 502
109. Nedwell, D.B., Brown, C.M. (eds.): Sediment Microbiology Academic Press London 1982
110. Billett, D.S.M., Lampitt, R.S., Rice, A.L., Mantoura, R.F.C.: Nature *302,* 520 (1983)
111. Malcolm, S.J., Stanley, S.O.: Sediment Microbiology (Nedwell, D.B., Brown, C.M., eds.). Academic Press London 1982, p. 1
112. Burns, R.G., Slater, J.H. (eds.): Experimental Microbial Ecology. Blackwell Scientific Publications, Oxford 1982
113. Burns, R.G.: Microbes in their Natural Environment (Slater, J.H., Whittenbury, R., Wimpenny, J.W.T., eds.) Cambridge University Press 1982, p. 249

114. Slater, J.H.: Microbial Ecology: A Conceptual Approach (Lynch, J.M., Poole, N.J., eds.). Blackwell Scientific Publications Oxford 1979, p. 45
115. Monod, J.: Ann. Rev. Microbiol. *3,* 371 (1949)
116. Gray, T.R.G., Hissett, R., Duxbury, T.: Rev. D'Ecol. Biol. Sol. *11,* 15 (1974)
117. Jenkinson, D.S., Ladd, J.N.: Soil Biochemistry. Volume 5 (Paul, E.A., Ladd, J.N., eds.). Marcel Dekker New York 1981, p. 415
118. Frankland, J.C.: Decomposer Basidiomycetes: their Biology and Ecology (Frankland, J.C., Hedger, J.N., Swift, M.J., eds.). Cambridge University Press 1982, p. 241
119. Hagstrom, Å., Larson, V., Horstedt, P., Normar, E.S.: Appl. Env. Microbiol. *32,* 805 (1979)
120. Cuhel, R.L., Taylor, C.D., Jannasch, H.W.: Appl. Env. Microbiol. *43,* 160 (1982)
121. Karl, D.M.: Appl. Env. Microbiol. *38,* 850 (1979)
122. van Es, F.B., Meyer-Reil, L.A.: Adv. Microb. Ecol. *6,* 111 (1982)
123. Pedros-Alio, C., Brock, T.D.: Appl. Env. Microbiol. *44,* 203 (1982)
124. Frankland, J.C.: Soil. Biol. Biochem. *7,* 339 (1975)
125. Warcup, J.H.: The Ecology of Soil Fungi (Parkinson, D., Waid, J.S., eds.). Liverpool University Press 1960, p. 3
126. Kuznetsov,S.I., Dubinina, G.A., Lapteva, W.A.: Ann. Rev. Microbiol. *33,* 377 (1979)
127. Poindexter, J.S.: Adv. Microb. Ecol. *5,* 63 (1981)
128. Morita, R.Y.: Adv. Microb. Ecol. *6,* 171 (1982)
129. Dow, C.S., Whittenbury, R., Carr, N.G.: Microbes in their Natural Environments (Slater, J.H., Whittenbury, R., Wimpenny, J.W.T., eds.). Cambridge University Press 1983, p. 187
130. Poindexter, J.S.: Microbiol. Rev. *45,* 123 (1981)
131. Hirsch, P.: Strategies of Life in Extreme Environments: (Shilo, M., ed.). Dahlem Conference Report Verlag Chemie Weinheim New York 1979, p. 357
132. Jannasch, H.W.: Limnol. Oceanog. *12,* 264 (1967)
133. Jannasch, H.W., Mateles, R.I.: Adv. Microb. Physiol. *11,* 165 (1974)
134. Harder, W., Kuenen, J.G., Martin, A.: J. Appl. Bact. *43,* 1 (1977)
135. Sieburth, J.M.: Sea Microbes. Oxford University Press New York 1979
136. Torrella, F., Morita, R.Y.: Appl. Env. Microbiol. *41,* 518 (1981)
137. MacDowell, M.T., Hood, M.A.: Appl. Env. Microbiol. *43,* 566 (1982)
138. Ohta, H, Hattori, T.: Soil Biol. Biochem. *15,* 1 (1982)
139. Mirocha, C.J., De Vay, J.G.: Can. J. Microbiol. *17,* 1373 (1971)
140. Friedmann, E.I.: Science *215,* 1045 (1982)
141. Staley, J.T., Palmer, F.J., Adams, J.B.: Science *215,* 1093 (1982)
142. Reanney, D.C., Gowland, P.C., Slater, J.H.: Microbes in their Natural Environments (Slater, J.H., Whittenbury, R., Wimpenny, J.W.T., eds.). Cambridge University Press 1983, p. 379
143. Alexander, M.: Introduction to Soil Microbiology 2nd Edition. John Wiley and Sons New York 1977
144. Atlas, R.M., Bartha, R.: Microbial Ecology: Fundamentals and Applications. Addison Wesley Massachusetts 1981
145. Domsch, K.H., Gams, W.: Fungi in Agricultural Soils. Longman London 1972
146. Christensen, M.: The Fungal Community (Wicklow, D.T., Carroll, G.C., eds.). Marcel Dekker New York 1981, p. 201
147. Winogradsky, S.N.: Microbiologie du Sol; Problems et Methodes. Masson et Cie Paris 1949
148. Ko, W.H., Chow, F.L.: J. Gen. Microbiol. *102,* 295 (1977)
149. Lockwood, J.L.: Biol. Rev. *52,* 1 (1977)
150. Lockwood, J.L.: Filonow, A.B.: Adv. Microb. Ecol. *5,* 1 (1981)
151. Rheinheimer, G.: Aquatic Microbiology. 2nd Edition John Wiley Chichester 1980
152. Caldwell, D.E.: CRC Crit. Rev. Microbiol. *5,* 305 (1977)
153. Tindall, B.J., Grant, W.D.: Anaerobic Bacteria in habitats other than Man. (Barnes, E.M., Mead, G., eds.). Academic Press London New York, in press
154. Padan, E., Cohen, V.: The Biology of the Cyanobacteria (Carr, N.G., Whitton, B.A., eds.). Blackwell Scientific Publications Oxford 1982, p. 215
155. Waterbury, B., Watson, S.W., Guillard, R.P.L., Brand, L.E.: Nature *277,* 293 (1979)

156. Rippka, R., Waterbury, J.B., Stanier, R.Y.: The Prokaryotes. Volume 1 (Starr, M.P., Stolp, H., Trüper, H.G., Balows, A., Schlegel, H.G., eds.). Springer-Verlag Berlin Heidelberg New York 1981, p. 247
157. Karl, D.M., Wirsen, C.O., Jannasch, H.W.: Science *207,* 1345 (1980)
158. Edmond, J.M., Von Damm, K.: Scientific American *248,* 70 (1983)
159. Powell, M.A., Somers, G.N.: Science *219,* 297 (1983)
160. Kepkay, P.E., Cooke, R.C., Novitsky, J.A.: Science *204,* 68 (1979)
161. Cavanaugh, C.M., Gardiner, S.L., Jones, M.L., Jannasch, H.W., Waterbury, J.B.: Science *213,* 340 (1981)
162. Felbeck, H., Childress, J.J., Somers, G.N.: Nature *293,* 291 (1981)
163. Cavanaugh, C.M.: Nature *302,* 58 (1983)
164. Williams, P.J. le B.: Kiel. Meeres. Sond. *5,* 1 (1981)
165. Jones, J.G.: Sediment Microbiology (Nedwell, D.B., Brown, C.M., eds.). Academic Press London 1982, p. 107
166. Novitsky, J.A.: Appl. Env. Microbiol. *45,* 1753 (1983)
167. Novitsky, J.A.: Appl. Env. Microbiol. *45,* 1761 (1983)
168. Joint, I.R., Morris, R.J.: Oceanog. Mar. Biol. Ann. Rev. *20,* 65 (1982)
169. Maksimov, V.N., Rudykh, N.R.: Microbiology *51,* 293 (1982)
170. Allen, D.A., Austin, B., Colwell, R.R.: J. Gen. Microbiol. *129,* 2043 (1983)
171. Austin, B., Garges, S., Conrad, B., Harding, E.E., Colwell, R.R., Simidu, U., Taga, N.: Appl. Env. Microbiol. *37,* 704 (1979)
172. Austin, B., Allen, D.A., Zachary, A., Belas, M.R., Colwell, R.R.: Can. J. Microbiol. *25,* 447 (1979)
173. Austin, B.: J. Appl. Bact. *53,* 253 (1982)
174. Hauxhurst, J.D., Krichevsky, M.I., Atlas, R.M.: J. Gen. Microbiol. *120,* 131 (1980)
175. Stewart, W.D.P.: Microbes in their Natural Environments (Slater, J.H., Whittenbury R., Wimpenny, J.W.T., eds.). Cambridge University Press 1983, p. 1
176. MacLeod, R.M.: Adv. Microbiol. Sea *1,* 95 (1968)
177. Baumann, P., Baumann, L.: The Prokaryotes. Volume II. (Starr, M.P., Stolp, H., Trüper, H.G., Balows, A., Schlegel, H.G., eds.). Springer-Verlag Berlin Heidelberg New York 1981, p. 1302
178. Witzel, K.P., Moaledj, K.O., Overbeek, H.J.: Arch. Hydrobiol. *95,* 507 (1982)
179. Bell, C.R., Albright, L.J.: Appl. Env. Microbiol. *43,* 1227 (1982)
180. Kirchman, D., Mitchell, R.: Appl. Env. Microbiol. *43,* 200 (1982)
181. Walsby, A.C.: Relations between Structure and Function in the Prokaryotic Cell (Stanier, R.Y., Rogers, H.J., Ward, B.J., eds.). Cambridge University Press 1978, p. 327
182. Blakemore, R.P.: Ann. Rev. Microbiol. *36,* 217 (1982)
183. Tabor, P.S., Deming, J.W., Ohwada, K., Colwell, R.R.: Appl. Env. Microbiol. *44,* 413 (1982)
184. Gareth Jones, E.B.: Recent Advances in Aquatic Mycology. Elek Science London 1976
185. Kohlmeyer, J., Kohlmeyer, G.: Marine Mycology: The Higher Fungi. Academic Press New York 1979
186. Emerson, R., Natvig, D.O.: The Fungal Community (Wicklow, D.T., Carroll, G.L., eds.). Marcel Dekker New York 1981, p. 109
187. Finlay, B.J.: Experimental Microbial Ecology (Burns, R.G., Slater, J.H., eds.). Blackwell Scientific Oxford 1982, p. 44
188. Stout, J.D.: Adv. Microb. Ecol. *4,* 1 (1980)
189. Sieburth, J.M., Davis, P.G.: Ann. Inst. Oceanog. *58,* 285 (1982)
190. Fenchel, T.M., Jørgensen, B.B.: Adv. Microb. Ecol. *2,* 1 (1978)
191. Reanney, D.C., Gowland, P.C., Slater, J.H.: Microbes in their Natural Environments (Slater, J.H., Whittenbury, R., Wimpenny, J.W.T., eds.). Cambridge University Press 1983, p. 379
192. Reanney, D.C., Roberts, W.P., Kelly, W.J. (1982) Microbial Interactions and Communities Volume I (Bull, A.T., Slater, J.H., eds.). Academic Press London 1982, p. 287
193. Zeikus, J.G.: Microbes in their Natural Environments (Slater, J.H., Whittenbury, R., Wimpenny, J.W.T., eds.). Cambridge University Press 1982, p. 423
194. Slater, J.H., Bull, A.T.: Phil. Trans. R. Soc. Lond. *B297,* 575 (1982)

195. Slater, J.H.: Mixed Culture Fermentations (Bushell, M.E., Slater, J.H., eds.). Academic Press London 1981, p. 1
196. Wolin, M.J.: Microbial Interactions and Communities. Volume 1 (Bull, A.T., Slater, J.H., eds.). Academic Press London 1982, p. 323
197. Mah, R.A.: Phil. Trans. R. Soc. Lond. *B297,* 599 (1982)
198. Biebl, H., Pfennig, N.: Arch. Microbiol. *117,* (1978)
199. Martinez, L., Silver, M.W., King, J.M., Alldredge, A.L.: Science *221,* 152 (1983)
200. Vincent, J.M.: The Prokaryotes. Volume I (Starr, M.P., Stolp, H., Trüper, H.G., Balows, A., Schlegel, H.G., eds.). Springer-Verlag Berlin Heidelberg New York 1981, p. 818
201. Trinick, M.J.: Nitrogen Fixation. Volume II (Broughton, W.J., ed.). Oxford University Press 1982, p. 76
202. Beringer, J.E., Bewin, W., Johnston, A.W.B., Schulman, H.M., Hopwood, D.A.: Proc. R. Soc. Lond. *B204,* 219 (1979)
203. Callahan, D., Tredici, P., Torrey, J.G.: Science *199,* 899 (1978)
204. Becking, J.H.: Int. J. Syst. Bact. *20,* 201 (1970)
205. Beringer, J.E., Brewin, N.J., Johnston, A.W.B.: Nitrogen Fixation. Volume 2 (Broughton, W.J., ed.). Oxford University Press 1982, p. 167
206. Dazzo, F.B., Hubbell, D.H.: Nitrogen Fixation. Volume 2 (Broughton, W.J., ed.). Oxford University Press 1982, p. 274
207. Urban, J.G., Dazzo, F.B.: Appl. Env. Microbiol. *44,* 219 (1982)
208. Tjepkema, J.D., Ormerod, W., Torrey, J.L.: Nature *287,* 633 (1980)
209. Appleby, C.A., Tjepkema, J.D., Trinick, M.J.: Science *220,* 951 (1983)
210. Harley, J.L., Smith, S.E.: Mycorrhizal Symbiosis. Academic Press London 1983
211. Mosse, B., Stribley, D.P., Le Tacon, F.: Adv. Microb. Ecol. *5,* 137 (1981)
212. Smith, S.E.: Biol. Rev. *55,* 475 (1980)
213. Hayman, D.S.: Can. J. Bot. *61,* 944 (1983)
214. Duddridge, J.A., Malibar, A., Read, D.J.: Nature *287,* 834 (1980)
215. Marks, G.C., Kozlowski, T.T.: Ectomycorrhizae. Academic Press New York 1973
216. Chilvers, G.A., Harley, J.L.: New Phytol. *84,* 319 (1980)
217. Callow, J.A., Capaccio, L.C.M., Parish, G., Tinker, P.B.: New Phytol. *80,* 125 (1978)
218. Read, P.J.: Can. J. Bot. *61,* 985 (1983)
219. Smith, D.C., Musatine, L., Lewis, D.H.: Biol. Rev. *44,* 17 (1969)
220. Cooper, K.M., Lösel, D.: New Phytol. *80,* 143 (1978)
221. Hale, M.E.: The Biology of Lichens. 3rd Edition Edward Arnold London 1983
222. Ahmadjian, V.: Prog. Phycol. Res. *1,* 179 (1982)
223. Collins, C.R., Farrar, J.F.: New Phytol. *81,* 71 (1978)
224. Farrar, J.F.: Lichenology: Progress and Problems (Brown, D.H., Hawkesworth, D.L., Bailey, R.H., eds.). Academic Press London New York 1976, p. 383
225. Hill, D.J.: Lichenology: Problems and Progress (Brown, D.H., Hawkesworth, D.L., Bailey, R.H., eds.). Academic Press London New York 1976, p. 457
226. Farrar, J.F.: New Phytol. *81,* 65 (1978)
227. Smith, D.C.: Strategies of Microbial Life in Extreme Environments (Shilo, M., ed.). Dahlem Conference Report Verlag-Chemie Weinheim New York 1979, p. 291
228. Smith, D.C.: Symbiosis (Jennings, D.H., Lee, D.L., eds.). Cambridge University Press 1975, p. 633
229. Tuominen, Y., Jaakkula, T.: The Lichens (Ahmadjian, V., Hale, M.E., eds.). Academic Press New York 1973, p. 185
230. Syers, J.K., Iskandar, I.K.: The Lichens (Ahmadjian, V., Hale, M.E., eds.). Academic Press New York 1973, p. 225
231. Puckett, K.J., Nieboer, E., Gorzynski, M.J., Richardson, D.H.S.: New Phytol. *72,* 329 (1978)
232. Goyal, R., Seaward, M.R.D.: New Phytol. *90,* 85 (1982)
233. Brown, D.H.: Lichenology: Problems and Progress (Brown, D.H., Hawkesworth, D.L., Bailey, R.H., eds.). Academic Press London 1976, p. 419
234. Ferry, B.W., Baddeley, M.S., Hawkesworth, D.L.: Air Pollution and Lichens. University of London Athlone Press 1973

235. Isenberg, H.D., Balows, A.: The Prokaryotes. Volume 1 (Starr, M.P., Stolp, H., Trüper, H.G., Balows, A., Schlegel, H.G., eds.). Springer-Verlag Berlin Heidelberg New York 1981, p. 83
236. Linton, A.H.: Microbes, Man and Animals. John Wiley and Sons London 1982
237. Breznak, J.A.: Ann. Rev. Microbiol. *36*, 323 (1982)
238. Orpin, C.G.: J. Gen. Microbiol. *99* (1977)
239. Bauchop, T.: Appl. Env. Microbiol. *38*, 148 (1979)
240. Bolin, B., Degens, E.T., Kempe, S., Ketner, P. (eds.): SCOPE Rep. 13. John Wiley and Sons, Chichester 1979
241. Golubić, S., Krumbein, W., Schneider, J.: Biogeochemical Cycling of Mineral-Forming Elements (Trudinger, P.A., Swaine, D.J., eds.). Elsevier Scientific Publishing Co., Amsterdam 1979, p. 29
242. Krumbein, W.E., Swart, P.K.: Microbial Geochemistry (Krumbein, W.E., ed.). Blackwell Scientific Publications, Oxford 1983, p. 5
243. Zehnder, A.J.B.: Handbook of Environmental Chemistry. Volume 2B (Hutzinger, O., ed.). Springer-Verlag Berlin Heidelberg New York 1982, p. 83
244. Fenchel, T., Blackburn, T.H.: Bacteria and Mineral Cycling. Academic Press, London New York 1979
245. Pfennig, N.: The Photosynthetic Bacteria. (Clayton, R.K., Sistrom, W.R., eds.). Plenum Press, New York London 1978, p. 3
246. Swift, M.J., Heal, D.W., Anderson, J.M.: Decomposition in terrestrial environments. Blackwell Scientific Publications, Oxford 1979
247. Swift, M.J.: The Role of Terrestrial and Aquatic Organisms in Decomposition Processes. (Anderson, J.M., Macfadyen, A., eds.). Blackwell Scientific Publications, Oxford 1976, p. 185
248. Zeikus, J.G.: Adv. Microb. Ecol. *5*, 211 (1981)
249. Eriksson, K-E.: Trends in the Biology of Fermentations for Fuels and Chemicals (Hollaender, A., ed.). Plenum Press, New York London 1981, p. 19
250. Burns, R.G.: Microbial Interactions and Communities. Volume 1 (Bull, A.T., Slater, J.H., eds.). Academic Press, London 1982, p. 475
251. Kirk, T.K., Higuchi, T., Chang, H.: Lignin Biodegradation: Microbiology, Chemistry and Potential Applications. Volume 1 and 2. CRC Press Inc., Boca Raton Florida 1980
252. Crawford, R.L.: Lignin biodegradation and transformation. John Wiley and Sons, New York 1981
253. Koenigs, J.W.: Wood and Fiber *6*, 66 (1974)
254. Highley, T.L.: Appl. Env. Microbiol. *40*, 1145 (1980)
255. Kirk, T.K., Chang, H.: Holzforschung *29*, 56 (1975)
256. Tien, M., Kirk, T.K.: Science *221*, 661 (1983)
257. Kaarik, A.A.: Biology of Plant Litter Decomposition. Volume 1. (Dickinson, C.H., Pugh, G.J.F., eds.). Academic Press, London New York 1974, p. 129
258. Orpin, C.G.: J. Gen. Microbiol. *123*, 287 (1981)
259. Stirling, D.I., Dalton, H.: FEMS Microbiol. Lett. *5*, 315 (1979)
260. Dalton, H., Stirling, D.I.: Phil. Trans. R. Soc. Lond. *B297*, 481 (1982)
261. Bartholomew, G.W., Alexander, M.: Appl. Env. Microbiol. *37*, 932 (1979)
262. Conrad, R., Meyer, D., Seiler, W.: Appl. Env. Microbiol. *42*, 211 (1981)
263. Leadbetter, E.R., Foster, J.W.: Arch. Mikrobiol. *30*, 91 (1958)
264. Perry, J.J.: Microbiol. Rev. *43*, 59 (1979)
265. Beam, H.W., Perry, J.J.: Arch. Mikrobiol. *91*, 87 (1973)
266. Beam, H.W., Perry, J.J.: J. Gen. Microbiol. *82*, 163 (1974)
267. Higgins, I.J., Best, D.J., Hammond, R.C.: Nature *286*, 561 (1980)
268. Degens, E.T., Mopper, K.: Chemical Oceanography. Volume 6. (Riley, J.P., Chester, R., eds.). Academic Press, New York London 1976, p. 59
269. Stout, J.D., Goh, K.M., Rafter, T.A.: Soil Biochem. *5*, 1 (1981)
270. Brooks, J.: Organic Maturation Studies and Fossil Fuel Exploration (Brooks, J., ed.). Academic Press, London New York 1981, p. 1
271. Philp, R.P., Calvin, M.: Environmental Biogeochemistry. Volume 1 (Nriagu, J.O., ed.). Ann Arbor Science Publishers, Ann Arbor 1976, p. 131

272. Hunt, J.M.: Petroleum Geochemistry and Geology. W.H. Freeman and Company, San Fransisco 1979
273. Degens, E.T., Paluska, A.: Nature *281,* 666 (1979)
274. Edmunds, K.L.H., Brassell, S.C., Eglinton, G.: Mineral Deposits and the Evolution of the Biosphere (Holland, H.D., Schildowski, M., eds.). Springer-Verlag Berlin Heidelberg New York 1982, p. 31
275. Given, P.E., Dickinson, C.H.: Soil Biochem. *3,* 123 (1975)
276. Tate, R.L.: Adv. Microb. Ecol. *4,* 169 (1980)
277. Vogels, G.D.: Antonie van Leeuwenhoek *45,* 347 (1979)
278. Anthony, C.: The Biochemistry of Methylotrophs. Academic Press, London New York 1982
279. Large, P.J.: Methylotrophy and Methanogenesis. Van Norstrand Reinhold (U.K.) Co. Ltd., Wokingham 1983
280. Zeikus, J.G.: Adv. Microb. Physiol. *24,* 215 (1983)
281. Bowien, B., Schlegel, H.G.: Ann. Rev. Microbiol. *35,* 405 (1981)
282. Stupperich, E., Hammel, K.E., Fuchs, G., Thauer, R.K.: FEBS Lett. *152,* 21 (1983)
283. Kenealy, W.R., Zeikus, J.G.: J. Bact. *146,* 133 (1981)
284. Archer, D.B., Harris, J.: Anaerobic Bacteria in Habitats other than Man (Barnes, E.M., Mead, G., eds.). Academic Press, London, in press
285. Stetter, K.O., Thomm, M., Winter, J., Wildgruber, G., Huber, H., Zillig, W., Janecović, D., König, H., Palm, P., Wunderl, C.: Zbl. Bakt. Hyg. I. Abt. Orig. *C2,* 166 (1981)
286. Belyaev, S.S., Ivanov, M.V.: Ecol. Bull. (Stockholm) *35,* 273 (1983)
287. Whittenbury, R., Krieg, N.R.: Bergey's Manual of Determinative Bacteriology 9th edition. Volume 1. (Krieg, N.R., Holt, J.G., eds.). Williams and Wilkins Co., Baltimore, 1984
288. Hanson, R.S.: Adv. Appl. Microbiol. *26,* 3 (1980)
289. Seiler, W.: Environmental Biogeochemistry and Geomicrobiology. Volume 3 (Krumbein, W.E., ed.). Ann Arbor Scientific Publishers, Ann Arbor 1978, p. 773
290. Tornabene, T.G.: Trends in the Biology of Fermentations for Fuels and Chemicals (Hollaender, A., ed.). Plenum Press, New York London 1981, p. 421
291. Gibson, D.T.: The Handbook of Environmental Chemistry. Volume 2A (Hutzinger, O., ed.). Springer-Verlag Berlin Heidelberg New York 1980, p. 162
292. Higgins, I.J., Gilbert, P.D.: The Oil Industry and Microbial Ecosystems (Chater, K.W.A., Somerville, H.J., eds.). Heyden and Sons Ltd., London 1978, p. 80
293. Atlas, R.M.: Microbiol. Rev. *45,* 180 (1981)
294. Cooper, D.G., Zajic, J.E.: Adv. Appl. Microbiol. *26,* 229 (1980)
295. Watkinson, R.J.: Hydrocarbons in Biotechnology (Harrison, D.E.F., Higgins, I.J., Watkinson, R.J., eds.). Heyden and Son Ltd., London 1980, p. 11
296. Bartha, R., Atlas, R.M.: Adv. Appl. Microbiol. *22,* 225 (1977)
297. McGill, W.B., Rowell, M.J., Westlake, D.W.S.: Soil Biochem. *5,* 229 (1981)
298. Chakrabarty, A.M.: Ann. Rev. Genet. *10,* 7 (1976)
299. Farrell, R., Chakrabarty, A.M.: Plasmids of Medical, Environmental and Commercial Importance (Timmis, K.W., Puhler, A., eds.). Elsevier/North-Holland Biomedical Press, Amsterdam 1979, p. 97
300. Davis, J.B.: Petroleum Microbiology. Elsevier, London Amsterdam New York 1967
301. Perry, J.J.: Adv. Appl. Microbiol. *26,* 89 (1980)
302. Söderlund, R., Svennson, B.H.: Nitrogen, Phosphorus and Sulphur – Global Cycles Scope Report 7. Ecological Bulletins (Stockholm) 1976, p. 23
303. Söderlund, R., Rosswall, T.: Handbook of Environmental Chemistry. Volume IB (Hutzinger, O., ed.). Springer-Verlag Berlin Heidelberg New York 1982, p. 61
304. Wallace, W., Nicholas, D.J.D.: Biol. Rev. *49,* 359 (1969)
305. Focht, D.D., Verstraete, W.: Adv. Microb. Ecol. *1,* 135 (1977)
306. Doxtader, K.G., Alexander, M.: J. Bacteriol. *91,* 1186 (1966)
307. Garcia, J.L., Tiedje, J.M.: Microbiology of Tropical Soils and Plant Productivity (Dommergues, Y.R., Diem, H.G., eds.). Martinus Nijhoff/Dr. W. Junk Publishers, The Hague, Boston, London 1982, p. 187
308. Watson, S.W., Valois, F.W., Waterbury, J.B.: The Prokaryotes. Volume 1 (Starr, M.P., Stolp, H., Trüper, H.G., Balows, A., Schlegel, H.G., eds.). Springer-Verlag Berlin Heidelberg New York 1981, p. 1005

309. Belser, L.W.: Ann. Rev. Microbiol. *33*, 309 (1979)
310. Suzuki, I., Kwok, S.C., Dular, V., Tsang, D.C.Y.: Can. J. Biochem. *59*, 477 (1981)
311. Vollenweider, R.A.: Scientific fundamentals of the eutrophication of lakes and flowing waters with particular reference to nitrogen and phosphorus as factors in eutrophication. OECD Paris 1968
312. Payne, W.J.: Bacteriol. Rev. *37*, 407 (1973)
313. Brown, C.M., McDonald-Brown, O.S., Meers, J.L.: Adv. Microb. Physiol. *11*, 1 (1974)
314. Stouthamer, A.H.: Adv. Microb. Physiol. *14*, 315 (1976)
315. Delwiche, C.C., Bryan, B.A.: Ann. Rev. Microbiol. *30*, 241 (1976)
316. Jeter, R.M., Ingraham, J.L.: The Prokaryotes. Volume 1. (Starr, M.P., Stolp, H., Trüper, H.G., Balows, A., Schlegel, H.G., eds.). Springer-Verlag Berlin Heidelberg New York 1981, p. 913
317. Knowles, R.: Soil Biochemistry. Volume 5 (Paul, G.A., Ladd, J.N., eds.) Marcel Dekker Inc. New York 1981, p. 323
318. Herbert, R.A.: Sediment Microbiology. (Nedwell, D.B., Brown, C.M., eds.). Academic Press New York London 1982, p. 53
319. Payne, W.J.: Bacteriol. Rev. *37*, 409 (1973)
320. Koike, I., Hattori, A.: J. Gen. Microbiol. *88*, 11 (1975)
321. Garber, E.A.E., Hollocher, T.C.: J. Biol. Chem. *256*, 5459 (1981)
322. Hall, J.B.: Microbiology – 1978 (Schlessinger, D., ed.). American Society for Microbiology – Washington 1978, p. 296
323. van Hartingsveldt, J., Marinus, M.G., Stouthamer, A.H.: Genetics *67*, 469 (1971)
324. Woldendorp, J.W.: Bull. Soc. Fr., Coll. Rhizosphere *122*, 89 (1975)
325. McCarty, P.L., Haug, R.T.: Microbial Aspects of Pollution (Sykes, G., Skinner, F.A., eds.). Academic Press London New York 1971, p. 215
326. Yamada, T., Sakaguchi, K.: Arch. Microbiol. *124*, 161 (1980)
327. Evans, H.J., Emerich, D.W., Ruiz-Argueso, T., Maier, R.J., Albrecht, S.L.: Nitrogen Fixation. Volume 2 (Newton, W.E., Orme-Johnson, W.H., eds.). University Park Press Baltimore 1980, p. 69
328. Hardy, R.W.F., Burns, R.C.: Ann. Rev. Biochem. *37*, 331 (1968)
329. Postgate, J.R.: Evolution in the Microbial World (Carlile, M., Skehel, J., eds.). Cambridge University Press 1974, p. 263
330. Becking, J.H.: Microbiology of Tropical Soils and Plant Productivity (Dommergues, Y.R., Diem, H.G., eds.). Martinus Nijhoff/Dr. W. Junk Publishers 1982, p. 109
331. Gauthier, D., Diem, H.G., Dommergues, Y.: App. Env. Microbiol. *41*, 306 (1981)
332. Kidd, G.H., Baker, D., Torrey, J.G.: Nature *281*, 76 (1979)
333. Capone, D.G., Carpenter, E.G.: Science *217*, 1140 (1982)
334. Martinez, L., Silver, M.W., King, J.M., Alldredge, A.L.: Science *221*, 152 (1983)
335. Watanabe, I., Furusaka, C.: Adv. Microb. Ecol. *4*, 125 (1980)
336. Watanabe, I.: Microbiology of Tropical Soils and Plant Productivity (Dommergues, Y.R., Diem, H.G., eds.) Martinus Nijhoff/Dr. W. Junk Publishers 1982, p. 169
337. Dobereiner, J., De-Polli, H.: Nitrogen Fixation (Stewart, W.D.P., Gallon, J.R., eds.). Academic Press London 1980, p. 301
338. Brown, M.E.: Bacteria and Plants (Skinner, F.A., Rhodes-Roberts, M.G., eds.). Academic Press London 1982, p. 25
339. Brill, W.J.: Microbiol. Rev. *44*, 449 (1980)
340. Old, R.W., Primrose, S.P.: Principles of Gene Manipulation. 2nd Edition. Blackwell Scientific 1981
341. Grawat, L., Rodhe, H.: Nitrogen, Phosphorus and Sulphur – Gobal Cycles Scope Report 7, Ecol. Bull. (Stockholm) *22*, 89 (1976)
342. Zehnder, A.J.B., Zinder, S.H.: Handbook of Environmental Chemistry. Volume 1A (Hutzinger, O., ed.). Springer-Verlag Berlin Heidelberg New York 1980, p. 105
343. Goldhaber, M.B., Kaplan, I.R.: The Sea, Volume 5 (Goldberg, G., ed.). John Wiley and Son 1974, p. 569
344. Freney, J.R., Jacq, V.A., Baldensberger, J.F.: Microbiology of Tropical Soils and Plant Productivity (Dommergues, Y.R., Diem, H.G., eds.). Martinus Nijhoff/Dr. W. Junk Publishers 1982, p. 271

345. Nedwell, D.B.: Sediment Microbiology (Nedwell, D.B., Brown, C.M., eds.). Academic Press London New York 1982, p. 72
346. Pfennig, N., Widdel, F.: Phil. Trans. R. Soc. Lond. *B298*, 433 (1982)
347. Peck, H.O., Le Gall, J.: Phil. Trans. R. Soc. Lond. *B298*, 443 (1982)
348. Siegel, L.M.: Metabolism of Sulphur Compounds (Greenberg, D.M., ed.). Academic Press 1975, p. 217
349. Pfennig, N., Biebl, H.: Arch. Microbiol. *110*, 3 (1976)
350. Zillig, W., Stetter, K.O., Schäfer, W., Janeković, D., Wunderl, S., Holz, I., Palm, P. Zentbl. Bakt. Hyg. I. Abt. Orig. *C2*, 205 (1981)
351. Stetter, K.O., Gaag, G.: Nature *305*, 309 (1983)
352. Pfennig, N., Widdel, F., Trüper, H.G.: The Prokaryotes. Volume 1 (Starr, M.P., Stolp, S., Trüper, H.G., Balows, A., Schlegel, H.G., eds.). Springer-Verlag Berlin Heidelberg New York 1981, p. 926
353. Pfennig, N., Biebl, H.: The Prokaryotes. Volume 1 (Starr, M.P., Stolp, S., Trüper, H.G., Balows, A., Schlegel, H.G., eds.). Springer-Verlag Berlin Heidelberg New York 1981, p. 941
354. Malcolm, S.J., Stanley, S.O.: Sediment Microbiology (Nedwell, D.B., Brown, C.M., eds.). Academic Press London New York 1982, p. 1
355. Postgate, J.R.: The Sulphate-reducing Bacteria. Cambridge University Press 1979
356. Postgate, J.R.: Phil. Trans. R. Soc. Lond. *B298*, 583 (1982)
357. Krouse, H.R., McCready, R.G.L.: Biochemical Cycling of Mineral-forming Elements. (Trudinger, P.A., Swaine, D.J., eds.). Elsevier 1979, p. 315
358. Trudinger, P.A., Williams, N.: Mineral Deposits and the Evolution of the Biosphere. (Holland, H.D., Schidlowski, M., eds.). Springer-Verlag Berlin Heidelberg New York 1982, p. 177
359. Miller, J.D.A.: Microbial Biodeterioration (Rose, A.H., ed.). Academic Press London New York 1981, p. 150
360. Thauer, R.: Microbial Energy Conversion (Schlegel, H.G., Barnea, J., eds.). E. Golze Göttingen 1976, p. 201
361. Fenchel, T.M., Jørgensen, B.B.: Adv. Microb. Ecol. *1*, 1 (1977)
362. Thauer, R.K.: Phil. Trans. R. Soc. Lond. *B298*, 467 (1982)
363. Garlick, S., Oren, A., Pada, E.: J. Bact. *129*, 623 (1977)
364. Robertson, L.A., Kuenen, J.G.: J. Gen. Microbiol. *129*, 2847 (1983)
365. Jannasch, H.W., Wirsen, C.O.: App. Env. Microbiol. *41*, 528 (1981)
366. Ruby, E.G., Wirsen, C.O., Jannasch, H.W.: App. Env. Microbiol. *42*, 317 (1981)
367. Felbeck, H., Childress, J.J., Somero, G.N.: Nature *293*, 291 (1981)
368. Killham, K., Lindley, L.D., Wainwright, M.: App. Env. Microbiol. *42*, 629 (1981)
369. Tuttle, J.U.: App. Env. Microbiol. *40*, 516 (1980)
370. Trüper, H.G., Fischer, U.: Phil. Trans. R. Soc. Lond. *B298*, 529 (1982)
371. Kelly, D.P.: Phil. Trans. R. Soc. Lond. *B298*, 499 (1982)
372. Trüper, H.G., Pfennig, N.: The Prokaryotes. Volume 1 (Starr, M.P., Stolp, H., Trüper, H.G., Balows, A., Schlegel, H.G., eds.). Springer-Verlag Berlin Heidelberg New York 1981, p. 299
373. Trudinger, P.A.: Phil. Trans. R. Soc. Lond. *B298*, 563 (1982)
374. Trüper, H.G.: Biology of Inorganic Nitrogen and Sulphur (Bothe, H., Trebst, A., eds.). Springer-Verlag Berlin Heidelberg New York 1981, p. 199
375. Brierley, C.L.: Scientific American *247*, 44 (1982)
376. Brierley, C.L.: CRC Crit. Rev. Microbiol. *6*, 207 (1978)
377. Lundgren, D.G., Silver, M.: Ann. Rev. Microbiol. *34*, 263 (1980)
378. Vinogradov, A.P., Grinenko, V.A., Ustinov, V.I.: Geokhimiya *5*, 973 (1962)
379. Burnie, S.W., Schwartz, H.P., Crocket, J.H.: Econ. Geol. *67*, 895 (1972)
380. Oehler, J.H., Logan, R.G.: Econ. Geol. *72*, 1393 (1977)
381. Kaplan, I.R., Rafter, T.A., Hulston, J.R.: N.Z. J. Sci. *3*, 338 (1960)
382. Harrison, A.G., Thode, H.G.: Trans. Faraday Soc. *54*, 84 (1958)
383. Iya, K.K., Sreenivasayi, M.: Curr. Sci. *14*, 243 (1945)
384. Dessau, G., Jensen, M.L., Nakai, N.: Econ. Geol. *57*, 410 (1962)

385. Ivanov, M.V.: Microbiological Processes in the Formation of Sulphur Deposits. Jerusalem Israel Programme for Scientific Translations 1968
386. Ruckmick, J.C., Wimbery, B.H., Edwards, A.F.: Econ. Geol. *74*, 469 (1979)
387. Thode, H.G., Wanless, R.K., Wallouch, R.: Geochima. Geophys. Acta *5*, 286 (1954)
388. Vinogradov, A.P., Grinenko, V.A., Ustinov, V.I.: Geokhimiya *11*, 1075 (1964)
389. Al-Sawaf, F.D.S.: Northern Iraq Econ. Geol. *72*, 608 (1977)
390. Marshall, K.C.: Biogeochemical Cycles of Mineral-Forming Elements (Trudinger, P.A., Swaine, D.J., eds.). Elsevier 1979, p. 253
391. Lundgren, D.G., Dean, W.: Biogeochemical Cycles of Mineral-Forming Elements (Trudinger, P.A., Swaine, D.J., eds.). Elsevier 1979, p. 211
392. Nealson, K.H.: Mineral Deposits and the Evolution of the Biosphere (Holland, H.D., Schidlowski, M., eds.). Springer-Verlag Berlin Heidelberg New York 1982, p. 51
393. Nealson, K.H.: Microbial Geochemistry (Krumbein, W.E., ed.). Blackwell Scientific Publications Oxford London Edinburgh Boston Melbourne 1983, p. 159
394. Nealson, K.H.: Microbial Geochemistry (Krumbein, W.E., ed.). Blackwell Scientific Publications Oxford London Edinburgh Boston Melbourne 1983, p. 191
395. James, H.L., Trendall, A.F.: Mineral Deposits and the Evolution of the Biosphere (Holland, H.D., Schidlowski, M., eds.). Springer-Verlag Berlin Heidelberg New York 1982, p. 199
396. Lundgren, D.G., Vestal, J.R., Tabita, F.R.: Microbial Iron Metabolism (Neilands, J.B., ed.). Academic Press New York 1974, p. 457
397. Kelly, D.P.: Companion to Microbiology (Bull, A.T., Meadow, P.N., eds.). Longmans 1978, p. 363
398. Brierley, C.L., Murr, L.E.: Science *179*, 488 (1975)
399. Wolfe, R.S.: Principles and Applications in Aquatic Microbiology (Heukelekian, W., Dondero, N.C., eds.). John Wiley New York 1963, p. 82
400. Van Veen, W.L., Mulder, G.C., Dienema, M.H.: Microbiol. Rev. *42*, 328 (1978)
401. Kuznetsov, S.I., Ivanov, M.V., Lyalikova, N.N.: Introduction to Geological Microbiology (Broneer, P.T., Translator) McGraw-Hill New York 1963
402. Aristovskaya, T.V., Zavarsin, G.A.: Soil Biochemistry. Volume 2 (McLaren, A.D., Skujins, J., eds.). Marcel Dekker 1971, p. 385
403. Ehrlich, H.L.: Soil Sci. *119*, 36 (1975)
404. Ehrlich, H.L.: Geomicrobiology. Marcel Dekker 1981
405. Cronan, D.S.: Endeavour (New Series) *2*, 80 (1978)
406. Cronan, D.S., Varentsov, I.M. (eds.): Monograph on the Geology and Geochemistry of Manganese, 25th International Geological Congress Sydney 1978
407. Glasby, G.P. (ed.): Marine Manganese Deposits. Elsevier Amsterdam 1977
408. Ehrlich, H.L.: Geomicrobiol. J. *1*, 65 (1979)
409. Summers, A.C., Silver, S.: Ann. Rev. Microbiol. *32*, 637 (1978)
410. Iverson, W.P., Brinkman, F.S.: Water Pollution Microbiology. Volume 2 (Mitchell, R., ed.). John Wiley and Son 1978, p. 201
411. Doran, J.W.: Adv. Microb. Ecol. *6*, 1 (1982)
412. Challenger, F.: Organometals and Organometalloids (Brinkman, F.E., Bellama, J.M., eds.). ACS Symp. Series No. 82 1978, p. 1
413. Craig, P.J.: Handbook of Environmental chemistry. Volume 1A (Hutzinger, O., ed.). Springer-Verlag Berlin Heidelberg New York 1980, p. 169
414. Halstead, R.L., McKercher, R.B.: Soil Biochemistry. Volume 4 (Paul, E.A., McLaren, A.D., eds.). Marcel Dekker 1975, p. 31
415. Henderson, M.E.K., Duff, R.B.: J. Soil Sci. *14*, 236 (1963)
416. Berthelin, J.: Microbial Geochemistry (Krumbein, W.E., ed.). Blackwell Scientific Publications Oxford London Edinburgh Boston Melbourne 1983, p. 223
417. Golubić, S., Krumbein, W.E., Schneider, J.: Biochemical Cycling of Mineral-Forming Elements (Trudinger, P.A., Swaine, D.J., eds.). Elsevier 1979, p. 29
418. Zehnder, A.J.B.: Handbook of Environmental Chemistry. Volume 1B. (Hutzinger, O., ed.). Springer-Verlag Berlin Heidelberg New York 1982, p. 83
419. Wilbur, K.M., Simkiss, K.: Biogeochemical Cycling of Mineral-Forming Elements (Trudinger, P.A., Swaine, D.J., eds.). Elsevier 1979, p. 69

420. Walter, M.R. (ed.): Stromatolites. Developments in Sedimentology *20*. Elsevier 1976
421. Krumbein, W.E.: Biogeochemical Cycling of Mineral-Forming Elements (Trudinger, P.A., Swaine, D.J., eds.). Elsevier 1979, p. 47
422. Kinsey, D.W., Davies, P.J.: Biogeochemical Cycling of Mineral-Forming Elements. (Trudinger, P.A., Swaine, D.J., eds.). Elsevier 1979, p. 131
423. Cook, P.J.: Handbook of Stratabound and Stratiform Ores. (Wolf, K.H., ed.). Elsevier 1976, p. 505
424. Cook, P.J., McElhinny, M.W.: Econ. Geol. *74*, 315 (1979)
425. Pierrou, U.: Biogeochemical Cycling of Mineral-Forming Elements (Trudinger, P.A., Swaine, D.J., eds.). Elsevier 1979, p. 163
426. Golubić, S., Schneider, J.: Biogeochemical Cycling of Mineral-Forming Elements (Trudinger, P.A., Swaine, D.J., eds.). Elsevier 1979, p. 107
427. Pierrou, U.: Biogeochemical Cycling of Mineral-Forming Elements (Trudinger, P.A., Swaine, D.J., eds.). Elsevier 1979, p. 205
428. Cosgrave, D.J.: Adv. Microb. Ecol. *1*, 95 (1977)
429. Pierrou, U.: Nitrogen, Phosphorus and Sulphur – Global Cycles. Scope Report *7*, Ecol. Bull. (Stockholm) *22*, 1976, p. 75
430. Emsley, J.: Handbook of Environmental Geochemistry. Volume 1A (Hutzinger, O., ed.). Springer-Verlag Berlin Heidelberg New York 1980, p. 147
431. Gieseking, J.E. (ed.): Soil Components. Volume 1. Springer-Verlag New York 1975
432. Haider, K., Martin, J.P., Filip, Z.: Soil Biochemistry. Volume 4 (Paul, E.A., McLaren, A.D., eds.). Marcel Dekker 1975, p. 195
433. Felbeck, G.T.: Soil Biochemistry. Volume 2 (McLaren, A.D., Skujins, D., eds.). Marcel Dekker 1971, p. 36
434. Stout, J.D., Goh, K.M., Rafter, T.A.: Soil Biochemistry. Volume 5 (Paul, E.A., Ladd, J.N., eds.). Marcel Dekker 1982, p. 1
435. Kirk, T.K.: Ann. Rev. Phytopathol. *9*, 185 (1971)
436. Zeikus, J.G.: Adv. Microb. Ecol. *5*, 211 (1981)
437. Riley, G.A.: Adv. Mar. Biol. *8*, 1 (1970)
438. Menzel, D.W.: The Sea. Volume 5 (Goldberg, E.D., ed.). John Wiley and Sons New York 1974, p. 659
439. Curtis, C.D.: Microbial Geochemistry (Krumbein, W.E., ed.). Blackwell Scientific Publications Oxford London Edinburgh Boston Melbourne 1983, p. 263
440. Foster, R.C.: Science *214*, 665 (1981)
441. Krumbein, W.E.: Deutsche Kunst- und Denkmalpflege, Deutscher Kunstverlag München Berlin 1973, p. 724
442. Myers, G.E., McCready, R.G.L.: Can. J. Microbiol. *12*, 477 (1966)
443. Webley, D.M., Henderson, M.E.K., Taylor, J.K.: J. Soil Sci. *14*, 102 (1963)
444. Hueck van der Plas, E.: Int. Biodeter. Bull. *4*, 11 (1968)
445. Strzelczyk, A.B.: Microbial Biodeterioration (Rose, A.W., ed.). Academic Press London New York 1981, p. 61
446. Silverman, M.D.: Biogeochemical Cycling of Mineral-Forming Elements (Trudinger, P.A., Swaine, D.J., eds.). Elsevier p. 445
447. Krumbein, W.E., Warner, D.: Microbial Geochemistry (Krumbein, W.E., ed.). Blackwell Scientific Publications Oxford London Edinburgh Boston Melbourne 1983, p. 125
448. Oehler, J.H.: Biogeochemical Cycling of Mineral-Forming Elements (Trudinger, P.A., Swaine, D.J., eds.). Elsevier 1979, p. 467
449. Wollast, R.: The Sea. Volume 5 (Goldberg, E.D., ed.). John Wiley and Sons New York 1974, p. 359
450. Wilding, L.P., Smeck, N.E., Drees, L.R.: Minerals in Soil Environments (Dixon, J.B., Weed, S.B., eds.). Soil Science Society of America Special Publication 1977, p. 471

Subject Index

The Handbook of Environmental Chemistry

Editor: **O. Hutzinger**

Volume 1

Part A

The Natural Environment and the Biogeochemical Cycles

With contributions by numerous experts
1980. 54 figures, 59 tables.
XII, 258 pages
ISBN 3-540-09688-4

Contents:

Springer-Verlag
Berlin
Heidelberg
New York
Tokyo

The Handbook of Environmental Chemistry

Editor: **O. Hutzinger**

Volume 1

The Natural Environment and the Biogeochemical Cycles

Part B
With contributions by numerous experts
1982. 84 figures. XV, 317 pages
ISBN 3-540-11106-9

Contents:
Basic Concepts of Ecology. – Natural Radionuclides in the Environment. – The Nitrogen Cycles. – The Carbon Cycle. – Molecular Organic Geochemistry. – Radiation and Energy Transport in the Earth Atmosphere System. – Subject Index.

Part C
With contributions by numerous experts
1984. 55 figures. XIII, 220 pages
ISBN 3-540-13226-0

Contents:
Humic Substances. Structural Aspects, and Photophysical, Photochemical and Free Radical Characteristics. – Organic Material in Sea Water. – Marine Gelbstoff. – The Surface of the Ocean. – Atmospheric Nitrogen. Chemistry, Nitrification, Denitrification, and their Interrelationships. – Carbon Dioxide: A Biogeochemical Portrait.

Springer-Verlag
Berlin
Heidelberg
New York
Tokyo

Made in the USA
Monee, IL
13 September 2026

62973769R00149